SURFACES, INTERFACES, AND COLLOIDS

MICHIGAN MOLECULAR INSTITUTE
1910 WEST ST. ANDREWS ROAD
MIDLAND, MICHIGAN 48640

SURFACES, INTERFACES, AND COLLOIDS

Principles and Applications

SECOND EDITION

Drew Myers

WILEY-VCH

New York · Chichester · Weinheim · Brisbane · Singapore · Toronto

For ordering and customer service call 1-800-CALL-WILEY.

Library of Congress Cataloging-in-Publication Data:

Myers, Drew, 1946–
    Surfaces, interfaces, and colloids : principles and applications /
Drew Myers.—2nd ed.
        p.   cm.   '
    Includes bibliographical references and index.
    ISBN 0-471-33060-4 (cloth : alk. paper)
    1. Surface chemistry.   2. Interfaces (Physical sciences)
3. Colloids   I. Title.
QD506.M39   1999
541.3′3—dc21   98-38906

Printed in the United States of America.

10 9 8 7 6 5 4 3 2

This work is dedicated to:
Christine
Betty
Jimmy
Linda

CONTENTS

PREFACE TO THE SECOND EDITION

Continuing the basic philosophy of the first edition, namely, the "generalization" of surface and colloid science in the overall context of chemistry, physics, biology, medicine, and other disciplines, this second edition is designed more in the style of a textbook. A number of errors and omissions pointed out by reviewers and readers have (hopefully) been corrected. Several important topics that were neglected in the first edition have been inserted and material that seemed *needlessly* redundant or of less importance has been deleted in an effort to maintain the size of the work at a manageable level. It is likely that new (and perhaps old) readers will still find errors and have questions about the material presented. Paraphrasing Abraham Lincoln: You can't please all of the people all of the time.

Because of the complex interrelationships among the various topics comprising surface and colloid science, it is difficult to break the subject up into "clean" units, especially when one wants to avoid writing an encyclopedic volume. As a result, the second edition, possibly more than the first, presents a number of topics in what seems to be an "out of order" manner or in a slightly different way depending on the context. At the time of writing, that seemed to be the best course to follow and the "winds" of time did not allow for much backtracking. Hopefully, the repetition will be helpful in the assimilation of the material in question.

At the request of several readers (and the publisher), I have added a number of questions and problems at the end of each chapter. The number involved is not large, and the difficulty ranges from ridiculously easy to pretty darned hard. In some cases students will (to their disgust) have to do some searching in order to find the needed clue to the solution. In others, the answer would bite them if it were a snake. It is hoped that the problems provided will help clarify the concepts involved without undue pain. The problems provided are intended to be indicative of the kinds of situations that might be addressed by the material presented in the chapter, and to serve as a stimulus for further application of those principles in diverse areas of science and technology.

It may be noticed that the book contains no specific literature references. It is the author's opinion that a general text or monograph of this type should be as free as possible of complicating factors in order to leave the mind open to the concepts presented. For those wishing to pursue a topic in more depth or to find specific reference material, general and specific chapter bibliographies are provided. A specific volume cited in the Bibliography at the end of the book may not be the latest edition of that work. Further, the edition of a work cited in the General Readings section of the Bibliography may not be the same as that given in a specific Chapter section of the Bibliography. Such occurrences simply reflect the fact that the author did not have access to the latest edition to confirm any organizational differences in the work. At the technical level of this book any differences in content should be of no significance.

The author wishes to express his thanks to José and Hugo Martina and the organization at ALPHA C.I.S.A. in Río Tercero, Córdoba, Argentina for allowing him the freedom to take the time to complete this work.

PREFACE TO THE FIRST EDITION

After much neglect over the years as a "stepsister" of physical chemistry and physics, the "twilight zone" of surface and interfacial science is now coming to be generally recognized as a vital, if not the vital, component of many of our most important and complex technological and biological processes. Surface effects, including colloidal systems, have been recognized for thousands of years as being of great potential use in many areas of the lives of humankind.

Historically, improvements in systems involving interfacial and colloidal phenomena, and the development of new applications, have been a result of repetitive processes of trial and error and the application of empirical rules developed over long periods of time—in other words, the area has been more art than science. Current economic and social conditions, however, tend to make the old "artistic" approaches to invention, development, and production much less desirable and tolerable. In today's technological climate, a small advantage or disadvantage can be the difference between survival and extinction in the economic jungle.

True to historical form, our educational institutions in general continue to gloss over or ignore completely the subject of surface and colloid science. Probably greater than 95% of graduates in chemistry, physics, biology, engineering, materials science, and other subjects, enter their careers totally ignorant of even the most basic concepts of surfaces, interfaces, and colloids—this, despite the fact that literally trillions of dollars in economic capacity are directly or indirectly involved with this scientific area. As a result, scientists and technologists (and their employers) risk losing large amounts of time and material (i.e., money!) in search of solutions that, in some cases, may be obvious or at least more easily found by the application of the fundamental principles of surface and colloid science.

Although there are a number of excellent standard reference texts available in the area of surface and colloid science, they are often found to be overly daunting to temporary visitors who hesitate to wade into the jungle of theory in order to find the basic concepts they seek. Intended primarily as texts for the training of surface and colloid specialists, those standard works often pose a significant barrier for someone with a limited familiarity with the finer points of thermodynamics, quantum mechanics, solution theory, electrical phenomena, and the like.

This book is intended to serve as a narrow footbridge for scientists, technologists, and students who may use or need to use some aspect of surface and colloid science in their work, or who want to attain some familiarity with the area during their training process. It is designed to provide a general introduction to concepts, rather than a strong theoretical background. While some theory must be included for clarity, details are left for the interested (or needy) reader to pursue in the cited references. In most cases, discussions center primarily on conceptualization, with semiquantitative or qualitative illustrations serving to highlight the principles involved.

Although a minimum amount of space has been dedicated to theory, the quantitative nature of the subject requires that certain mathematical formulas be introduced. However, formulae are basically tools to be used when needed and stored away when not needed. For that reason, large portions of this work will be found to be devoid of mathematics.

Likewise, while extensive references to the original literature may occasionally be useful, most readers probably prefer not to chase down such details. For that reason, very few such references are provided here. Instead, various works that do contain original references are cited in the Bibliography for each chapter.

I wish to thank Drs. G. H. Pearson, B. W. Rossiter, and D. A. Smith for their strong professional support over many years, and Drs. H. D. Bier and W. P. Reeves for that first important push. In addition, I want to salute the faculty and staff of the Surface and Coilloid Group in the School of Chemistry, the University of Bristol, England, who, along with a few other groups throughout the world, strive to maintain a high standard of excellence in this field.

DREW MYERS

Río Tervero (Córdoba)
Argentina

SURFACES, INTERFACES, AND COLLOIDS

1 Surfaces and Colloids: The Twilight Zone

1.1. INTRODUCTION: THE WORLD OF NEGLECTED DIMENSIONS

In 1915 Wolfgang Ostwald described the subject matter of colloid and surface science as a "world of neglected dimensions." The reason for such a description stemmed from the unique nature of interfaces and related colloidal phenomena—they could not be readily interpreted based on "classical" atomic or solution theories, and the regions of space involved were beyond the reach of existent experimental techniques. Science has since taken a firm theoretical and experimental hold on the nature of matter at its two extremes: at the molecular, atomic, and subatomic levels, and in the area of bulk materials, including their physical strengths and weaknesses and their chemical and electrical properties. Legions of chemists, physicists, materials scientists, engineers, and others are continuously striving to improve on that knowledge in academic and industrial laboratories around the world. Between those two extremes still lies the world referred to by Ostwald, and even with the latest advanced techniques for studying the region between phases, a great many mysteries remain to be solved. For that reason, I like to refer to the study of interfaces and colloids as entering the "twilight zone." That "region" of the physical world represents a bridge not only between chemical and physical phases, but also plays a vital but often unrecognized role in other areas of chemistry, physics, biology, medicine, engineering, and other disciplines.

Our understanding of the nature of the interfacial region and the changes and transformations that occur in going from one chemical (or physical) phase to another has historically lagged behind that in many other scientific areas in terms of the development and implementation of both theoretical and practical concepts. That is not to say, however, that we are particularly ignorant when it comes to interfacial and colloidal phenomena. Great strides were made in the theoretical understanding of interactions at interfaces in the late nineteenth and early twentieth centuries. Modern computational and analytical techniques made available in the last few years have led to significant advances toward a more complete understanding of the unique nature of interfaces and the interactions that result from their unique nature. However, because of the unusual and sometimes complex character of interfaces and

1

associated phenomena, the development of fully satisfying theoretical models has been slow. By "fully satisfying" is meant a theory that produces good agreement between theory and experiment in situations that are less than "ideal" or "model" systems.

The degree of "satisfaction" one obtains from a given theory is quite subjective, of course, so there exists a great deal of controversy in many areas related to colloids and interfaces. For the surface and colloid scientist (as in all science), such controversy is not bad, since it represents the fuel for continued fundamental and practical research. However, for the practitioner who needs to apply the fruits of fundamental research, such uncertainty can sometimes complicate attempts to solve practical interfacial and colloidal problems.

It is likely that for every trained surface and colloid scientist in academia and industry, there are hundreds of scientists, engineers, and technicians whose work directly or indirectly involves some surface and/or colloidal phenomena. And very probably, of those hundreds, a relatively small percentage have been formally introduced to the subject in more than a cursory way during their scientific training. It therefore becomes necessary for them to learn "on the fly" enough of the subject to allow them to attack their problems in a coherent way. This book has been designed in a way and at a level that will (hopefully) provide a useful introduction to surface and colloid science at an undergraduate or graduate level while at the same time serving as an accessible reference for those already trained in other fields of science but needing some initial guidance into the twilight zone.

It would be practically impossible to list all of the human activities (both technological and physiological) that involve surface and colloidal phenomena, but a few examples have been listed in Table 1.1. For purposes of illustration, the examples in Table 1.1 have been divided into four main categories, each of which is further divided (somewhat arbitrarily, in some cases) according to whether the main principle involved is "colloidal" or "interfacial." More exact definitions of what those two terms imply will be given in the appropriate chapters; however, for present purposes one can think of "colloidal" as being a state of subdivision of matter in which the particle (or molecular) size of the basic unit involved varies from just larger than that of "true" molecular solutions to that of coarse suspensions—that is, between 10 and 10,000 nm (Fig. 1.1a). "Interfacial" phenomena may be defined, in this context, as those related to the interaction of at least one bulk phase (solid or liquid) with another phase (solid, liquid, or gas) or a vacuum in the narrow region in which the transition from one phase to the other occurs (Fig. 1.1b). As will quickly become apparent, the two classes of phenomena are intimately related and often cannot be distinguished. For present purposes (and according to this author's preference) the examples have been divided according to those definitions based on the principle phenomenon involved. By examining each subdivision in Table 1.1, one can quickly see that interfacial and colloidal

TABLE 1.1. Some Common Examples of Surface and Colloidal Phenomena in Industry and Nature

Surface Phenomena	Colloidal Phenomena
Products Manufactured as Colloids or Surface Active Materials	
Soaps and detergents (surfactants)	Latex paints
Emulsifiers and stabilizers	Aerosols
(nonsurfactant)	Foods, (ice cream, butter, mayonnaise, etc.)
Herbicides and pesticides	Cosmetics and topical ointments
Fabric Softeners	Pharmaceuticals
	Inks
	Lacquers, oil-based paints
	Oil and gas additives
	Adhesives
Direct Application of Surface and Colloidal Phenomena	
Lubrication	Control of rheological properties
Adhesion	Emulsions
Foams	Emulsion and dispersion polymerization
Wetting and waterproofing	Drilling muds
	Electrophoretic deposition
Use for the Purification and/or Improvement of Natural or Synthetic Materials	
Tertiary oil recovery	Mineral ore separation by flotation
Sugar refining	Grinding and communition
Sintering	Sewage and wastewater treatment
Physiological Applications	
Respiration	Blood transport
Joint lubrication	Emulsification of nutrients
Capillary phenomena in liquid	Enzymes
transport	Cell membranes
Arteriosclerosis	

phenomena are ubiquitous. We and our world simply would not function or even exist as we know it in their absence.

Somewhat surprisingly, considering its inherent importance, this "neglected dimension" has historically been shorthanded in terms of scientists and technicians formally trained in the theoretical and experimental aspects of the discipline. As a result, one can speculate that large amounts of time, money, and other resources have been wasted over the years simply because chemists, physicists, biologists, engineers, and technical operators were ignorant of certain basic ideas about interfaces and colloids that could have solved or helped solve many practical and theoretical problems.

(a)

(b)

FIGURE 1.1. Colloids and interfaces represent distinct, but related "zones" that span the size range between atoms or molecules and bulk phases: (*a*) a colloid may be defined as a unit (particle, drop, bubble, molecule, etc.) larger than a "normal" (i.e., nonpolymeric) molecule (<10 nm in diameter), but smaller than what may be considered a "true" bulk phase (>10,000 nm diameter); (*b*) an interface is a region of space through which a system changes from one phase (liquid or solid) to another (solid, liquid, or gas), that change usually occurring over a distance of one to several tens of unit (atomic or molecular) diameters.

1.2. A HISTORICAL PERSPECTIVE

While the fields of interface science and colloid science are now known to be intimately related, such was not always the case. In the history of their development, the two branches evolved from somewhat different sources and slowly grew together as it became obvious that the basic laws controlling phenomena related to the two were, in fact, the same.

Although the formal studies of interface and colloid science began in the early nineteenth century, humans observed and made use of such phenomena thousands of years earlier. The Bible and other early religious writings refer to strange clouds and fogs, that were colloidal in nature (aerosols). Ancient Egyptian hieroglyphic paintings show scenes of slaves lubricating great stones being moved to build pyramids and other monuments. Hebrew slaves made bricks of clay, a classic colloid, while many ancient seafaring cultures recognized the beneficial effect of spreading oil on storm-tossed waters in order to help protect their fragile crafts. The preparation of inks and pigments, baked bread, butter, cheeses, glues, and other substances all represent interfacial and colloidal phenomena of great practical importance to ancient cultures.

In more modern times, such notables as Benjamin Franklin began to take formal notice of interfacial and colloidal phenomena in philosophical discussions of, for example, the amount of oil required to cover a small pond in London completely with the thinnest possible layer (monolayer coverage). The first important quantitative analyses of surface phenomena were probably the works of Young, Laplace, Gauss, and Poisson to be discussed later.

From the middle of the nineteenth century on, understanding of the phenomenology of interfaces became better at the molecular level, although the nature of the forces involved remained uncertain until the advent of quantum mechanical theory in the 1930s. The study of colloidal phenomena followed a similar track in that certain characteristics of colloidal systems were recognized and studied in the last century (and before), but a good quantitative understanding of the principles and processes involved remained elusive.

The main reason for the delay in developing good quantitative theories for surfaces and colloids was the lack of good, well-characterized systems which gave consistent and reproducible results in the hands of different investigators (and sometimes those of the same one). System purity is a key requirement for understanding the nature of most surface and colloidal phenomena; contaminants at levels of small fractions of a percent can significantly affect the performance of a system. In the pioneering days of surface and colloid science, such low levels of impurities often went undetected, so that correlation between theory and experiment were sometimes less than optimal. Modern techniques of purification and analysis have made it possible for us to work with much better information about the systems involved, with the result that experimental data can be used with more confidence. Even now, however, the sensitivity and complexity of interfaces and colloids causes difficulties in developing complete understanding of many phenomena of academic and practical importance.

1.3. A VIEW TO THE FUTURE

Because the modern, quantitative approach to interface and colloid science is a relatively new development, a great deal remains to be done in terms of extending the basic ideas and concepts originating in the "classic" period of the science to include new information and theories. Not only the desire for improved theories, but technological innovation demands that our understand-

TABLE 1.2. Some Important Areas of Surface and Colloid Science Inviting Future Research

Theoretical studies	Surface energies of solids, surface and interfacial tensions and the interfacial region, thermodynamics of colloidal systems, improved electrical double layer theory, adsorbed polymer layers and steric stabilization, relationships between surface energies and bulk properties
Surface chemistry	Equilibrium and dynamic wetting and spreading processes, adhesion, physical adsorption, chemisorption and heterogeneous catalysis, spectroscopic and optical studies of surfaces, flow through porous media
Interparticle interactions	Measurement of forces between surfaces, effects of adsorbed layers, the role of solvation, attractive forces
Colloidal stability	Hydrodynamic and solvation factors, emulsion stability, microemulsions, multiple emulsions, coagulation and flocculation theory, foam stability, demulsification and defoaming, the effects of adsorbed polymers on stability and flocculation
Colloidal properties	Optical and spectroscopic properties of model colloids, rheological properties, electrophoretic properties
Chemical reactions	Heterogeneous catalysis by colloids, chemical at surfaces and biological reactions in colloidal systems, interfacial reactions, including those between body fluids and foreign surfaces
Lyophilic colloids	Association colloids, gels, studies of polymers in solution and adsorbed onto surfaces, microgels, liquid crystals
Aerosols	Methods of formation, stabilization, and destruction
Biocolloids	Membranes, cell and particle adhesion, cell–antibody interactions, drug delivery, transport phenomena

ing of surface and interfacial phenomena be improved. Just a few of the general areas that still warrant close, specific attention are listed in Table 1.2.

The list in Table 1.2 is obviously incomplete, but it may serve to lead the student or interested reader into new and important areas of research. It is hoped that the following material, as limited as it is in its coverage of a large subject, will assist the "needy" or just the interested in finding a door into the "neglected dimension" of interface and colloid science. Welcome to the "twilight zone!"

PROBLEMS

1.1. If Benjamin Franklin poured one standard tablespoon of oil onto the surface of a $\frac{1}{2}$-acre pond, would the amount of oil be sufficient to produce a complete monolayer of oil on the surface? Assume that the oil has a molecular weight of 310 and a density of 0.94 g cm^{-3}.

1.2. The Biblical reference to the effect of pouring oil on "troubled waters" actually has a strong basis in scientific principles. Explain in simple "mechanical" terms how one might explain the calming effect that such a process can have on turbulent seas.

1.3. Arteriosclerosis is an important medical problem in much of the developed world. Given the nature of the problem suggest, in general conceptual terms, how studies of surface and colloid science might be helpful in its prevention and treatment.

2 Surfaces and Interfaces: General Concepts

The subject matter to be covered in the following chapters is concerned with the regions of our physical world that lie between two distinct and identifiable phases of matter. The bulk characteristics of various phases will not, for the most part, be considered, except insofar as they affect interphase interactions. Our primary area of interest lies in that region of space in which the system as a whole undergoes a transition from one phase to another. For purposes of terminology, it is common practice to refer to that nebulous region as a "surface" or an "interface." As will become evident, the exact definition of what constitutes a surface or an interface is not always unequivocal. While the two terms are often used to indicate distinct situations, they are in practice interchangeable, exact usage depending more on personal preference than on any physically definable differences. In general, however, one usually finds that the term "surface" is applied to the region between a condensed phase (liquid or solid) and a gas phase or vacuum, while "interface" is normally applied to systems involving two condensed phases. Where complete generality is implied, "interface" is probably the better term. That convention will generally be employed in the material to follow. However, no guarantee of complete consistency is given or implied.

There are several types of interfaces that are of great practical importance and that will be discussed in turn. These general classifications include, solid–vacuum, liquid–vacuum, solid–gas, liquid–gas, solid–liquid, liquid–liquid, and solid–solid. From a practical standpoint, solid– and liquid–vacuum interfaces are of little concern. They are most often encountered in the context of theoretical derivations, since the absence of a second phase simplifies matters greatly, or in studies of high-vacuum processes such as deposition, and sputtering. The true two-phase systems (assuming that a vacuum is not considered to be a true "phase") are the ones which are of most importance in practical applications and that are addressed in most detail here. A list of commonly encountered examples of these interfaces is given in Table 2.1.

2.1. THE NATURE OF INTERFACES

In order for two phases to exist in contact, there must be a region through which the intensive properties of the system change from those of one phase

8

TABLE 2.1. Common Interfaces of Vital Natural and Technological Importance

Interface Type	Occurrence or Application
Solid–vapor	Adsorption, catalysis, contamination, gas–liquid chromatography
Solid–liquid	Cleaning and detergency, adhesion, lubrication, colloids
Liquid–vapor	Coating, wetting, foams
Liquid–liquid	Emulsions, detergency, tertiary oil recovery

to those of the other, as for example in the boundary between a solid and a liquid. In order for such a boundary to be stable it must possess an interfacial free energy such that work must be done to extend or enlarge the boundary or interface. If such is not the case, and if no other external forces such as gravity act to separate the phases by density, and other factors, then no energy will be required to increase the interfacial area and random forces, including the uncertainty principle, Brownian motion or the chaotic butterfly, will distort, fold, and convolute the interface until the phases become mixed. In other words, if the interface does not have a positive free energy, it cannot exist as a stable boundary between two phases.

In surface and colloid science "stable" is or can be a relative term (as will become apparent later). For that reason, one should always have clearly in mind just what is intended by the term in a given situation. Our "chemical" world is one of both thermodynamics and kinetics, so that even if a system is thermodynamically unstable (i.e., diamond), it may require a rather long time for it to reach its most stable configuration. Such systems may be considered kinetically stable. They are also sometimes referred to as being "metastable." While thermodynamics is an essentially irresistible drive to a final lower energetic state, we can sometimes use kinetics as a tool to slow that drive for periods of time sufficient to achieve a particular goal. As will be seen in later chapters, that tool is of vital importance in many surface and colloidal phenomena.

In order to define an interface and show in chemical and physical terms that it exists, it is necessary to think in terms of energy, keeping in mind that nature will always act so as to attain a situation of minimum total free energy. In the case of a two-phase system, if the presence of the interface results in a higher (positive) free energy, the interface will spontaneously be reduced to a minimum—the two phases will tend to separate to the greatest extent possible within the constraints imposed by the container, gravitational forces, mechanical motion, and other factors. If the condition or composition of the system is altered, the energetic situation at the interface may also be altered, possibly producing a lower interfacial energy or some other effect that results in an increase in the time required for complete phase separation. That is, the change may alter the energetic drive to phase separation or it may alter the rate at which the phase separation occurs (i.e., its kinetics), or both. Overall, the interfacial energy will still be positive, but the changes caused by the alteration may prolong the "life" of any "excess" interfacial area. Such an effect may be beneficial, as in the case of a cosmetic emulsion, or detrimental, as in a petroleum–seawater emulsion. The important point is that although thermodynamics is almost always working to reduce interfacial area, we have access to tools that allow us to control, to some extent, the rate at which area changes occur.

There exist innumerable practical situations in which the energetic balance of interfacial regions must be controlled in order to make use of the unique characteristics of a system. The primary purpose of this work is to present in a "bare bones" way the fundamental natures of various interfaces and illustrate how those characteristics can (in principle) be manipulated to the best advantage of ourselves and our environment.

In the discussions to follow, the concept of the interfacial region will be presented from a molecular (or atomic) perspective and from the viewpoint of the thermodynamics involved. In this way one can obtain an idea of the situations and events occurring at interfaces and have at hand a set of basic mathematical tools for understanding the processes involved and to aid in manipulating the events to best advantage. The concepts presented are intended to be primarily qualitative in nature and do not necessarily represent "reality" in every detail. Similarly, the mathematical tools will be, for the most part, the basic elements necessary for accomplishing the purpose, with little or no derivation presented. More elaborate and sophisticated treatments of the subjects will be referenced but left for the more adventurous reader to pursue as needed.

2.2. SURFACE FREE ENERGY

Before beginning any discussion of surfaces and interfaces, it is important to have a clear concept of just what is meant by surface free energy. As will be seen throughout, the unique characters of interfaces and interfacial phenomena arise from the fact that atoms and molecules at interfaces, because of their special environment, often possess energies and reactivities significantly different from those of the same species in a bulk or solution situation. If one visualizes a unit (an atom or molecule) of a substance in a bulk phase, it can be seen that, on average, the unit experiences a uniform force field due to its interaction with neighboring units (Fig. 2.1a). If the bulk phase is cleaved in vacuum, isothermally and reversibly, along a plane that just touches the unit in question (Fig. 2.1b), and the two new faces are separated by a distance H, it can be seen that the forces acting on the unit are no longer uniform. Instead, it will continue to "feel" the presence of the adjacent units in the adjoining bulk phase, while having less interaction with those units being removed in the separated section.

Because the unit at the new surface is in a different energetic environment relative to its nearest neighbors, its total free energy must change. In this

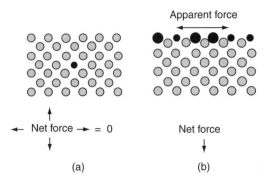

FIGURE 2.1. As seen in a two-dimensional representation, the net force acting on a unit (atom or molecule shown in black) in the interior of a bulk phase (*a*) will essentially be zero, assuming the absence of "faults" in a solid that would alter interactions with neighboring units. The net forces acting on a unit at the surface (*b*) will be unbalanced in the direction of the bulk phase.

case, since the interactions in the bulk phase produce a net lowering of the free energy of the units, the removal of those interactions by the formation of new surface results in an increase in the free energy of the units at or near the interface.

The net increase in free energy of the system as a whole resulting from the new situation will be proportional to the area, *A,* of new surface formed and the density (i.e., number) of interfacial units. The actual change in system free energy will also depend on the distance of separation, since unit interactions will generally fall off by some inverse power law. When the two new interfaces are separated by what can be termed practical infinity, the free energy of the system becomes constant. The "additional" energy is termed the "surface free energy" or more accurately the "excess" surface free energy. When the term "specific" excess surface free energy is used it refers to energy per unit area, usually in mJ m^{-2}. It should be remembered that the excess free energy is not equal to the total free energy of the system, but only that part resulting from the units location at the surface.

While the quantitative details will be explained in more detail in later chapters, it should be intuitively clear that atoms or molecules at a surface will experience a net positive inward (i.e., into the bulk phase) attraction normal to the surface, the resultant of which will be a state of lateral tension along the surface, giving rise to the concept of "surface tension." For a flat surface, the surface tension may be defined as a force acting parallel to the surface and perpendicular to a line of unit length anywhere in the surface (Fig. 2.2). The definition for a curved surface is somewhat more complex, but the difference becomes significant only for a surface of very small radius of curvature.

The units for surface tension and specific excess surface free energy are dimensionally equivalent and, for a pure liquid in equilibrium with its vapor,

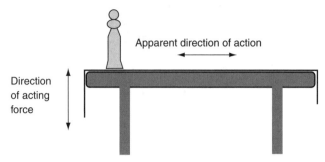

FIGURE 2.2. Although the force acting on an individual unit at the surface is directed inward, the resultant force on all surface units gives rise to the apparent existence of a tangential tension. Mechanically, one can visualize the situation as being analogous to a child pulling down on a table cloth, perpendicular to the table surface, with a resultant horizontal movement of the items on the surface (until they reach the edge, of course).

numerically equal. They represent intensive thermodynamic properties of the system, unlike the excess surface free energy, which is extensive (i.e., it depends on the area of surface being considered). The specific thermodynamic definition of surface tension for a pure liquid is given by

$$\Delta A_H = \Delta W = 2\sigma A \tag{2.1}$$

where A_H is the Helmholtz free energy of the system, W is the amount of reversible work necessary to overcome the attractive forces between the units at the new surface, and A is the area of new surface formed. The proportionality constant σ, termed the "surface tension," is numerically equal to the specific excess surface free energy for pure liquids at equilibrium; that is, when no adsorption of a different material occurs at the surface. The importance of that qualification will be made more clear in Chapter 9.

For two pure, mutually immiscible liquids having a common flat interface we can define the terms "interfacial tension" and "excess interfacial free energy," based on the same concepts used for the liquid–vapor systems. However, because unlike atoms or molecules at a liquid–liquid interface experience mutual attractions from units in the adjacent phase, those interactions become important in determining the properties of the system. The specific excess interfacial free energy will be dimensionally equivalent to and numerically equal to the interfacial tension.

When one ventures into the realm of solid surfaces, the situation becomes less clear cut than that given for liquids. In principle, the same concepts of surface formation and surface energetics should apply. However, the special nature of solids—specifically the lack of mobility of the atoms or molecules composing the material—means that in reality units at a freshly formed surface cannot re-accommodate themselves to their new situation and true equilibrium will not be obtained (at least over a reasonable time period), unlike in liquids where equilibrium is attained rapidly. The surface tension of a solid, therefore, will not be numerically equal to the specific excess surface free energy. In the case of solid surfaces a new definition of surface tension is commonly employed: it is the force needed to bring the freshly formed surface to its equilibrium state.

The SI (International System of Units) units of surface tension are mN m^{-1}, which can be interpreted as a two-dimensional analog of pressure (mN m^{-2}). As a concept, then, surface (and interfacial) tension may be viewed as a two-dimensional negative pressure acting along the surface as opposed to the usual positive pressures encountered in our normal experience. In liquid–vapor and liquid–liquid systems, the measurement of surface tension is a relatively easy task (with proper precautions, of course). For systems involving solid surfaces, life becomes much more difficult and the determination (or estimation) of surface tension and other thermodynamic quantities becomes very difficult and often very ambiguous.

2.2.1. The Work of Cohesion and Adhesion

At this point it is convenient to introduce two terms related to Equation (2.1) that will appear in various contexts in later chapters, namely, the "work of cohesion" and the "work of adhesion." The work of cohesion, W_c, is defined as the reversible work required to separate two surfaces of unit area of a single material with surface tension σ (Fig. 2.3a). Based on the distinction between solid and liquid surfaces explained above, the definition applies

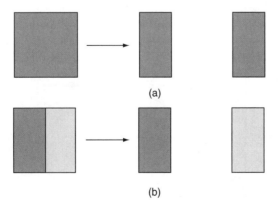

(a)

(b)

FIGURE 2.3. (a) When new surface is formed by dividing a homogeneous material, a certain amount of work is required. That work, the work of cohesion, is related to the surface tension of the material by Equation (2.2). (b) If the new surface results from the separation of two different materials, the resulting work of adhesion is given by Equation (2.3).

strictly to liquid surfaces, although the concept is useful for solid surfaces as well. Since the process involves the creation of two unit areas of fresh surface, and since the work required for that process is the surface tension, the work of cohesion is simply

$$W_c = 2\sigma \tag{2.2}$$

It must be remembered that W_c is a reversible thermodynamic function and represents a minimum amount of work for carrying out the process. Additional work may be expended in associated irreversible processes such as heat generation. Related to W_c is the work of adhesion, $W_{a(12)}$, defined as the reversible work required to separate unit area of interface between two different materials (1 and 2) to leave two "bare" surfaces of unit area (Fig. 2.3*b*). The work is given by

$$W_{a(12)} = \sigma_1 + \sigma_2 - \sigma_{12} \tag{2.3}$$

where the subscripts refer to the two phases being separated and the σ values are the respective surface tensions.

The nature of the immediate environment of freshly formed surfaces will affect the actual excess surface free energy of the system, as illustrated in Figure 2.4. If a vacuum separates the surfaces, there are, naturally, no atoms or molecules present to interact with the "exposed" units. Those units, therefore, can be considered to have "bare" areas that represent a high energy situation relative to the bulk (Fig. 2.4, curve *a*). When in contact with an adjacent fluid phase (liquid or gas), surface units can interact to some extent with the fluid phase and thereby lose some of the excess energy they have gained by virtue of their position (Fig. 2.4, curve *b*). The greater the interaction between surface units and the adjacent phase, the greater the reduction in

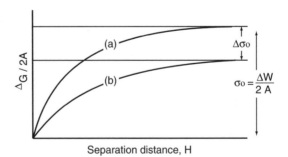

FIGURE 2.4. The excess surface free energy of newly formed surface will depend on the nature of the new phase it contacts. (*a*) If the new surface contacts a vacuum, the excess free energy will be maximized. (*b*) If another phase is present (liquid or gas) the excess surface energy will be reduced by an amount depending on the new interactions.

excess surface energy. The net result is that the excess energy of a surface will be greater in a vacuum than in the presence of a fluid. In some cases, such as liquid surface tensions, the difference between vacuum and vapor may be negligible. For high-energy solid surfaces, however, the difference can be significant, as will be seen in later chapters. Conversely, for liquid–liquid and solid–liquid interfaces where significant interactions take place, the interfacial tension can be quite low. Even solid–solid interfaces can, over time, show the results of mutual attraction across an interface in the form of sintering or spontaneous weld or joint formation.

Since the result of the formation of new interface is an increase in the free energy of the system, it should not be surprising that most systems will be thermodynamically driven to minimize interfacial area. A vivid illustration of the effect of surface thermodynamics is the picture of a blob of liquid forming itself into an almost perfect sphere when left to its own devices—that is, when no mechanical agitation, gravitational effects, and other forces are present. The technological consequences of interfacial thermodynamics are far reaching. It is our ability (or inability) to control interfacial effects that makes the study of surface and interface science such a technologically and economically important matter.

In addition to the tendency for liquids to form spherical droplets in order to minimize their surface area, it can be easily demonstrated that liquid surfaces have other properties that can be traced back to the concept of the work necessary to form new surface area. For example, if one takes a clean needle and carefully places it on the surface of pure distilled water, the needle will float, even though it has a density many times that of the water. In order for the needle to sink, it must penetrate the surface of the water. Penetration of the liquid surface involves increasing the water interfacial area with respect to both the vapor phase and the needle. The force causing the needle to sink, of course, is its mass times the acceleration due to gravity. Opposing it is the surface tension of the water.

A classical way to visualize the resultant phenomenon is to think of the liquid surface as having a membrane under tension stretched across it and supporting the needle. The concept of the stretched membrane gave rise to the picture of a "surface of tension" running parallel to the interface along the bulk phase. In fact the operative phenomenon is really an energy term, so that the surface tension is more correctly a surface energy. The two terms are often interchanged and for liquids are, as we have seen, numerically equal. The units employed are different although dimensionally equivalent—millinewtons per meter ($mN\ m^{-1}$) in SI units ($dyn\ cm^{-1}$ in older publications) for surface tension and millijoules per meter ($mJ\ m^{-2}$) (or $ergs\ cm^{-2}$) for energy.

If one thinks of the forces acting between molecules as being springs, one can visualize the situation as follows. In the bulk phase, molecules are being pulled and pushed from all sides by vibrating springs of equal strength. The time-average result is some equilibrium position for a specific molecule (Fig. 2.5). At an interface, the pull of springs into the bulk phase is stronger (in

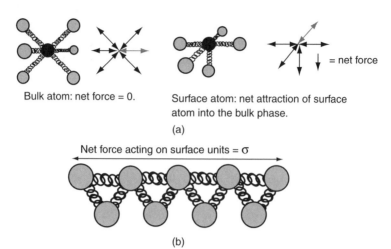

Bulk atom: net force = 0.   Surface atom: net attraction of surface atom into the bulk phase.

(a)

Net force acting on surface units = σ

(b)

FIGURE 2.5. Schematic representation of a "spring" model of surface energy: (*a*) for the individual atom, location at the surface results in an unbalanced force pulling it into the bulk; (*b*) for the surface in general, the summation of the individual attraction for the units produces the net effect of surface tension or surface energy.

general) than that due to springs in the adjacent phase. As a result, the interfacial molecules are pulled into the bulk (to the extent allowed by their finite size and repulsive interactions) and the net density of molecules in the surface region is decreased. There is more space between surface molecules and the springs acting between them are therefore stretched beyond their equilibrium length, creating a tension pulling along the surface working to keep the molecules together. The force of the springs pulling along the surface, then, is the surface tension or surface energy.

The application of the above concept to solid surfaces is not quite as straightforward. While it is certainly true that the forces and stresses experienced by atoms or molecules at the solid surface differ greatly from those in the bulk, those stresses will not usually be isotropic, as is (or is assumed to be) the case for more mobile liquid systems. If one defines the surface tension of a solid in the same way as that of a liquid, the tension must be expected, at least for a crystalline material, to depend on the direction in the surface being considered as well as the exact crystal structure of the surface. It should be immediately obvious that for a solid the idea of a homogeneous surface tension (in the sense of the spring analogy) can become quite complicated, and a completely satisfactory definition in those terms difficult to achieve. It is therefore more convenient (and more accurate) when talking about solid interfaces to speak directly in terms of energy and to avoid completely the concept of tensions. In that way many of the various conceptual problems associated with the normally heterogeneous nature of solid surfaces can be avoided.

In summary, the surface "energy" and "tension" for solids are not necessarily equivalent, and the energy term is most often used. The concept of "tension" is best applied to the interface between two fluid phases, while "energy" is most appropriate with respect to systems involving at least one solid phase. In addition, for solid systems, the actual surface will not generally be molecularly smooth. Rather, it will be irregular with different surface units being located in distinct environments relative to their neighbors (Fig. 2.6a). As a result, the free energies of the surface units will vary and the total excess surface free energy will be history-dependent and not uniform over the entire surface.

The presence of an asymmetric force field at the phase boundary is manifested in liquids by the apparent presence of a tension at the interface acting tangent to that region at the point of interest. It is conventional to consider that this surface tension resides in a narrow monomolecular region between the two phases. However, experimental evidence indicates that contributions can arise from second, third, and possibly even deeper molecular layers. For that reason it is most convenient at times to refer to a surface or interfacial "region" with the implication that more than one molecular layer must be considered. That can be even more true for solid surfaces in which unit dislocations from equilibrium may be evidenced tens or hundreds of unit lengths into the "bulk" phase. Such an approach can sometimes cause "philosophical" problems in the discussion of an interface using certain models and mathematical approaches. In reality, however, since we still do not fully understand all aspects of molecular interactions in interfacial regions, it is best not to dwell too much upon such apparent contradictions. Nature is full of apparent contradictions resulting from our own ignorance of the true situation. For the time being we must use what tools we have that seem to work and hope for further enlightenment in the future.

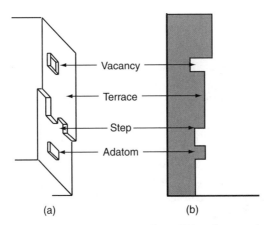

FIGURE 2.6. A "typical" solid interfacial region will be characterized by an irregular physical profile (a) and a parallel variable concentration profile (b).

2.2.2. Standard Reference States

The simplified description of surface energy given above is far from sufficient to fully explain all the surface and interfacial phenomena such as wetting, adhesion, and colloidal stability, that are of theoretical and practical importance. In fact, depending on the specific situation, it is often necessary, or at least convenient, to approach the question of surface interactions from completely opposite points of view. For example, when one is considering a question of colloidal stability, in which the desired effect is to prevent two surfaces from interacting in an attractive way (or at least reduce such interactions to a significant extent), it is convenient to think in terms of imposing a barrier, either energetic or physical, between the two interacting species that prevents or inhibits the dispersed state from passing to the energetically more favorable state of phase separation. For the case of adhesion, on the other hand, it is convenient to think in terms of increasing the net attractive interactions between the interfaces to be joined, so that it may be conceptually easier to consider the situation in terms of decreasing the interfacial energy between the surfaces.

In chemistry and physics it is customary to discuss energies with reference to some specified state. In other words, instead of stating an absolute energy (which may be difficult or impossible to determine) for a system, the change in energy relative to a standard state is measured. For example, the preceding discussion of surface energy was given in terms of an initial state of zero separation distance between two surfaces, going to a state of some "infinite" separation distance, H. It may be more useful, however, to think in terms of an initial state of infinite separation and measure energy changes as a function of the approach of two surfaces. Because each situation carries its own requirements there can be no set rules governing the choice of standard reference for all interfacial interactions. In each specific area of interest, it is important to define the starting point and be consistent throughout further operations.

2.2.3. The Molecular Nature of the Interfacial Region

It has been stated that the free energy of an interface results from asymmetric forces acting on atoms or molecules at or in the boundary region between phases. While the quantitative nature of those forces will be addressed in Chapters 4 and 5, it will be useful to develop the qualitative picture of the situation a bit more at this point. To begin with, let us assume that there are only three phases with which we need be concerned: solid, liquid, and vapor. We will for the moment neglect the vacuum "phase" and ignore the existence of the various classes of solids, including crystalline, quasi-crystalline, liquid crystalline, glass, and amorphous. In a practical context, the differences between the classes of solid surfaces cannot be ignored because that nature may greatly affect it surface properties. For now, however, we will attempt to keep life simple.

When two phases are in contact, there is a transition region of molecular dimensions in which the composition of the system changes from that of one phase to that of the other. In the case of a nonvolatile molecularly smooth solid surface in contact with an inert gas, the transition region will be essentially one molecule in thickness. Thus, there will be a sharp boundary at which the composition will change abruptly from molecules of the solid to molecules of the gas. For a more common irregular surface, the transition region will reflect the physical irregularities of the surface (Fig. 2.6*a*). A concentration profile of the region will reflect the presence or absence of solid phase units resulting from vacancies, steps, adatoms, etc., present on the surface (Fig. 2.6*b*).

For a pure liquid in contact with its vapor, the transition will be much less abrupt, going from a molecular density corresponding to the bulk material, through a zone where the unit concentration gradually decreases until the density reaches that of the pure vapor. In such a case, the transition region may be found to be several unit diameters thick (Fig. 2.7). At a mixed liquid–vapor interface, each component will have its own concentration profile depending on such factors as volatility and miscibility. For example, the vapor region directly adjacent to the liquid phase may have a higher concentration of liquid phase units which then decreases (relative to other vapor-phase components) with distance from the interface. A similar situation would hold for the interface between two liquid phases with some finite mutual solubility. In fluid systems, critical phenomena require that the interfacial region become thicker as the temperature of the system is increased, until the point where the critical temperature is reached and the two phases cease to exist as such. Thus, the interfacial transition region becomes less distinct as the physical differences between the two phases lessen until a single phase is obtained. Solid–liquid systems will also exhibit the concentration profiles similar to those noted above, although the details will depend on the solubility of the solid in the liquid (and vice versa).

Quantitative details of the concepts introduced above will be given in later chapters. As a beginning, however, it is important that one begins to get a "feel" for the nature of the beast one is to confront as a first step into the twilight zone.

FIGURE 2.7. For a liquid–vapor system, the interfacial region will be smooth with a narrow transition region and smooth concentration profile.

PROBLEMS

2.1. Calculate the total reversible thermodynamic work required to produce a spray of water of droplet diameter 2000 nm from 40 liters of water at 25°C. Take the surface tension of water as 72 mN m^{-1}. How many water drops would be produced, theoretically, if the droplets are all of equal size?

2.2. Explain why the experimentally determined work of cohesion for the cleavage of a solid crystalline material in vacuum will always be greater than that obtained by the same process in the presence of nitrogen.

2.3. It is generally found that the surface tension of a liquid decreases with an increase in temperature. Give a qualitative explanation for that phenomenon based on the general rules of molecular motion and the "spring" concepts presented in the chapter.

2.4. On the basis of the molecular "spring" concept presented above, would one expect to find a correlation among the surface tensions of typical liquids and other intensive characteristics such as boiling point, and freezing point? If so, support your point with some representative values from the literature.

3 Surface Activity and Surfactant Structures

Throughout the wide range of topics related to surfaces and colloids one encounters reference to chemical species that have a special propensity to concentrate (i.e., adsorb) at interfaces, or to form colloidal aggregates in solution at very low molar concentrations. Such materials are given the general name of surface active agents or surfactants. The physical chemistry of surfactants, in the specific context of interfaces and colloids, will be covered in subsequent chapters. This chapter will be devoted to a description of the structural aspects of surfactant molecules, that is, the atomic compositions and groupings which produce the observed physicochemical characteristics of such materials. For a short time, then, readers will need to put on recall their (for some) long filed away organic chemistry.

In the following discussion reference will often be made to the process of "adsorption." While the details of adsorption processes will be discussed in some detail in Chapter 9, the idea should be sufficiently obvious so that the concept of surface activity will be understandable in the present context.

3.1. BASIC STRUCTURAL REQUIREMENTS FOR SURFACE ACTIVITY

Surface-active materials (surfactants) possess a characteristic chemical structure that consists of (1) molecular components that will have little attraction for one surrounding (i.e., the solvent) phase, normally called the lyophobic group, and (2) chemical units that have a strong attraction for that phase—the lyophilic group (Fig. 3.1). Although, in principle, surface activity and related concepts are applicable to any system composed of at least one condensed phase, the bulk of the scientific and technological literature is concerned with aqueous solvents and their interaction with a second phase. As a result, the term hydrophobic will quite often be employed in place of the more general lyophobic; analogously, hydrophilic will be employed instead of lyophilic. It should always be kept in mind, however, that generality is implied in most discussions, even when the specific terms applicable to water-based systems are used.

Hydrophilic   Hydrophobic tail
head group

FIGURE 3.1. The basic molecular structure of a surface-active material includes the hydrophobic (or lyophobic) group having little attraction for water (or the solvent) and the hydrophilic (or lyophilic) group having strong interactions with water (or the solvent).

Materials that possess chemical groups leading to surface activity are generally referred to as being amphiphilic ("liking both"), indicating that they have some affinity for two essentially immiscible phases. When a surface-active material is dissolved in a solvent (whether water or an organic liquid), the presence of the lyophobic group causes an unfavorable distortion of the liquid structure, increasing the overall free energy of the system. In an aqueous surfactant solution, for example, such a distortion (in this case ordering) of the water structure by the hydrophobic group decreases the overall entropy of the system (Fig. 3.2). That entropy is regained when surfactant molecules are transported to an interface and the associated water molecules released. The surfactant will therefore preferentially adsorb at interfaces, or it may undergo some other process to lower the energy of the system (e.g., micelle formation). Since less work is required to bring surfactant molecules to an interface relative to solvent molecules, the presence of the surfactant decreases the work required to increase the interfacial area resulting in a decrease in interfacial tension.

The amphiphilic structure of surfactant molecules not only results in the adsorption of surfactant molecules at interfaces and the consequent alteration

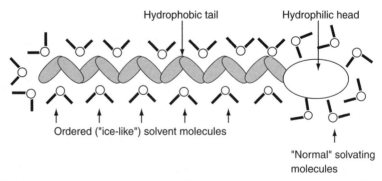

Hydrophobic tail   Hydrophilic head

Ordered ("ice-like") solvent molecules

"Normal" solvating
molecules

FIGURE 3.2. For a surfactant molecule in water, the hydrophobic tail will be "solvated" with an icelike structure of associated solvent molecules. The hydrophilic head will be solvated in the usual way.

FIGURE 3.3. When adsorption occurs at the interface, the adsorbed molecules will have a preferred orientation that tends to minimize unfavorable interactions between the aqueous phase and the surfactant molecular sections.

of the corresponding interfacial energies, but it will often result in the preferential orientation of the adsorbed molecules such that the lyophobic groups are directed away from the bulk solvent phase (Fig. 3.3). The resulting molecular orientation produces some of the most important macroscopic effects observed for surface active materials. Energetic considerations aside for the moment, it is important to understand the qualitative relationships between the nature of interfaces and the general chemical structures required for a molecule to exhibit significant surface activity.

The chemical structures having suitable solubility properties for surfactant activity vary with the nature of the solvent system to be employed and the conditions of use. In water, the hydrophobic group (the "tail") may be, for example, a hydrocarbon, fluorocarbon, or siloxane chain of sufficient length to produce the desired solubility characteristics when bound to a suitable hydrophilic group. The hydrophilic (or "head") group will be ionic or highly polar, so that it can act as a solubilizing functionality. In a nonpolar solvent such as hexane the same groups may function in the opposite sense. As the temperature, pressure, or solvent environment of a surfactant varies, significant alterations in the solution and interfacial properties of the system may occur. As a result, changes in conditions may require modifications in the chemical structure of the surfactant to maintain a desired degree of surface activity.

3.2. SURFACTANT STRUCTURES AND SOURCES

In order to understand the relationship between the surface activity of a given material and its chemical structure, it is important to understand the chemistry of the individual chemical components that in concert produce the observed phenomena. The following discussion will introduce some of the structural aspects of surfactants, ranging from basic raw materials and sources to the chemical group combinations that result in the observed surface activity. Since the chemical compositions and synthetic pathways leading to surface active molecules are limited primarily by the creativity and ingenuity of the synthetic chemist and production engineer, it is not possible to discuss all chemical classes, their preparation, and subtle variations. However, the majority of surfactants of academic and technological interest can be grouped into a rather limited number of basic chemical types and synthetic processes.

The chemical reactions that produce most surfactants are rather simple, understandable to anyone surviving the first year of organic chemistry. The challenge to the producer lies in the implementation of those reactions on a scale of thousands of kilograms, reproducibly, with high yield and high purity (or at least known levels and types of impurity), and at the lowest cost possible. With very few exceptions, there will always be a necessity to balance the best surfactant activity in a given application with the cost of the material that can be borne by the added value of the final product or process.

3.2.1. The Classification of Surfactants

Considering all of the possible chemical structures available to the synthetic chemist for surfactant design, it is necessary to have some system of classification to guide the user to the material best suited to immediate and future needs. It therefore seems reasonable to have clearly in mind where one wants to go before looking for the best route to get there.

Surfactants may be classified in several ways, depending on the intentions and preferences of the interested party (e.g., the author). One of the more common schemes relies on classification by the application under consideration, so that surfactants may be classified as emulsifiers, foaming agents, wetting agents, dispersants, or similar. For the user whose work is confined to one type of application, such a classification scheme has certain obvious advantages. It does not, however, say much about the specific chemical nature of a surfactant, nor does it give much guidance as to other possible uses of a material.

Surfactants may also be generally classified according to some physical characteristic such as it degree of water or oil solubility, or its stability in harsh environments. Alternatively, some specific aspect of the chemical structure of the materials in question may serve as the primary basis for classification; an example would be the type of linking group (oxygen, nitrogen, amide, etc.) between the hydrophile and the hydrophobe. Perhaps the most useful scheme from a general point of view, however, is that based on the overall chemical structure of the materials in question, in particular its ionic character. In such a classification system, it is easier to correlate chemical structures with interfacial activity, and thereby develop some general rules of surfactant structure–performance relationships.

The simplest structural classification procedure is that in which the primary type is determined by the nature of the solubilizing functionality (the lyophilic group or the hydrophile in aqueous systems). Within each primary classification by solubilizer, there will exist subgroups according to the nature of the lyophobic moiety. It is possible to construct a classification system as complex as one might like, breaking down the lyophobic groups by their finest structural details such as branching and unsaturation. Such extremes, however, can introduce unnecessary complications in any discussion of structure–performance relationships, especially since industrially important surfactant systems often consist of several isomers or homologues, or other complex mixtures.

In aqueous systems, which constitute by far the largest number of surfactant applications, the hydrophobic group generally includes a long-chain hydrocarbon radical, although there are useful examples using fluorinated or oxygenated hydrocarbon or siloxane chains (designated as R below). The hydrophilic group will be an ionic or highly polar group that gives some water solubility to the molecule. The most useful chemical classification of surface-active agents is based on the nature of the hydrophile; subgroups are defined by the nature of the hydrophobe. The four general groups of surfactants are defined as follows:

1. *Anionic,* with the hydrophilic group carrying a negative charge such as carboxyl ($RCOO^-\ M^+$), sulfonate ($RSO_3^-\ M^+$), or sulfate ($ROSO_3^-\ M^+$).

2. *Cationic,* with the hydrophile bearing a positive charge, as for example, the quaternary ammonium halides ($R_4N^+\ X^-$).

3. *Nonionic,* where the hydrophile has no charge but derives its water solubility from highly polar groups such as polyoxyethylene ($-OCH_2CH_2O-$), sugars or similar groups.

4. *Amphoteric* (and zwitterionic), in which the molecule has, or can have, a negative and a positive charge on the principal chain (as opposed to a counterion, M^+ or X^-) such as the sulfobetaines, $RN^+(CH_3)_2CH_2CH_2SO_3^-$

3.2.2. Building Blocks for Surfactant Molecules

Synthetic surfactants and the natural fatty acid soaps are amphiphilic materials that tend to exhibit some solubility in water as well as some affinity for nonaqueous solvents. As a basis for understanding the relationship between surfactant structures and surface activity, it is useful to work through the simple example of how changes in the polarity (ie, the head group) for a specified hydrocarbon chain affects its solubility and surface activity. As an illustration, consider the simple, straight-chain hydrocarbon dodecane,

$$CH_3(CH_2)_{10}CH_3$$

a material that is, for all practical purposes, insoluble in water.

If a terminal hydrogen in dodecane is exchanged for a hydroxyl group ($-OH$), the new material, n-dodecanol,

$$CH_3(CH_2)_{10}CH_2OH$$

still has very low solubility in water, but the tendency toward solubility has been increased substantially and the material begins to exhibit characteristics of surface activity. If the alcohol functionality is placed internally on the dodecane chain, as in 3-dodecanol,

$$CH_3(CH_2)_8CH(OH)CH_2CH_3$$

the resulting material will be similar to the primary alcohol but will have slightly different solubility characteristics (slightly more soluble in water). Those differences will generally be carried over in other functional modifications. The effects of the position of substitution on surfactant properties can be quite large and will be discussed in more detail later.

If the original dodecanol is oxidized to dodecanoic acid (lauric acid)

$$CH_3(CH_2)_{10}COOH$$

the compound still has limited solubility in water; however, when the acid is neutralized with alkali it becomes water soluble—a classic soap. The alkali carboxylate will be a reasonably good surfactant.

If the hydrocarbon chain length is increased to 16 carbons, its solubility will decrease, but many of the surfactant properties (e.g., foaming and detergency) will improve significantly. By altering the balance between the hydrophobic group—the hydrocarbon chain or the "tail"—and the hydrophilic group—the carboxylate or "head"—one can gain effective control over the surfactant characteristics of a molecule and, in theory at least, design a molecule to meet the specific requirements of a system. From a purely structural standpoint, then, the family of carboxylate soaps represents an almost ideal series of surfactants in that the hydrocarbon chain length of the tail can be easily varied over a wide range to produce the characteristics one needs for any application.

An important drawback to the use of the carboxylate soaps, however, has always been their great sensitivity to their aqueous environment. The main ionic components of "hard water" are calcium, magnesium, and other di- and trivalent salts. In the presence of such materials, the carboxylic acid soaps form salts of very low water solubility, that loose a good deal of their surface activity and precipitate to produce scum deposits, often described as "bathtub ring." On a molecular basis, the effectiveness as surfactants of the polyvalent carboxylate salts may not be greatly reduced from that of the alkali salts; however, their solubility in water is simply too low for the system to attain a high enough concentration to produce optimum results. An additional problem is that of the working temperature of a system. The solubility of carboxylate soaps is very temperature-dependent. As a result, they tend to lose solubility, and therefore concentration related surface activity, at low temperatures and are not generally well suited to cold-water applications.

Similar solubility problems are encountered in acidic media where the soap will be neutralized to produce the free carboxylic acid, which will have little surface activity and probably precipitate from solution. In nonaqueous solvents, the polyvalent salts of carboxylate soaps show a significantly enhanced solubility and perform admirably in many surfactant functions.

The inability of the carboxylate soaps to tolerate the presence of commonly encountered cations, as well as their sensitivity to pH changes, was one of the major driving forces for the development of synthetic surfactants (detergents) that are not so adversely affected by the common circumstances of hard water and cool washing temperatures, each detrimental to the effectiveness of the soaps.

The alcohol dodecanol already encountered can be sulfated by a relatively simple process to produce the dodecane sulfuric acid ester, having a carbon–oxygen–sulfur linkage

$$CH_3(CH_2)_{10}CH_2OSO_3H$$

a compound with relatively high water solubility. However, the ester retains strong acid characteristics that severely limit its potential utility as a surfactant. When the sulfuric acid ester is neutralized with alkali, organic amines, or other basic reagents the resulting salt

$$CH_3(CH_2)_{10}CH_2OSO_3^- \; M^+$$

is highly soluble in water and will be an excellent surfactant for many applications.

By modifying the synthetic process, the parent hydrocarbon can be sulfated to yield dodecane sulfonic acid, having a direct carbon–sulfur bond

$$CH_3(CH_2)_{10}CH_2SO_3H$$

which closely resembles the sulfuric acid ester and has similar miscibility with water. When neutralized with the proper base, the resulting material

$$CH_3(CH_2)_{10}CH_2SO_3^- \; M^+$$

is an excellent surfactant. It should be noted that while the sulfonic acid is related to the sulfate ester, their chemical, solution, and surfactant properties are not identical, so that their potential applications may be different as well. The organic sulfate and related sulfonate salts are probably the most extensively studied and best understood class of synthetic surfactants.

As a branch of the surfactant family tree, if the tried-and-true dodecanol is treated with ethylene oxide (OE) and base under rigid anhydrous conditions, the material obtained is an alkyl polyoxyethylene (POE) polyether.

$$C_{11}H_{23}CH_2-O-CH_2CH_2(OCH_2CH_2)_nOCH_2CH_2OH$$

The polyether surfactants have widely varying solubility characteristics, depending on the value of n, the number of OE groups added to the molecule. For the dodecanol derivative, if $n = 10$, the material will exhibit significant solubility in water and will have good surfactant properties. If n is as little as 5, its water solubility will decrease significantly, as will its usefulness as a surfactant. If n is increased to 20 or higher, high water solubility will be maintained, but most of the good surfactant qualities will be lost.

If our original dodecane molecule is chlorinated to produce dodecyl chloride

$$CH_3(CH_2)_{10}CH_2Cl$$

and subsequently reacted with an amine such as trimethylamine, $(CH_3)_3N$, the resulting compound, dodecyltrimethylammonium chloride

$$CH_3(CH_2)_{10}CH_2N^+(CH_3)_3Cl^-$$

is a water-soluble compound exhibiting some surfactant properties, although its properties are not generally as useful as the anionic analogs. The utility of such compounds is limited not so much by their surface activity, however, as by their interaction with various oppositely charged components found in practical systems. That topic will be discussed in more detail below.

To this point we have covered the first three of the four general classes of surfactants defined previously. To produce an example of the fourth class, an amphoteric or zwitterionic surfactant, it is only necessary to react dodecyl chloride with a difunctional material such as N,N-dimethyl-3-aminopropane-1-sulfonic acid

$$(CH_3)_2NCH_2CH_2CH_2SO_3H$$

The resulting material

$$CH_3(CH_2)_{10}CH_2N^+(CH_3)_2CH_2CH_2CH_2SO_3^-$$

is just one of several possible chemical types that possess the amphoteric or zwitterionic character.

As one can see, the number of modifications of the simple dodecane molecule that can lead to materials with good surfactant characteristics is limited primarily by the imagination and skill of the organic chemist—and by the time and money available for indulgence in creative molecular architecture. In each example discussed, a solubilizing group has been added to the basic hydrophobe to produce materials with varying amounts of useful surfactant characteristics. When one considers the wide variety of hydrophobic groups (instead of dodecane) that can be employed with the hydrophiles discussed, the number of possible combinations becomes impressive. For that reason, listings of commercially available surfactants number in the thousands.

3.2.3. Surfactant Solubilizing Groups

To expand on the ideas presented above, one can say that the solubilizing groups of modern surfactants fall into two general categories: those that ionize in aqueous solution (or highly polar solvents) and those that do not. Obviously, the definition of what part of a molecule is the solubilizing group depends on the solvent system being employed. For example, in water the solubility will be determined by the presence of an ionic or highly polar group, while in organic systems the active group (in terms of solubility) will be the organic "tail." It is important, therefore, to define the complete system under consideration before discussing surfactant types.

The functionality of ionic hydrophiles derives from a strongly acidic or basic character, which, when neutralized, leads to the formation of true, highly ionizing salts. In this context, the carboxylic acid group, while not generally considered as such, is classified as a strong acid. A weak acid would be an alcohol or phenol. The nonionic hydrophilic groups, on the other hand, have functional groups that are individually rather weak hydrophiles but have an additive effect so that increasing their number in a molecule increases the magnitude of their solubilizing effect.

The most common hydrophilic groups encountered in surfactants today are illustrated in Table 3.1, where R designates some suitable hydrophobic

TABLE 3.1. The Most Commonly Encountered Hydrophilic Groups in Commercially Available Surfactants

Class	General Structure
Sulfonate	$R-SO_3^- \, M^+$
Sulfate	$R-OSO_3^- \, M^+$
Carboxylate	$R-COO^- \, M^+$
Phosphate	$R-OPO_3^- \, M^+$
Ammonium	$R_x H_y N^+ X^-$ $(x = 1-3, \ y = 4-x)$
Quaternary ammonium	$R_4 N^+ X^-$
Betaines	$RN^+(CH_3)_2 CH_2 COO^-$
Sulfobetaines	$RN^+(CH_3)_2 CH_2 CH_2 SO_3^-$
Polyoxyethylene (POE)	$R-OCH_2 CH_2 (OCH_2 CH_2)_n OH$
Polyols	Sucrose, sorbitan, glycerol, ethylene glycol, etc
Polypeptide	$R-NH-CHR-CO-NH-CHR'-CO-$ $\ldots -CO_2H$
Polyglycidyl	$R-(OCH_2 CH[CH_2 OH]CH_2)_n - \ldots -$ $OCH_2 CH[CH_2 OH]CH_2 OH$

group that imparts surface activity, M is an inorganic or organic cation, and X is an anion (halide, acetate, etc.). The list is in no way complete, but the great majority of surfactants available commercially fall into one of the classes.

It is possible, and sometimes even advantageous, to combine two or more functionalities to produce materials with properties superior to a monofunctional analog. Prime examples of that would be the alcohol ether sulfates in which a POE nonionic material is terminally sulfated

$$R(OCH_2CH_2)_n OSO_3^- \ M^+$$

and, of course, the zwitterionic and amphoteric materials already mentioned, which often exhibit the advantages of both ionic and nonionic surfactants while having fewer of their potential drawbacks. The "hybrid" classes of surfactants, while not yet composing a large fraction of total surfactant use, can be particularly useful because of their flexibility and, especially in personal care items such as shampoos, because of their low level of eye and skin irritation.

3.2.4. Common Surfactant Hydrophobic Groups

By far the most common hydrophobic group used in surfactants is the hydrocarbon radical having a total of 8–20 carbon atoms. Commercially there are two main sources for such materials that are both inexpensive enough and available in sufficient quantity to be economically feasible: biological sources such as agriculture and fishing, and the petroleum industry (which is, of course, ultimately biological). Listed below and illustrated structurally in Figure 3.4 are the most important commercial sources of hydrophobic groups, along with some relevant comments about each. There are, of course, alternative synthetic routes to the same basic molecular types, as well as other surfactant types that require more elaborate synthetic schemes. Those shown, however, constitute the bulk of the synthetic materials used today. Each source of raw materials may have its own local geographic or economic advantage, so that nominally identical surfactants may exhibit slight differences in activity due to the subtle influences of raw materials variations and processing parameters. Such small variations in surface activity may not be important for most applications but should be kept in mind in critical situations.

Natural Fatty Acids. Obtained primarily from the hydrolysis of triglycerides such as animal and vegetable fats and oils, the most useful members of the group have from 12 to 18 carbon atoms. The most commonly employed members of the family have hydrocarbon chains that are fully saturated, although some unsaturated examples are employed (especially oleic acid). Because most vegetable oils contain high proportions of unsaturated fatty acids, they usually require extensive processing in order to separate the useful

FIGURE 3.4. The most commonly encountered hydrophobic materials used in the commercial manufacture of surfactants.

isomers for soap and detergent production. The saturated fatty acid content can be increased, however, by hydrogenation. The alkyl chains usually have an even number of carbons and very little branching or heteroatom (not carbon or hydrogen) substitution except for the carboxyl group itself. An important exception is recinoleic acid (from castor oil) which is unsaturated and contains a hydroxyl group near the middle of the chain. Other hydroxy acids, such as lactic, tartaric, malic, and citric, are becoming more important as components in surfactant structures for use in foods, pharmaceuticals, and cosmetics, although their function is more in the hydrophile than the hydrophobic group.

Paraffins. These hydrocarbon materials are obtained from petroleum distillates boiling higher than gasoline. They are generally saturated materials with

10–20 carbon atoms. The mixture will normally contain branched isomers, some cyclic materials, and aromatic groups (benzene, toluene, naphthalene, etc.). Since saturated hydrocarbon materials in this family are relatively unreactive, their conversion to surfactants usually involves a preliminary reaction such as chlorination that, by its somewhat random nature, results in the production of a more complex mixture of isomers. The production of a relatively pure product requires further purification and fractionation, increasing the cost of such materials. Paraffin-based materials are therefore commonly encountered as complex mixtures of isomers and homologues.

Olefins. The surfactant range olefins (C_{10}–C_{20}) are generally mono-unsaturated, with the exact point of unsaturation being determined randomly or, in some cases, by the specific process employed. When the unsaturation occurs predominantly at the C-1 position they are commonly referred to as α-olefins. They are conveniently prepared by oligomerization of ethylene or propene, or by cracking of higher-molecular-weight petroleum fractions. Because of their more uniform isomeric content, relative to the "natural" paraffin fractions, the α-olefins are a more desirable starting material for surfactant synthesis, where cost and availability allow. They are also good raw materials for the synthesis of the next class of materials discussed.

Alkylbenzenes. These materials are industrially prepared by the Friedel–Crafts reaction between olefins and benzene. The reaction may be carried out using surfactant-range olefins, or by using olefins such as propene or butene under conditions where a limited amount of oligomerization accompanies the alkylation. Generally the final alkyl group will contain an average of 8–12 carbon atoms. The products will be highly branched, with the benzene being substituted randomly along the hydrocarbon chain. Once the basic hydrocarbon group is prepared (e.g., dodecylbenzene, $C_{12}H_{25}C_6H_5$) the aromatic ring is further functionalized by sulfonation, etc., to produce a "detergent" such as dodecylbenzene sulfonate.

Alkyl Aromatics. Related to the alkyl benzenes discussed above are materials based on the so-called polynuclear aromatics such as naphthalene and anthracene. Of this class, the primary surfactant samples prepared and used industrially are the alkylnaphthalene sulfonates. They usually consist of mixtures of mono-, di-, and trialkyl naphthalene sulfonates with the alkyl group usually being in the C_2–C_4 range.

Alcohols. Long-chain alcohols (C_8–C_{18}) are found in nature, but generally in such limited quantities that their direct use is economically prohibitive. They may be readily prepared in industrial quantities, however, by the catalytic reduction of fatty acids or fatty acid esters, by the oxidation of oligomers of ethylene or other light olefins, or by oxidation of α-olefins. They may have an even or odd number of carbon atoms, depending on the raw materials source, with significant amounts of secondary alcohols usually being present.

Alkylphenols. These materials are produced by the reaction of phenol with olefins. The products are mixtures of linear and branched alkyl chains with random substitution of the ring along the chain and mixed-ring substitution with respect to the hydroxyl group (*ortho-* , *meta-* , and *para-*).

Polyoxypropylenes. This is the most important class of hydrophobic groups containing noncarbon atoms in the primary chain. The most important members of the class are prepared by the base-catalyzed oligomerization of propylene oxide (OP). The degree of hydrophobicity of the chain is controlled by the degree of oligomerization, that is, the average number of propylene oxide groups combined. The nature of the reaction makes it relatively easy to control the chain length and produce materials with well-defined characteristics and narrow distributions of isomers. The polyoxypropylene (OP_n) materials are particularly important in the preparation of block copolymer surfactants with ethylene oxide (OE) with the basic structure

$$OP_nOE_mOP_n \quad \text{or} \quad OE_mOP_nOE_m$$

where m and n refer to the average number of oxide units incorporated in each section. The flexibility of the reactions involved makes it possible to tailor the characteristics of the product to suit specific needs while using one set of raw materials, one basic reaction type, and maintaining the same general chemical nature.

Fluorocarbons. Fluorocarbons are prepared primarily by the electrolytic substitution of fluorine for hydrogen on the carbon chain of carboxylic acid fluorides or sulfonyl fluorides. They may be completely fluorinated (perfluoro-) or have a terminal hydrogen atom. In that respect, it is important to know which type of chain is present in a material, since the properties of the two may differ significantly in critical applications. They are also prepared by the oligomerization of tetrafluoroethylene. Linkage to many hydrophilic groups is accomplished through a short-chain hydrocarbon unit.

The fluorocarbon surfactants, while expensive, have much greater surface activity than hydrocarbon analogues, produce greater surface tension reductions, and usually withstand harsher environmental conditions without significant degradation. They also sometimes exhibit other unusual and useful characteristics that justify their added expense.

Silicones. These are generally oligomers of dimethylsiloxane attached to an appropriate solubilizing group. While they are expensive relative to the hydrocarbon materials, they often show surface activity in organic systems as well as other advantages similar to the fluorocarbons mentioned above, thereby justifying their additional cost.

3.3. THE ECONOMIC IMPORTANCE OF SURFACTANTS

Economic considerations can often be almost as important as surface activity in selecting a surfactant for a given application. Unless the cost of the surfactant is insignificant compared to the rest of the system, the least expensive material producing the desired effect will usually be chosen. Economics, however, cannot be the only factor in the choice, since the final performance of the system may well be of crucial importance. To make a rational selection, without resorting to an expensive and time-consuming trial-and-error approach, the formulator should have some knowledge of (1) the surface and interfacial phenomena that must be controlled; (2) the characteristic chemical and physical properties of the available surfactant choices; (3) the relationships between the structural properties of the available surfactants and their effects on the pertinent interfacial phenomena; (4) any restrictions to the use of available materials, as in, for example, foods, cosmetics, or pharmaceuticals; and (5) economic constraints on the choice of surfactant.

For example, a developer who requires a surfactant that has good detergent action, but low foaming characteristics and produces little or no skin irritation for application in a hospital environment must look beyond the classical alkylbenzene sulfonate detergents (usually high foaming with relatively high possibilities of producing skin irritation) and perhaps consider a nonionic material that costs a little more, but produces little foam and little or no irritation. In such an application, cost should have less importance than function.

As another example, the early synthetic detergents were found to produce serious environmental problems (i.e., foaming in rivers and effluent streams). The residual effects were traced to the low biodegradability of the highly branched alkyl chains in the hydrophobic group, derived from the oligomerization of propene and 2-butene. After extensive study, linear chain alkylbenzene sulfonates prepared from α-olefins were found to be much more easily degraded, produced fewer residue problems, and have largely replaced the branched analogues in commercial detergents.

The applications of surfactants in science and industry are legion, ranging from primary processes such as the recovery and purification of raw materials in the mining and petroleum industries, to enhancing the quality of finished products such as paints, cosmetics, pharmaceuticals, and foods. Table 3.2 lists some of their major areas of application. As the technological and legal demands placed on products and process additives such as surfactants increase, it seems obvious that our need to understand the relationships between the chemical structures of those materials and their physical manifestations in particular circumstances becomes more important.

TABLE 3.2. Some of the Major Modern Applications of Surfactants

Industrial	Consumer
Agricultural crop applications	Adhesives
Building materials	Cleaning fluids
Cement additives	Cosmetics
Coal fluidization	Disinfectants
Coating and leveling additives	Foods and beverages
Electroplating	Household cleaning and laundering
Emulsion polymerization	Paints
Graphic arts	Pharmaceuticals
Industrial cleaning	Photographic products
Leather processing	Soaps, shampoos, creams
Lubrication	Waxes and polishes
Mold release agents	
Ore flotation	
Paper manufacture	
Petroleum recovery	
Printing and printing inks	
Surface preparations	
Textiles	
Waterproofing	

For many of the applications noted in Table 3.2, the desired properties will vary significantly. For that reason, such characteristics as solubility, surface tension reducing capability, critical micelle concentration (cmc), detergency power, wetting control, and foaming capacity may make a given surfactant perform well in some applications and less well in others. The "universal" surfactant that meets all of the varied needs of surfactant applications has yet to emerge from the industrial or academic laboratory.

There have been developed over the years a number of useful generalizations relating surfactant structures to their activity in a given application. Some of those generalizations are pointed out in the appropriate context in later chapters. For now, it is enough to remember that each application may have specific requirements that determine the utility of a particular structure. Some of the fundamental characteristics that must be evaluated for a surfactant proposed for some specific applications are given in Table 3.3.

When discussing the commercial aspects of surfactant technology, especially with regard to raw-materials sources, it is common to refer to materials on the basis of their original starting materials. While such classifications may be useful from economic and technological points of view, the complex natures of most materials, especially the hydrophobic groups, make it very difficult to illustrate the role of chemical structures in determining surfactant properties. It should always be kept in mind, therefore, that nominally identical surfactants derived from different raw materials sources may exhibit significant differences in activity due to different isomer distributions.

TABLE 3.3. Typical (But Not All) Characteristics for Surfactants that Must Be Evaluated for Various Applications

Application	Characteristics
Detergency	Low cmc, good salt and pH stability, biodergradability, good foaming properties
Emulsification	Proper HLB,[a] environmental and biological (safety) aspects for application
Foods	Biocompatibility, functionality, sensory perception
Lubrication	Chemical stability, absorption at surfaces
Mineral flotation	Proper adsorption characteristics on the ore(s) of interest, low cost
Petroleum recovery	Proper wetting of oil-bearing formations, microemulsion formation and solubilization properties, ease of emulsion breaking after oil recovery
Pharmaceuticals	Biocompatibility, toxicity

[a] Hydrophile–lipophile balance.

3.4. SURFACTANTS IN THE ENVIRONMENT

The use of surfactants is increasing at a rate in excess of the population growth because of generally improved living conditions and processed material availability in the less industrially developed Third World countries. Hand in hand with increased surfactant use go the problems of surfactant disposal. As the more developed nations have learned by painful and expensive experience, the ability of an ecosystem to absorb and degrade waste products such as surfactants can significantly affect the potential usefulness of a given material.

Of particular importance are the effects of surfactants on groundwater and waste treatment operations. Although it may be technologically possible to remove all detectable residual surfactants physically or chemically from effluent streams, the economic costs would undoubtedly be totally unacceptable. When possible, the preferred way to address the problem is to allow nature to take its course and solve the problem by natural biodegradation mechanisms.

3.4.1. Biodegradation of Surfactants

Biodegradation may be defined as the removal or destruction of chemicals through the biological action of living organisms. For surfactants, such degradation may be divided into two stages: (1) primary degradation, leading to modification of the chemical structure of the material sufficient to eliminate any surface-active properties and (2) ultimate degradation, in which the material is completely removed from the environment as carbon dioxide, water, inorganic salts, or other materials that are the normal waste byproducts of biological activity. Years of research indicate that it is at the first stage of primary degradation that the chemical structure of a surfactant molecule most heavily impacts biodegradability.

As already mentioned, some of the earliest observations on the biodegradability of synthetic surfactants indicated that linear secondary alkyl sulfates (LAS) were biodegradable, while the branched alkylbenzene sulfonates (ABS) in extensive use at the time were much more resistant to biological action. Continued investigation showed that the distinction between the LAS and ABS surfactants was not nearly as clear as first thought; that is, the observed differences in biodegradability did not stem directly from the presence of the benzene ring in ABS systems. It was found, rather, that the biodegradability of a particular ABS sample depended on the source, and therefore the chemical structure, of the sample. Early producers of ABS surfactants used either petroleum-derived kerosene (largely linear) or tetrapropylene (highly branched) as their basic raw material, without great consideration for the structural differences between the two. As a result, great variability was found in the assay of materials for determination of biodegradability. In fact, those materials derived from tetrapropylene showed little degradation while the nominally identical materials based on the kerosene feedstocks were much more acceptable in that respect. The difference lay in the degree of branching in the respective alkyl chains.

It was subsequently shown conclusively that the resistance of tetrapropylene ABS surfactants to biodegradation was a result of the highly branched structure of the alkyl group relative to that of the kerosene-derived materials and the linear alkyl sulfates. As a result of extensive research on the best available model surfactant compounds it was concluded that it was the nature of the hydrophobic group on the surfactant that determined its relative susceptibility to biological action, and that the nature and mode of attachment of the hydrophile was of minor significance. Subsequent research using an increasingly diverse range of molecular types has continued to support those early conclusions.

3.4.2. Rules for Biodegradation

Over the years, the following generalizations have been developed to explain the biodegradation of most surfactant types:

1. The chemical structure of the hydrophobic group is the primary factor controlling biodegradability; high degrees of branching, especially at the alkyl terminus, inhibit biodegradation.
2. The nature of the hydrophilic group has a minor effect on biodegradability.
3. The greater the distance between the hydrophilic group and the terminus of the hydrophobe, the greater is the rate of primary degradation.

The relationship between the chemical structure of a surfactant and its resultant surface-active properties is quite complex. Many of the following chapters will point out general rules relating the two types of information. In the end, it will usually be firsthand experience that leads to a final decision on the selection of a surfactant for a given end use. That "experience" can be made less painful, however, by the application of "rules of thumb" and chemical common sense.

PROBLEMS

3.1. Using basic principles of organic synthesis, suggest a process for the synthesis of sodium dodecylsulfate $(CH_3(CH_2)_{11}OSO_3^-Na^+)$ from lauric acid.

3.2. Detergents were originally developed to replace carboxylate soaps for use under conditions of "hard" water and low temperatures. Why do detergents work better under those conditions?

3.3. Most higher-molecular-weight carboxylic acids do not have sufficient water solubility to be effective surfactants and must be neutralized with alkali to produce classic soaps. Strongly acidic materials such as alkylsulfonic acids $(R-SO_3H)$ and sulfuric acid esters $(R-OSO_3H)$, however, are usually surface active as the free acid, although normally employed as the alkali salt (detergents). If R is taken as a C_{16} hydrocarbon chain, would you expect the following reaction product to be a good surfactant:

$$R\text{-}SO_3H + R_3N \rightarrow R\text{-}SO_3^- \ {}^+NHR_3$$

Explain your reasoning based on general concepts of solutions, solubility, and so on.

3.4. Suggest why highly branched alkyl chains in a surfactant molecule result in a longer persistence than a straight-chain analog in lakes, rivers, and underground water tables.

3.5. Explain why amphoteric surfactants, while having discrete electrical charges, tend to behave more like nonionic than ionic surfactants.

3.6. Alkylphosphonic acids and salts $(R-PO_3H/M^+)$ and phosphoric acid esters of long chain alcohols $(R-OPO_3H/M^+)$ can make good surfactants, but in fact represent a small percentage of commercial products. Suggest some reasons for their lack of "popularity."

3.7. Other things being equal, which would probably represent a more desirable source of raw materials for detergent synthesis—animal, vegetable, or mineral feedstocks?

3.8. Most commercial fluorocarbon surfactants are produced by the electrolytic substitution of fluorine for hydrogen on the carbon backbone of a carboxylic acid fluoride or sulfonic acid fluoride

$$H_3C-(CH_2)_n-COF + HF/NaF + e^- \rightarrow F_3C-(CF_2)_nCOF$$

$$H_3C-(CH_2)_n-SO_2F + HF/NaF + e^- \rightarrow F_3C-(CF_2)_nSO_2F$$

followed by hydrolysis to the acid and neutralization. Other materials may be prepared by the oligomerization of tetrafluoroethylene to produce the alcohol followed by further reaction

$$F_2C=CF_2 + OH^- \rightarrow HO(CF_2-CF_2)_nH$$

Would you expect two surfactants, one prepared by each process, having the same carbon chain length and hydrophilic group, to be equivalent in terms of surface activity? Explain.

4 Attractive Forces

4.1. CHEMICAL AND PHYSICAL INTERACTIONS

When two atoms bind to form a typical nonionized molecule, the forces involved in bond formation are referred to as covalent forces, and the resulting bonds, covalent bonds. In such bond formation, the electrons involved are shared by two (or more) atoms and the individual characters of the atoms are (so far as the shared electrons are concerned), to some extent, lost. The formation of metallic bonds also involves the sharing of electrons among a number of atoms so that the nature of the "bonded" atoms differs from that of a "free" atom of the same metal.

We know from the general principles of chemistry that the exact nature and number of covalent bonds formed by a given atom depends on its location in the periodic table, relative to the other atoms involved. In addition, covalent bonds have certain characteristic bond lengths and bond angles which depend on the atoms involved; that is, they are directional. The number of bonds an atom may form is a fundamental property of that atom and controls how it contributes to the overall structure of the resulting molecule. The number, length, and direction of bonds to an atom will control how the atoms and resulting molecules can arrange themselves in three-dimensional patterns or lattices. For example, the special bonding properties of the carbon atom results in its ability to form the perfectly regular three-dimensional lattice structure of diamond or the more two-dimensional structure of graphite, as well as determining the three-dimensional structure of all carbon-based organic compounds.

Obviously, covalent bonding is of primary importance to the nature of things as we know them. However, covalent bonds are very localized in the bonding regions between atoms, and are short range in the sense that they act over bond distances of 0.1–0.2 nm. Most covalent bond energies fall in the range of 150 to 900 kJ mol^{-1} ($\approx$100-300 kT), and generally decrease in strength as the bond length increases. Some typical covalent bond energies are given in Table 4.1.

While covalent bonds are strong, they are restricted in the range of their actions. They are, in effect, limited to the interactions between atoms involved in molecular formation and formal chemical reactions. They may be considered, as a first approximation, strictly chemical forces. In most systems involving surface and colloidal phenomena, on the other hand, one is not so much

TABLE 4.1. Characteristic Strengths of Covalent Bonds

Bond Type	Strength (kJ mol^{-1})	Bond Type	Strength (kJ mol^{-1})
F—F (F$_2$)	150	C—H (CH$_4$)	430
N—O (NH$_2$OH)	200	O—H (H$_2$O)	460
C—O (CH$_3$OH)	340	C=C (C$_2$H$_4$)	600
C—C (C$_2$H$_6$)	360	C=O (HCHO)	690
Si—O	370	C≡N (HCN)	870

concerned with intramolecular forces as with intermolecular (or interatomic) forces acting between discrete, nonbonded atoms or molecules over distances significantly greater than molecular bond dimensions (tens to thousands of nanometers)—they are therefore generally nondirectional, nonstochiometric, long-range forces. Interactions due to long-range forces are sometimes referred to as "physical" interactions, implying that no formal chemical reaction is involved. While physical interactions do not, in general, involve electronic transformations analogous to covalent bond formation, they can, under some circumstances, be equally strong. While physical interactions may perturb the electronic configurations of the atoms or molecules involved, the electrons themselves remain associated with their original system. (An exception to this would be heterogeneous catalytic reactions, to be addressed later.)

While physical interactions lack the qualifications to be considered classic chemical "bonds," it is through such interactions that all but the most elemental properties of most materials in our world are determined. They also exhibit themselves as the fundamental factors involved in all colloidal systems, all biological assemblies, and all natural phenomena not involving chemical interactions. That covers quite a lot of territory! Because of their importance, physical interactions will be covered in somewhat more theoretical detail below than most topics to be presented in the following chapters.

4.2. THE IMPORTANCE OF LONG-RANGE PHYSICAL FORCES

We have seen from the discussion in Chapter 2 that as two uncharged surfaces in a vacuum are separated to an effectively infinite distance, their free energies increase to some maximum value characteristic of the system involved. This means that the net force acting between the two surfaces must be attractive. That will always be the case for pure substances in vacuum, but in reality, many situations exist in which a maximum in the free energy-separation distance curve is encountered.

Many important systems and processes, especially biological assemblies such as cell walls and protein secondary and tertiary structures form as a result of physical, inter- and intramolecular interactions. Such assemblies and conformations exist because the physical forces binding them together operate over distances greater than those of covalent bonds yet hold the various molecules at the proper distance and with the proper strength so that they can successfully carry out their vital functions.

Under certain circumstances it is possible to utilize physical interactions to maintain surfaces at some minimum distances of separation as a result of an energy maximum in the interaction energy. The practical result of such long-range energy maxima is that, properly utilized, they can prevent or at least retard the natural tendency of surfaces to approach and join spontaneously, thereby reducing interfacial area. This effect is especially important in colloids, foams, emulsions, and similar systems.

Although our aim is to understand interactions at interfaces, it must be remembered that an interface is nothing more than a collection of individual atoms or molecules, and that its macroscopic properties will be a reflection of the interactions of all of the individual atomic or molecular interactions involved. Therefore, we will begin the discussion by addressing the question of the source and nature of the various types of interactions experienced by individual units (atoms or molecules), followed by an integration of those interactions over all of the units in the surface.

4.3. CLASSIFICATION OF PHYSICAL FORCES

The fundamental physical forces controlling the nonchemical or physical interactions among atoms and molecules are of two kinds—formal coulombic or electrostatic interactions, and those lumped together under the general term of van der Waals forces. The place of coulombic interactions can be somewhat unclear in terms of their classification as chemical or physical interactions because they are, of course, involved in chemical bonding in ionic molecules. However, they may also be classified as physical interactions because they function over distances much greater than those of covalent bonds and produce dramatic effects on the interactions among ions, among ions and polar molecules, and among ions and nonpolar molecules.

Coulombic interactions are by far the strongest (in absolute terms) of the physical interactions, equaling and exceeding the magnitude of covalent bonds. They are not, however, the most widely encountered type of interaction, since they are present only in systems containing charged species. Nor are they even always the most important, since many factors can greatly diminish or completely nullify their net effect in a system.

The term "van der Waals forces" is often encountered in contexts where it implies that only one type of interaction is involved. In fact, however, van der Waals forces include three separate types of atomic and molecular interactions, each of which has its own characteristics, its own theoretical basis, and its own limitations. Two of the three forces are reasonably easy to understand because they are based on relatively straightforward electrostatic principles similar to those used for the much stronger coulombic interactions. The third force is sometimes less clear because it is quantum mechanical in origin, and quantum mechanics, even is its simplest form, seems to adversely effect the sanity of many who venture into its realm. In any case, the following discussion will attempt to present the ideas involved as simply as possible while still conveying the essence of the subject.

For reasons of convenience, the discussion will begin with the "simplest" long range force, that of direct coulombic interaction between two charged species. The presentation of the material has been kept brief because the intention is to provide a conceptual understanding of the basis of charge–charge interactions, leaving the fine points and complications (of which there can be many) to more advanced texts on the subject referenced in the Bibliography. Other interactions involving ions will be addressed as well.

The interactions generally grouped together as van der Waals forces will then be covered more or less in order of increasing complexity—that is, interactions of permanent dipoles, induced dipolar interactions, and finally the quantum mechanical forces.

4.3.1. Coulombic or Electrostatic Interactions

The interaction between two charged atoms or molecules is potentially the strongest form of physical interaction to be considered at interfaces and in colloidal systems. The basic concepts and equations involved are fundamental to many areas of physics and chemistry and will not be developed in detail here. More will be said about them in Chapter 5 in the context of repulsive interactions between surfaces. A few basic points of review, however, will be useful in order to facilitate reference to them in later discussions of dipolar ininteractions.

For two point charges Q_1 and Q_2, the free energy of interaction $w(r)$ is given by

$$w(r) = \frac{Q_1 Q_2}{4\pi\varepsilon_0\varepsilon r} = \frac{z_1 z_2 e^2}{4\pi\varepsilon_0\varepsilon r} \tag{4.1}$$

where ε_0 is the permittivity of a vacuum or free space, e is the relative permittivity or dielectric constant of the medium, and r is the distance between the two charges. The right-hand form of the equation is commonly used, where the value of Q can be readily specified in terms of the sign and valency of each ion, z, and the elementary charge, e ($= 1.602 \times 10^{-19}$ C) (coulombs).

The force of the coulombic interaction, F_c, is the differential with respect to r of the free energy

$$F_c = \frac{dw(r)}{dr} = \frac{Q_1 Q_2}{4\pi\varepsilon_0\varepsilon r^2} = \frac{z_1 z_2 e^2}{4\pi\varepsilon_0\varepsilon r^2} \tag{4.2}$$

For two charges of the same sign, both $w(r)$ and F will be positive, which means that the interaction will be repulsive; for unlike charges they will be attractive. In terms of magnitude, the force (whether attractive or repulsive) is at a maximum when the distance of separation r is a minimum, that is, when the two ions are in contact and r equals the sum of the two ionic radii. For example, for a sodium and a chloride ion in contact, r will be ≈ 0.276 nm, and the binding energy will be

$$w(r) = \frac{(-1)(+1)(1.602 \times 10^{-19})^2}{4\pi(8.854 \times 10^{-12})(0.276 \times 10^{-9})} = -8.4 \times 10^{-19} \text{ J}$$

Throughout the following discussions, reference will be made to a standard unit of thermal energy, kT, where k is Boltzmann's constant and T is absolute temperature (K). The reference energy at room temperature (≈ 300 K) will be $kT = (1.38 \times 10^{-23})(300) = 4.1 \times 10^{-21}$ J. The energy of the sodium and chloride atom interaction, then, is approximately $200kT$. From Table 4.1, an average covalent bond energy may be estimated to be in the range of 9×10^{-19} J or $220\ kT$. Obviously, coulombic interactions must be considered to be at least equal in strength to covalent bonds.

According to Equation (4.2), the magnitude of the coulombic interaction falls off as the inverse square of the distance between the charges. A quick calculation shows that, in a vacuum, the interaction energy will fall to kT only as the separation distance approaches 60 nm, a large distance in the normal world of atoms, ions, and molecules.

Whenever we talk about the free energy of a system, we are usually talking about a change in free energy rather than an absolute energy. It is therefore necessary to keep in mind that such changes in free energy are normally compared to some reference state that is specified or understood. When considering intermolecular interactions, the usual reference state to be considered is that in which the ions or molecules involved start with a separation of $r = \infty$ and come together to form a condensed solid or liquid state. If the interaction occurs in the gaseous phase so that the surrounding medium is a vacuum, the values for the denominator in Equation (4.2) will be $\varepsilon_o = 8.854 \times 10^{-12}$ C^2 J^{-1} m^{-1} and $\varepsilon = 1$, the dielectric constant of a vacuum. If the interaction occurs in a condensed liquid, then ε becomes the dielectric constant of the liquid. Normally encountered liquid media range from hydrocarbons ($\varepsilon \approx 2$) to water ($\varepsilon = 80$). Obviously, in media other than vacuum, the coulombic interactions will vary significantly with the nature of the intervening medium.

It is important to point out that there exists an apparent discrepancy between Equation (4.2) and experimental measurements of such interactions. It is generally found in experiments that all intermolecular forces decay faster than the inverse 4th power of distance (r^{-4}), while Equation (4.2) indicates a much slower inverse square relationship. The apparent contradiction is removed, however, if one remembers that the relationship is derived for two isolated ions, while in reality, ions will interact to some extent with all ions of opposite charge in their vicinity. The presence of these "associated" ions, which are not considered in simple one-to-one calculations of the interaction potential, acts to "screen" interactions between the charges, reduces the strength of their interaction and, therefore, the range over which its effects are significant. Even with that, however, coulombic interactions remain strong and significant over distances much greater than covalent bonds and are therefore considered to be long-range forces.

4.3.2 Other Interactions Involving Ions

Many commonly encountered materials include ionic and molecular units that may interact by mechanisms that may be considered to be "hybrids" between strictly coulombic phenomena and the van der Waals interactions to be considered below. Although weaker and perhaps more subtle, interactions involving a charged species and a neutral (but usually polar) unit can make a significant contribution to the total interaction energy of a system.

Dipoles and Polarization Phenomena. Many molecules do not carry formal electrical charges, so that their mutual interactions do not involve the direct coulombic interactions discussed above. However, if one examines the structures of many useful chemical species, including polymers, proteins, and drugs, it is apparent that they often include bonds that can impart an overall polar nature to the molecule as permanent dipoles, or they can be polarized by the effect of neighboring electric fields producing induced dipoles. The presence of permanent or induced dipoles means that the molecules can become involved in specific interactions with charged species, other dipoles, or nonpolar molecules, and those interactions can significantly affect the physical characteristics of the system.

Some molecules have dipoles that result from differences in the electronegativity of the bonded atoms, an example being the commonly encountered carbonyl group ($C^{\delta+}=O^{\delta-}$) found in many organic molecules. Other important molecules, for example the amino acids (and therefore the proteins they make up), contain acid and base functionalities that usually exist as zwitterions in which the two functionalities form, in effect, an internal salt. For glycine, the molecule takes the form

$$CH_2(NH_3{}^+)COO^-$$

In water at the isoelectric point, the molecule is electrically neutral, but the charge separation in the zwitterion produces a strong dipole that to a great extent governs the nature of the interactions of the molecule. At pH values other than the isoelectric point, the molecule becomes formally charged and coulombic factors prevail. Obviously, the interactions of molecules of these types can become quite complex and may exhibit a significant sensitivity to their solvent environment.

The Dipole Moment. A dipole results from the presence of an unsymmetrical distribution of electron density within a molecule, due either to a formal charge separation, such as in amino acids, or due to differences in the electronegativities of the atoms forming a covalent bond, as in carbonyl compounds, water, and alcohols. An isolated, neutral atom, of course, cannot have an unsymmetrical electron distribution; therefore atoms cannot be dipolar in nature. That is not to say, of course, that they cannot be polarized, or have their electron cloud distorted by an external electric field, but that subject is considered later. The dipole moment, μ, of a molecule is defined as

$$\mu = ql \tag{4.3}$$

where l is the distance between the two charges, $+q$ and $-q$, of the dipole (or the positive and negative ends of the asymmetrical electron cloud in a covalent bond; Fig. 4.1). For a zwitterionic species, for example, where $q = \pm e$ and the charge separation is $l = 0.1$ nm, the dipole moment will be

$$\mu = (1.602 \times 10^{-19})(1 \times 10^{-10}) = 1.6 \times 10^{-29} \text{ C} \cdot \text{m} = 4.8 \text{ } D$$

where D is the Debye unit $= 3.336 \times 10^{-30}$ C $\cdot$ m.

The magnitudes of the dipole moment for commonly encountered bonds and molecules range from approximately 1 D for slightly polar organic molecules, such as chloroform, to closer to 2 for water, carboxylic acids, and similar species. The dipole moments of complex molecules can, in principle, be calculated by taking a vectorial sum of all the bond moments of the functionalities present. That can become quite complicated (and of marginal utility) in the case of complex organic molecules, especially since modern experimental techniques make their direct determination relatively easy. The dipole and bond moments of some common molecules and functionalities are given in Table 4.2.

The Polarization of Nonpolar Atoms and Molecules. Molecules having no permanent dipole can also take part in electrostatic interactions as a result of deformation or polarization of their electron clouds by the presence of an

FIGURE 4.1. Typical dipolar molecular structures.

TABLE 4.2. Characteristic Molecular and Bond Dipole Moments,
μ **(in Debye Units)**

Molecule	μ	Bond	μ
n-Alkanes	0	C—C	0
Bezene (C_6H_6)	0	C=C	0
Carbon tetrachloride (CCl_4)	0	C—H	0.22
Carbon dioxide (CO_2)	0	N—O	0.03
Chloroform ($CHCl_3$)	1.06	C—H	0.4
Hydrogen chloride (HCl)	1.08	C—O	0.74
Ammonia (NH_3)	1.47	N—H	1.31
Methanol (CH_3OH)	1.69	O—H	1.51
Acetic acid (CH_3COOH)	1.7	C—Cl	1.5–1.7
Water (H_2O)	1.85	F—H	1.94
Ethylene oxide (C_2H_4O)	1.9	N=O	2.0
Acetone (CH_3COCH_3)	2.85	C=O	2.3–2.7
N,N -Dimethyl formamide	3.82	—	—
Acetonitrile	3.92	—	—

external electric field. Polarization phenomena are referred to as "induced dipole interactions" and essentially involve the distortion of a normally symmetrical electron cloud as a result of the presence of some strongly polar unit in the vicinity. The process is illustrated schematically in Figure 4.2.

All atoms and molecules are, in principle, polarizable. The ease with which the electron cloud of a given species can be distorted, its "polarizability," α, is defined by the strength of the induced dipole formed when placed under the influence of an electric field of strength E

$$\mu_{ind} = \alpha E \tag{4.4}$$

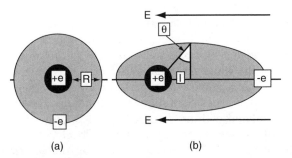

(a)   (b)

FIGURE 4.2. Schematic illustration of the polarization mechanism: (a) isolated atom with no external field, $\mu_{ind} = 0$; (b) with applied external field, E, the electrons are shifted by distance l, and $\mu_{ind} = le = \alpha_o E$.

For a nonpolar molecule, the induced dipole results simply from the displacement of the center of mass of the negative electron cloud with respect to that of the positive nuclei. In polar molecules, the situation becomes more complex because the net induced dipole, μ'_{ind}, will depend on the orientation of the induced dipole, μ_{ind}, with respect to the permanent dipole of the molecule.

To illustrate the point, imagine an atom with one associated electron of charge $-e$ in a spherically symmetric orbital of "radius" R, the atomic radius. If the electron is subjected to an external field E, the orbit of the electron will be shifted by a distance r away from the nucleus. The induced dipole moment μ_{ind} is determined by the external force, F_{ext}, acting on the electron(s) as a result of the field E

$$F_{ext} = eE \qquad (4.5)$$

balanced against the attractive force between the electron and the nucleus, the internal restoring force, F_{int}, where

$$F_{int} \approx \frac{e}{4\pi\varepsilon_0 R^3} \mu_{ind} \qquad (4.6)$$

At equilibrium $F_{int} = F_{ext}$ so that

$$\mu_{ind} = 4\pi\varepsilon_0 R^3 E = \alpha_0 E \qquad (4.7)$$

where the polarizability, α_0, is

$$\alpha_0 = 4\pi\varepsilon_0 R^3 \qquad (4.8)$$

The phenomenon related to the displacement of electrons around an atom or molecule is termed electronic polarizability and has the dimensions $C^2.m^2\ J^{-1}$. As a rule of thumb, it is usually found that the magnitude of the polarizability of a given atom or molecule will be of the order of, but slightly less than, its radius cubed multiplied by the term $4\pi\varepsilon_0$. For water, $\alpha_0/4\pi\varepsilon_0 = 1.48 \times 10^{-30}\ m^3$, which is approximately equal to $(0.114\ nm)^3$, some 15% less than the radius of the water molecule $(0.135\ nm)$.

The electronic polarizabilities of some characteristic atoms and molecules are given in Table 4.3. It has been found that the polarizability of a molecule can often be calculated with an accuracy of a few percent by summing the polarizabilities of isolated bonds within the molecule. While the procedure works well for isolated bonds, systems involving neighboring nonbonding electrons or delocalized structures, such as aromatic rings, are not so well behaved. In those circumstances, group polarizabilities have been assigned. Typical values of bond and group polarizabilities are also included in Table 4.3.

TABLE 4.3. Electronic Polarizabilities of Typical Atoms, Molecules, Bonds, and Molecular Groups, in Units of $\alpha_0/5\pi\varepsilon_0$ (Å^3)

Atoms and Molecules					
He	0.20	NH_3	2.3	$CH_2{=}CH_2$	4.3
H_2	0.81	CH_4	2.6	C_2H_6	4.5
H_2O	1.48	HCl	2.6	Cl_2	4.6
O_2	1.60	CO_2	2.6	$CHCl_3$	8.2
Ar	1.63	CH_3OH	3.2	C_6H_6	10.3
CO	1.95	Xe	4.0	CCl_4	10.5

Bond Polarizabilities					
Aliphatic C—C	0.48	C—H	0.65	C=C	1.65
Aromatic C=C	1.07	N—H	0.74	C—Cl	2.60

Molecular Groups			
C—O—H	1.28	CH_2	1.84
C—O—C	1.13	C=O	1.36

The Polarization of Polar Molecules. Equation (4.7) gives an expression for the electronic polarizability of a spherically symmetric atom or molecule. If a molecule has a freely rotating permanent dipole, which has a time-averaged dipole moment of zero, in the presence of an external field E, there may develop an induced orientational dipole. This would then be related to the orientational polarizability of the molecule. If at some instant, the permanent dipole μ of the molecule is at an angle θ to the applied field, its energy in the field will be given by

$$w(r,\theta) = -\mu E(r) \cos \theta \qquad (4.9)$$

so that the time-averaged induced dipole moment will be

$$\mu_{\text{ind}} = \frac{\mu^2 E}{kT} < \cos^2 \theta > = \left(\frac{\mu^2}{3kT}\right)E, \qquad \mu E \ll kT \qquad (4.10)$$

From this equation μ_{ind} is proportional to E, and $\mu^2/kT = \alpha_{\text{orient}}$ represents an additional contribution to the polarizability of the molecule, the orientational polarizability. The total polarizability of the molecule, α, will be the sum of the electronic and orientational polarizabilities

$$\alpha = \alpha_0 + \frac{\mu^2}{3kT} \qquad (4.11)$$

As an example, assume that a molecule has a permanent dipole moment of 1 D. At 300 K it will have an orientational polarizability of

$$\alpha_{orient} = \frac{(3.336 \times 10^{-30})^2}{3(1.38 \times 10^{-23})300}$$

$$= 9 \times 10^{-40} \text{ C}^2 \cdot \text{m}^2 \text{ J}^{-1} = (4\pi\varepsilon_0)8 \times 10^{-30} \text{ m}^3$$

Such a value is comparable to the electronic polarizabilities of molecules as shown in Table 4.3. In the case of a very polar molecule in a very high field, as for example a water molecule next to a lithium ion, or at very low temperatures where molecular rotation is greatly limited, the polar molecule may become completely aligned with the field. At that point, the induced dipole moment will no longer be proportional to E and the simple concept of the molecule's polarizability breaks down. Many of the unique, and sometimes perplexing, solvent characteristics of water can be traced to the ability of water molecules to be aligned and structured in solution by the presence of strong electric fields.

Ion–Dipole Interactions. Somewhere between the strong electrostatic and relatively weak dipole–dipole interactions to be discussed later lies an intermediate area involving ions and dipoles. If a charge Q lies at a distance r from the center of a polar molecule of moment μ and dipolar length l, with the orientation shown in Figure 4.3, the coulombic interaction between ion and dipole will be the sum of the interactions between Q and each end of the dipole with charges $\pm q$.

$$w(r) = -\frac{Qq}{4\pi\varepsilon_0\varepsilon[1/AB - 1/AC]} \tag{4.12}$$

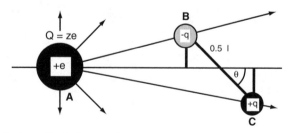

FIGURE 4.3. In ion–dipole interactions, the net force will be the sum of the interactions between the ion and each end of the dipole.

where

$$AB = \left[\left(\frac{r - \frac{1}{2}}{\cos \theta} \right)^2 + \left(\frac{\frac{1}{2}}{\sin \theta} \right)^2 \right]^{1/2} \tag{4.13}$$

and

$$AC = \left[\left(\frac{r - \frac{1}{2}}{\cos \theta} \right)^2 + \left(\frac{\frac{1}{2}}{\sin \theta} \right)^2 \right]^{1/2} \tag{4.14}$$

At distances greater than the dipole length the interaction energy can be estimated by

$$w(r,\theta) = - \frac{Q\mu \cos \theta}{4\pi\varepsilon_0 \varepsilon r^2} = - \frac{(ze)\mu \cos \theta}{4\pi\varepsilon_0 \varepsilon r^2} \tag{4.15}$$

Equation (4.15) gives the theoretical interaction between an ion of charge Q and a "point" dipole of moment μ, for which the length of the dipole is theoretically zero. In such a case, it is hard to visualize exactly what is meant by saying that the dipole is pointing "toward" or pointing "away from" the interacting charged species. That point aside for the moment, when the dipole is pointing away from the charge ($\theta = 0°$, in Fig. 4.3) the interaction will be at its most negative value (attractive). For $\theta = 180°$, the value will be its most positive (repulsive).

In fact, of course, point dipoles do not exist, so that some deviation from the simple theory presented here is expected. It is only when the value of r approaches $2l$ that significant problems arise, however. Thus, for all reasonable ion–molecule separations and a dipole length of 0.1 nm or so, the use of eq. 4.15 produces no significant error. In zwitterions, however, in which the dipole arises from a formal separation of charges on different atoms in a molecule, the length of the dipole may be greater than 0.1 nm and larger deviations from Equation (4.15) are found. In that case, the total interaction must be calculated from each separate interaction between the ion and the ends of the dipole according to (4.4). In those cases it is found that the total interaction is greater than that predicted by (4.15).

Another consequence of the ion–dipole interaction is that it is often strong enough to bind polar solvent molecules to the ion and establish a local aligned or "frozen" structure composed of the ion and several associated polar molecules within the matrix of the solution (see Fig. 3.2). For the small monovalent cation Li^+ in the presence of water molecules (as opposed to dissolved *in* water), for example, the following data may be used: μ (water) = 1.85 D, r_w (the molecular radius of water) ≈ 0.14 nm, and r_{Li^+} (the ionic radius of Li^+) = 0.068 nm. Application of Equation (4.15) gives

$$w(r,\theta=0) = - \frac{(1.602 \times 10^{-19})(1.85 \times 3.336 \times 10^{-30})}{4\pi(8.854 \times 10^{-12})[(0.14 + 0.068) \times 10^{-9}]^2} = -2.05 \times 10^{-19} \text{ J}$$

which corresponds to a value of 123 kJ mol^{-1}, compared to the experimental value of 142 kJ mol^{-1}. Good correspondence is found between theory [Eq. (4.15)] and experiment for most of the small mono- and polyvalent cations. Because of their strong interaction with dipolar molecules such as water, cations are important in a number of phenomena stemming from the structuring and aligning of water molecules, not the least of which are the nucleation of raindrops and snowflakes in cloud formations, cloud chamber experiments, and similar.

The preceding analysis assumes that the dielectric constant of the medium of interaction is unity (i.e., vacuum). In aqueous solution, the interaction will be reduced by a factor of 80, the dielectric constant of water. Even there, however, the strength of the interaction can be significant with respect to kT for small di- and polyvalent ions and cannot be ignored for small monovalent ions.

Ion Solvation. In aqueous solution, when a "bulk" water molecule approaches an ion, the situation is not the same as when a water molecule approaches an isolated ion. In the liquid phase, the water molecule in question must replace a molecule already associated with the ion, so that the net result is not a net change in free energy, but simply an exchange of two water molecules. However, even here, Equation (4.15) suggests some finite interaction energy between the ion and individual water molecules in bulk water. The equation contains the orientational term cos θ which becomes important at small distances of separation.

When a water molecule is far away from the ion, it will be randomly oriented with respect to the ion, producing a spatial average for cos θ = zero. If the water molecule in question maintains that random orientation right up to its closest approach to the ion, the net interaction energy would be zero. However, as the approach takes place, the water molecule is no longer randomly oriented, but begins to assume orientations that produce a favorable (more negative) interaction between the ionic charge and the water dipole; that is, the molecule becomes oriented relative to the ion and bulk solvent molecules.

For small monovalent and polyvalent ions, the effect will produce a "shell" of oriented water molecules bound to the ion, with the orientation favoring $\theta = 0°$ for cations and 180° for anions. The surrounding shell of water molecules thus formed constitutes the waters of solvation or hydration of the ion. The number of water molecules associated with an ion (its hydration number) is characteristic of that ion but normally ranges between 4 and 6. Waters of hydration are not completely and irreversibly bound to a given ion, of course. They are slowly (on a molecular timescale), but continually exchanged for other bulk water molecules.

Closely related to the hydration number is the hydrated radius of an ion in water, which is larger than its crystal lattice radius. Small ions, which may have a closer approach to the water molecules and thus a stronger interaction, tend to have relatively larger hydration radii than larger ions. On a molar basis they tend to structure the water more by orienting more molecules per ion. The effect of the hydration radius is manifested in a number of solution physical properties including viscosity, conductivity, compressibility, diffusion, and a number of thermodynamic and spectroscopic properties of ionic solutions.

The structuring of polar solvent molecules, especially water, around an ion is not limited to the molecules directly "in contact" with the ion. The interaction is transmitted, to an increasingly reduced extent, to molecules in a second, third (etc.) layer of surrounding solvent molecules. The structuring effect decays in an approximately exponential way but may extend several molecular diameters into the bulk liquid. This region of enhanced structuring in the solvent is referred to as the "solvation zone around the ion." This solvation or structuring effect can have a number of important consequences for the interactions between ions, molecules, colloidal particles, and interfaces. If the orientation and mobility of solvent molecules near an ion differ from those in the bulk, one might expect, as is the case, that many physical properties of the solvation zone differ from those of the bulk. In particular, there are found to be differences in density, dielectric constant, conductivity, and other parameters.

Of particular importance for present purposes is that the dielectric constant of the solvation zone may differ significantly from the bulk value because the molecules there cannot respond to an imposed electric field in the same way as in the bulk liquid. Since the equations for calculating ionic, dipolar, and van der Waals interactions include the dielectric constant of the medium, differences such as those encountered in the solvation zone may have important consequences. In particular, when two ions approach close enough that their respective zones touch, their interaction energy may be much different than than predicted using the bulk value of ε. If the effective dielectric constant in the solvation zone is less than the bulk, as is most often the case, then the interaction will be stronger; if e is increased (less common, but not unheard of) then the interaction will be decreased. Ion–solvent interactions are very specific and cannot be handled well by a general theory as yet. However, because of their potential importance in solvation-mediated processes (e.g., nucleation, crystal growth), in colloidal stabilization, and in interfacial interactions, solvation effects should be kept in mind to help explain seemingly anomalous experimental results.

Interactions Between Ions and Nonpolar Molecules. When a nonpolar molecule is near an ion generating an electric field **E,** where

$$\mathbf{E}(r) = \frac{(ze)}{4\pi\varepsilon_0\varepsilon r^2} \tag{4.16}$$

the ion will induce a dipole moment in the molecule of

$$\mu_{ind} = \alpha \mathbf{E}(r) = \frac{\alpha(ze)}{4\pi\varepsilon_0\varepsilon r^2} \tag{4.17}$$

For a methane molecule with an electronic polarizability $\alpha_o = 2.6 \times 10^{-29}$ $C^2 \cdot m^2 J^{-1}$, located 0.4 nm from a monovalent ion ($\mathbf{E} = 9 \times 10^9$ V m^{-1}), the induced dipole moment will be

$$\mu_{ind} = \alpha_0 \mathbf{E} = 4\pi\varepsilon_0(26 \times 10^{-30})(9 \times 10^9) = 2.6 \times 10^{-29} \text{ C·m} = 0.78 \text{ D}$$

Considering the above induced dipole in a neutral molecule due to the presence of the ion, what will be the net interaction between the two species? If the ion involved is a cation, the induced dipole on the molecule will point away from the ion (Fig. 4.4). An anion will induce a dipole pointing toward the ion. In either case, the overall interaction due to the induced dipole will be attractive. The induced dipole will interact with the ion with a "reactive" field given by

$$\mathbf{E}_r = \frac{-2\mu_{ind}}{4\pi\varepsilon_0\varepsilon r^3} = \frac{-2\alpha\mathbf{E}}{4\pi\varepsilon_0\varepsilon r^3} = \frac{-2\alpha(ze)}{(4\pi\varepsilon_0\varepsilon)^2 r^5} \tag{4.18}$$

The attractive force between ion and induced dipole will be

$$F = -\frac{2\alpha(ze)^2}{(4\pi\varepsilon_0\varepsilon)^2 r^5} \tag{4.19}$$

for an overall interaction free energy of

$$w(r) = \frac{-\alpha(ze)^2}{2(4\pi\varepsilon_0\varepsilon)2r^4} = -\frac{1}{2}\alpha\mathbf{E}^2 \tag{4.20}$$

where the energy is given as one-half that expected for the interaction between an ion and an aligned, permanent dipole. The reason is that, in the process

(a)   (b)

FIGURE 4.4. When a dipole is induced in a neutral atom or molecule, the orientation will depend on the sign and field of the ion inducing the dipole.

of inducing the dipole, some energy is used up in polarizing the nonpolar molecule. The energy used is that required to displace the center of the negative electron charge from the center of the positive nuclear charge.

The question of the full nature and consequences of coulombic interactions is, of course, much more complex than the material presented here. For excellent, more extensive discussions of the subject as it applies to surfaces and colloids the reader is referred to the works in the (end-of-book) Bibliography for this chapter. When the general topic of the stabilization of colloids is reached, the functional importance of coulombic interactions will become much more apparent. For now, however, our attention will turn to more subtle (but more universal) interactions classed together as van der Waals forces.

4.4. VAN DER WAALS FORCES

It was stated that there are four principal types of forces acting between atoms and molecules at the molecular and macroscopic levels. The first, coulombic interaction involving at least one formally charged species, was covered above. The remaining three forces make up what are commonly termed van der Waals forces and are comprised of three types of interaction. Of the three, two involving permanent and induced dipoles, are closely related to the forces discussed above, although they do not involve interactions of formally charged species. The third is the most fundamental and universal force, and although generally the weakest of the three in absolute terms, it is often the most important contributor to the total van der Waals interaction. This force is the so-called London–van der Waals force or London dispersion force first postulated by van der Waals in 1873 but actually explained by London in 1930. Beginning with the "simplest" dipole–dipole interactions, each of the three forces will be described in general terms below.

4.4.1. Dipole–Dipole Interactions

If two polar molecules with dipole moments μ_1 and μ_2 approach in a vacuum there will develop a dipole–dipole interaction between the ends of the dipoles analogous to the interaction between the ends of two magnets. If the dipoles are oriented with respect to each other at a distance r as shown in Figure 4.5a, the interaction energy for the two dipoles will be given by

$$w(r,\theta_1,\theta_2,\phi) = -\frac{\mu_1\mu_2}{4\pi\varepsilon_0\varepsilon r^3}(2\cos\,\theta_1\,\cos\,\theta_2 - \sin\,\theta_1\,\sin\,\theta_2\,\cos\,\phi) \quad (4.21)$$

Maximum interaction will occur when the two dipoles are lying in line ($\theta_1=\theta_2=0°$), so that

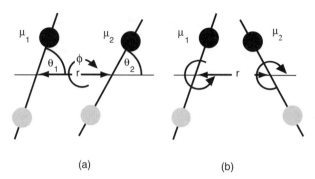

(a)   (b)

FIGURE 4.5. In dipole–dipole systems, the intensity of the interaction will depend on the strength of the dipoles and their relative orientations. For fixed dipoles (*a*) the interaction will be determined by the angle between the two. In freely rotating systems (*b*) the two may rotate to maximize attractive or minimize repulsive interactions.

$$w(r,0,0,\phi) = -\frac{2\mu_1\mu_2}{4\pi\varepsilon_0\varepsilon r^3} \tag{4.22}$$

Two like dipoles in a vacuum aligned for maximum interaction must, according to (4.22), approach to within ≈ 0.36 nm for their interaction energy to be equal to kT. If aligned parallel, the distance must be (0.29 nm. In a solvent medium, the interaction will be reduced so that the distance of separation for maximum interaction will be even less. The expression in Equation (4.21) is an idealized expression that under many circumstances gives results that differ significantly from reality. That is found to be the case especially when the distance of separation is less than three times the length of the dipole. Since the distance of approach for maximum dipole–dipole interaction is of the same order of magnitude as normal molecular separations in condensed materials, it can be seen that dipolar interactions are of little practical significance (acting alone) in aligning or binding molecules to produce liquid or solid systems, or for imparting significant molecular structure in a system. Dipole–dipole interactions are usually significant only in systems involving very polar molecules. The exceptions involve small molecules with very large dipole moments such as water (O—H), hydrogen fluoride (F—H), and ammonia (N—H).

In those cases, the dipoles involve the small electron-deficient hydrogen atom bonded to a very electronegative atom. Here, the interactions become much more complex and are given the more specific descriptive name of "hydrogen bonding" interactions. (They are also referred to in some instances as acid–base interactions, although that terminology has yet to "catch on.") In such cases the electron-deficient hydrogen atom has such a small size that the electronegative atoms can approach very closely and the hydrogen experiences a much stronger electric field. This results in an enhanced attractive interaction in the condensed state. Molecules undergoing this special hydrogen bonding interaction form a special and very important class of liquids called "associated" liquids. Their nature and the nature of their interactions with other species is of great importance in many practical areas of surface and colloid science, chemistry in general, biology, and so on. For now, we will be concerned with the more normal molecules exhibiting weak dipolar interactions.

4.4.2. Angle-Averaged Dipolar Interactions

To this point the discussion has been concerned with the interaction between two dipoles effectively fixed in space by their mutual interaction. However, at large separations, where the interaction energy falls below the value of kT, or in a medium of high dielectric constant, the thermal energy of the system is such that the dipoles can no longer be considered fixed; instead, they rotate or tumble with respect to one another. Due to the random motion of the molecules, the overall averages for the components of Equation (4.21) are zero. The angle-averaged values of the interaction potential, however, will never be zero because there will always exist a Boltzmann weighting factor that gives more importance to some angles (or orientations) than to others. In other words, those orientations which produce more negative (lower energy) interaction potentials will be favored over those with less negative or positive values. For the dipole–dipole interaction, the Boltzmann angle-averaged interaction will be given by the expression

$$w(r) = -\frac{(\mu_1\mu_2)}{3(4\pi\varepsilon_0\varepsilon)^2 kTr^6} \quad \text{for} \quad kT > \frac{\mu_1\mu_2}{4\pi\varepsilon_0\varepsilon r^3} \tag{4.23}$$

The Boltzmann angle-averaged interaction potential is generally referred to as the orientation or Keesom interaction and represents one of the three "6th power of distance" relationships involved in the total van der Waals interaction.

4.4.3 Dipole-Induced Dipole Interactions

The interaction between a polar molecule and a nonpolar molecule is similar to that between an ion and a nonpolar molecule except that the force field inducing the dipole arises as a result of a dipole rather than a formal charge center. For a dipole of moment μ oriented at an angle θ to the line joining the dipolar and nonpolar molecules, the strength of the electric field acting on the nonpolar (but polarizable) molecule is given by

$$E = \frac{\mu(1 + 3\cos^2\theta)^{1/2}}{4\pi\varepsilon_0\varepsilon r^3} \tag{4.24}$$

The interaction energy, therefore, will be

$$w(r, \theta) = -\tfrac{1}{2}\, \alpha_0 E^2 = \frac{-\mu\alpha_0(1 + 3\cos^2\theta)}{2(4\pi\varepsilon_0\varepsilon)^2 r^6} \tag{4.25}$$

For commonly encountered values of μ and α_0, the strength of the interaction is not sufficient to orient the two molecules completely, as sometimes occurs in the case of ion-induced dipole and dipole–dipole interactions. The net effective interaction, $w_{\text{eff}}(r, \theta)$, will be the angle-averaged energy. For the function $\cos^2\theta$, the angle average is $\tfrac{1}{3}$, so that the interaction becomes

$$w(r) = \frac{-\mu^2\alpha_0}{(4\pi\varepsilon_0\varepsilon)^2 r^6} \tag{4.26}$$

It may happen, of course, that a weakly polar molecule can be further polarized in the presence of another dipole. For such a situation in which two different molecules, each possessing a permanent dipole μ_1 and μ_2 and polarizabilities α_{o1} and α_{o2}, the net dipole-induced dipole interaction energy will be

$$w(r) = \frac{-(\mu_1^2\alpha_{01} + \mu_2^2\alpha_{02})}{(4\pi\varepsilon_0\varepsilon)^2 r^6} \tag{4.27}$$

The interaction energy given by Equation (4.27), often referred to as the Debye interaction, represents the second of the three "inverse 6th power" contributions to the total van der Waals interaction between molecules.

4.4.4. The London–van der Waals (Dispersion) Force

As mentioned above, the London–van der Waals or London dispersion force (hereafter generally referred to simply as "dispersion force") often makes the most important contribution to the total van der Waals interaction because of its universal nature, as contrasted with the dipolar and induced dipolar forces, which vary with the exact chemical natures of the species involved and may or may not make a significant contribution to the total energy. Dispersion forces are important in a wide variety of phenomena, including the condensation of nonpolar molecules to the liquid and solid states; the boiling points, surface tensions, and other physical properties of condensed states; adsorption, adhesion, and lubrication processes; the bulk physical strength of primarily covalent materials; the aggregation and flocculation of molecular and particulate systems; and the structures and interactions of synthetic polymers, proteins, and other complex biological systems.

Although arising from complex quantum-mechanical factors, dispersion forces have several easily understood characteristics, such as the following:

1. They have a relatively long range of action compared to covalent bonds, their effect in some cases extending to a range of 10 nm or more.
2. They may be attractive or repulsive, depending on the situation, and generally do not adhere to simple power laws with respect to their dependence on separation distances.
3. They are nonadditive, in that the interaction between any two atoms or molecules will be affected by the presence of other nearby atoms and molecules.

The dispersion force is basically quantum mechanical in nature because it involves interactions between rapidly fluctuating dipoles resulting from the movement of the outer-valence-shell electrons of an atom or molecule. Rigorous derivations, therefore, can become quite complex and will serve little useful purpose in the present discussion. The interested reader is referred to the works cited in the Bibliography for further enlightenment.

For a system of two isolated atoms (or molecules), one can visualize the electrons around one atom as being particles (much like in the Bohr atom) that, although traveling at close to the speed of light, can at any instant be located asymmetrically with respect to the nucleus with which they are associated (Fig. 4.6). The asymmetric charge "distribution" produces an instantaneous dipole in the atom or molecule. That dipole generates a short-lived electric field that can then polarize a neighboring atom or molecule, inducing a dipole in the neighbor (Fig. 4.6b). The result is a net coulombic attraction between the two species. Using the simple model of dispersion forces arising from fluctuating dipoles, their complex quantum-mechanical origin can, to a first approximation, be forgotten and they can be treated as simple electrostatic interactions.

The strength of the attraction, F_{att}, between two such instantaneous dipoles is found to be proportional to the inverse 7th power of the distance separating the two nuclei

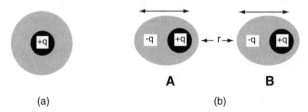

                                                A   **B**

          (a)   (b)

FIGURE 4.6. Dispersion interactions involve the distortion of an otherwise symmetrical electron cloud by a neighboring atom or molecule: (*a*) the isolated atom or molecule has a net symmetrical charge distribution; (*b*) in the presence of a neighboring electron cloud the "instantaneous" quantum mechanical dipole present at any instant in atom A induces a dipole in neighboring atom B.

$$F_{att} = \frac{-A'}{r^7} \tag{4.28}$$

where r is the distance between nuclei or, for approximately spherical molecules, the centers of mass, and A' is a quantum-mechanical constant related to the structure of the atom or molecule.

The amount of work required to separate reversibly a pair of atoms or molecules from a distance r to infinity is

$$\Delta W = -\int_r^\infty F_{att}\, dr = A' \int_r^\infty \frac{1}{r^7}\, dr = \frac{A'}{6r^6} = \frac{A}{r^6} \tag{4.29}$$

If it is assumed that the interaction energy at infinite separation is zero, then the free energy of attraction will be

$$\Delta G_{att} = \frac{-A}{r^6} \tag{4.30}$$

The constant A ($= A'/6$) was given by London, for two identical units, as

$$A = \frac{3}{4} h\nu \alpha_0^2 \tag{4.31}$$

where h is Planck's constant, α_o is the electronic polarizability of the atom or molecule, and n is a characteristic frequency identified with the first ionization potential of the atom or molecule, usually falling in the ultraviolet region. For two different interacting units, 1 and 2, the expression is

$$A_{12} = \frac{3}{2} h \left(\frac{\nu_1 \nu_2}{\nu_1 + \nu_2} \right) \alpha_{o_1} \alpha_{o_2} \tag{4.32}$$

The shape of the force–distance curve for Equation (4.30) is given schematically in Figure 4.7a. The free energy and attractive force between units becomes more negative as the separation distance decreases until the electron clouds of the respective units begin to interact. If no bonding interactions between the two units are possible, the interaction becomes repulsive (Born repulsion) and rises to infinity as the electron clouds begin to overlap.

Theoretically, the Born repulsion function takes the form

$$F_{rep} = Be^{-a\,r} \tag{4.33}$$

where a and B are constants. This produces a contribution to the total interaction potential at distance r of

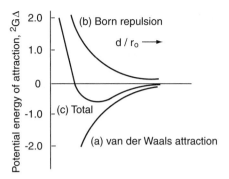

FIGURE 4.7. A plot of the attractive (*a*) and repulsive (*b*) interactions produce a force–distance relationship for the total dispersion interaction between atoms and molecules.

$$\Delta G_{\text{rep}} = \frac{B}{a} e^{-a r} \tag{4.34}$$

For ease of manipulation, it is common practice to approximate (4.34) by

$$\Delta G_{\text{rep}} = \frac{B'}{r^{12}} \tag{4.35}$$

which produces a curve similar to (*b*) in Figure 4.7.

The total potential energy of interaction between two units will be the sum of the attractive and repulsive interactions and is illustrated by curve (*c*):

$$\Delta G = \Delta G_{\text{rep}} + \Delta G_{\text{att}} = \frac{B'}{r^{12}} - \frac{A}{r^6} \tag{4.36}$$

This equation is commonly referred to as the Lennard-Jones 6–12 potential. It may be noted that the attractive term in (4.36) is an inverse 6th-power relationship, the third such term in the overall van der Waals attraction.

One may estimate the strength of dispersion interactions between two small identical spherical atoms or molecules using the London equation as

$$w(r) = \frac{-(\frac{3}{4})\alpha_o^2 h\nu}{(4\pi\varepsilon_o)^2 r^6} \tag{4.37}$$

The ionization potential of the atom or molecule (typically in the range of 2×10^{-18} J) can be substituted for $h\nu$. An approximate value for the term $\alpha_o/4\pi\varepsilon_o$ is 1.5×10^{-30} m^3. For two atoms in contact ($r \approx 0.3$ nm) at room temperature, the interaction potential will be $w(r) = -4.6 \times 10^{-21}$ J, which is approximately equal to 1 kT. It can be seen, then, that dispersion forces can be quite significant in magnitude considering that they arise from little more than fleeting variations in the balancing act between electrons and nuclei in the atom or molecule.

It is important to emphasize that Equation (4.37), while extremely useful for the type of systems specified, has several limitations when applied to larger and nonspherical molecules and to many-bodied systems such as condensed liquids and solids. These limitations include the following:

1. The analysis neglects the effect of short-range repulsive forces at small r that tend to lower the net attractive interaction between the atoms or molecules.
2. Higher level quadrapole interactions, which tend to increase the attractive interaction, are not considered.
3. Interactions with distant neighbors in condensed systems, which also tend to increase the attraction, are neglected.

By lucky chance, it appears that limitation 1 is pretty much canceled out by 2 and 3 for simple systems, so that theoretical calculation based on Equation (4.37) are found to agree surprisingly well with experiment.

4.4.5. Total van der Waals Interactions between Polar Molecules

We have seen, now, that there are three types of interactions that can be involved in the total van der Waals interaction between atoms or molecules: dipole–dipole (orientational or Keesome), dipole-induced dipole (induced or Debye), and dispersion (London) interactions. The theories for all three interactions are found (to a first approximation) to involve an inverse 6th power of the distance separating the two interacting centers. The total van der Waals interaction potential, $w_{vdw}(r)$, can then be written as

$$w_{vdw}(r) = \frac{-C_{vdw}}{r^6} \qquad (4.38)$$

where C_{vdw} is the overall van der Waals constant for which

$$C_{vdw} = (C_{disp} + C_{ind} + C_{orient}) \qquad (4.39)$$

The individual constants are simply the coefficients of $1/r^6$ for each contributing factor [Eqs. (4.23), (4.27), and (4.37)].

For two identical molecules the full equation takes the form

$$w_{vdw}(r) = \frac{-[(3\alpha_o^2 h\nu_1/4) + 2\mu^2\alpha_o + (\mu^4/3kT)]}{(4\pi\varepsilon_o)^2 r^6} \qquad (4.40)$$

For two unlike polar molecules, this equation becomes

$$w_{vdw}(r) = \frac{-\{[3\alpha_{o1}\alpha_o{}^2 h\nu_1\nu_2/2(\nu_1 + \nu_2)] + (\mu_1^2\alpha_{o_2} + \mu_2^2\alpha_{o_1}) + (\mu_1^2\mu_2^2 / 3kT)\}}{(4\pi\varepsilon_o)^2 r^6} \quad (4.41)$$

Some representative values for total van der Waals interactions are given in Table 4.4.

One can see from Table 4.4, the dispersion force contribution to the total van der Waals interaction can be quite significant. Except for very small and very polar molecules such as water, the dispersion force will exceed the Keesom and Debye contributions and dominate the character of the interaction, even in the case of two dissimilar molecules.

In many cases the interactions between dissimilar molecules cannot be calculated, so that some form of estimate must be employed. Experience has shown that in most cases, the experimentally determined value for the van der Waals interaction, and therefore the van der Waals constant C_{vdw}, falls somewhere between that for the two identical molecules. In practice, it is usually found that the geometric mean of the two values produces a theoretical result that agrees reasonably well with experiment. For example, the geometric mean of C_{vdw} for Ne ($= 4$) and methane ($= 102$) is

$$C_{vdw} = (4 \times 102)^{1/2} = 20$$

TABLE 4.4. Typical Values for Various Partial Contributions to Total van der Waals Interaction in Vacuum (10^{-79} J m^{-6}) According to Equations (4.40) and (4.41)

Interacting Units	C_{ind}	C_{oriet}	C_{disp}	Experimental Theoretical	From Gas Law	C_{disp} as % Total
Ne/Ne	0	0	4	4	4	100
CH$_4$/CH$_4$	0	0	102	102	101	100
HCl/HCl	6	11	106	123	157	86
HBr/HBr	4	3	182	189	207	96
HI/HI	2	0.2	370	372	350	99
CH$_3$Cl/CH$_3$Cl	32	101	282	415	509	68
NH$_3$/NH$_3$	10	38	63	111	162	57
H$_2$O/H$_2$O	10	96	33	139	175	24
Ne/CH$_4$	0	0	19	19	—	100
HCl/HI	7	1	197	205	—	96
H$_2$O/CH$_4$	9	0	58	67	—	87

compared to the experimental value of 19. For HCl–HI (123 and 372) the geometric mean of 214 also compares reasonably well with the measured value of 205.

Just as the van der Waals force theory breaks down somewhat for the interaction between two water molecules, so the geometric mean approximation begins to lose its usefulness for interactions between water and dissimilar molecules. For the interaction between water and methane, for example, the geometric mean approach predicts a van der Waals constant of 120 while experimental values lie more in the range of 60–70. What that tells us is that methane molecules and water molecules greatly prefer to interact with their "own kind." And it helps explain the observed fact that methane (and other hydrocarbons) are immiscible with water. The same applies for other molecules such as fats and oils which may have some polar nature, but are predominantly hydrocarbon in makeup, as well as fluorocarbons, silicones, and other compounds. Conventionally, this mutual "dislike" has come to be called hydrophobicity and the phenomenological effect the hydrophobic (water fearing) effect. We will see in the coming chapters that this "xenophobic" aspect of van der Waals forces has many important ramifications in real-world applications.

The simplified theory of van der Waals forces presented above is useful as a foundation for understanding the general principles involved in the physical interactions between atoms and molecules. As pointed out, it works well for many simple systems, due in part to the theory and in part to good fortune, but begins to fail when the system of interest begins to deviate from one of small, spherically symmetrical units. Unfortunately, almost all practical systems deviate significantly from those restrictions. In addition, the theory assumes that atoms and molecules have only one ionization potential, and it cannot readily handle the presence of an intervening medium other than a vacuum between the interacting units. Theories that attempt to take such limitations into consideration have been developed but are presently of little practical use, except, of course, to the theoretician.

4.4.6. Effects of a Nonvacuum Medium

The equation for the total van der Waals interaction between two atoms or molecules [Eq. (4.40)] includes a factor for corrections due to changes in the dielectric characteristics of an intervening medium other than vacuum. That aspect of the theory can be of great importance both quantitatively and qualitatively and has significant ramifications in practical systems. A full discussion of the theoretical aspects of the effects of medium on van der Waals interactions is beyond the scope of this book, but the reader is referred to the work by Israelachvili for further enlightenment. From a practical standpoint, however, several important points arise from an analysis of the dispersion force equation for media of differing dielectric constants. The relevant points include the following:

1. Since, for interactions in a vacuum, the ionization potential I (or $h\nu$) for most materials is much greater than kT for $\nu > 0$, the dispersion contribution to the total interaction is usually greater than the dipolar contributions for $\nu = 0$. This agrees with the results given in Table 4.4.

2. The magnitudes of the van der Waals interactions are greatly reduced in the presence of a medium other than vacuum. For example, in the case of two nonpolar molecules, the magnitude of their interaction in vacuum may be reduced by an order of magnitude in the presence of an intervening medium.

3. Dispersion force contributions in a medium other than vacuum may be either attractive or repulsive, depending on the relative ionization potentials of the materials involved.

4. In some cases where the interaction between two molecules in a solvent is very small, such as the lower molecular weight alkanes in water, the interaction becomes dominated, not by the dispersion force, but by an entropic term of the form

$$w(r)_{\nu = 0} \approx \frac{-kT\, a_1^6}{r^6} \tag{4.42}$$

where a_1 is the radius of the small interacting molecules. In general terms, this means that there is an overall increase in entropy (of the water molecules) as the distance between two alkane molecules decreases. This agrees qualitatively with the observation that the interaction between "hydrophobic" species such as alkanes in water is primarily entropic in nature.

5. For all but spherically symmetrical molecules, van der Waals forces are anisotropic. The polarizabilities of most molecules are different in different molecular directions because the response of electrons in a bond to an external field will usually be anisotropic. A consequence of this effect is that the dispersion force between two molecules will depend on their relative molecular orientation. In nonpolar liquids, the effect is of minor importance because the molecules are essentially free to tumble and attain whatever orientation is energetically favorable. However, in solids, liquid crystals, and polar media, the effect can be important in determining the relative fixed orientation between molecules, thereby affecting or controlling specific conformations of polymers or proteins in solution, critical transition temperatures in liquid crystals and membranes, and so on. Repulsive forces in polar molecules are also orientation dependent, and are often of greater importance in controlling conformations and orientations.

6. Van der Waals forces are nonadditive and are affected by the presence of other interacting bodies in the vicinity. What this means is that the total interaction among a group of molecules or particles will not be a simple sum of the individual pairwise interactions. In fact, in most cases, a molecule interacting with a second molecule in a group not only will experience the force of interaction directly, but will also feel a "reflected" force due to the polarization of other neighboring molecules polarized by the first. The net result will be a total interaction somewhat greater than that which would be predicted by a simple summation. Although the effect is usually small (perhaps 30% of the "normal" interaction) it can be significant, especially in the case of relatively large particles interacting with a surface.

7. Over relatively large distances, dispersion forces experience a retardation effect that results from the nature of the fluctuating dipoles which give rise to the interaction. When two molecules are an appreciable distance apart, the electric field generated by a fluctuating dipole takes a finite amount of time to reach a neighboring molecule. By the time the second "induced" dipole can retransmit its effect back to the first, the first has had time to reorient itself and may no longer have an orientation suitable for maximum interaction. Therefore the total interaction will be reduced in magnitude. Thus, at large distances, the magnitude of the dispersion interaction is found to fall off faster than predicted by the r^{-6} relationship. In a vacuum, retardation effects begin to appear at distances of approximately 5 nm, which makes them of little practical importance. However, in a liquid medium they can begin to be observed at shorter distances and become important, as in the interaction between particles and surfaces in liquid media. That is the case because in condensed media, the speed of light is reduced, thereby allowing more time between cause and effect, weakening the net interaction. Since it is only dispersion forces that suffer from retardation effects, induced and orientational dipolar contributions may become more important to the total van der Waals interaction.

4.5. INTERACTIONS BETWEEN SURFACES AND PARTICLES

The above discussion centered on the forces controlling the interactions between two isolated atoms or molecules. For multiunit systems it is assumed that the units interact mutually according to the Lennard–Jones potential, and that the total interaction is the sum of all individual interactions. For the repulsive term, it is common to neglect the repulsive component for units in one bulk phase and consider only repulsion between opposing surfaces.

Mathematically, the simplest situation to analyze is that involving two hard, flat, effectively infinite surfaces separated by a distance, H, in a vacuum. The free energy of attraction per unit area in such a case is approximated by

$$\Delta G_{att} = \frac{-A_H}{(12\pi H^2)} \qquad (4.43)$$

where A_H is the Hamaker constant. The value of A_H is related to A of Equation (4.31) by

$$A_{\mathrm{H}} = \tfrac{3}{4}h\nu\alpha^2\pi^2 n^2 = A\pi^2 n^2 \qquad (4.44)$$

where n is the number of atoms or molecules in unit volume of the phase. For two identical spheres of radius a, where $H/a \ll 1$, a similar type of approximate equation is

$$\Delta G_{\mathrm{att}} = \left(\frac{-A_{\mathrm{H}}a}{12H}\right)\left[\frac{1 + (\tfrac{3}{4})H}{a + \text{higher terms}}\right] \qquad (4.45)$$

In most practical instances, it is safe to neglect all of the higher terms.

A comparison of Equations (4.30) and (4.43) shows that the free energy of attraction between two surfaces falls off much more slowly than that between individual atoms or molecules. This extended range of bulk interactions plays an important role in determining the properties of systems involving surfaces and interfaces. A combination of the attractive and repulsive forces between surfaces leads to a curve such as in Figure 4.7a.

Because interactions between surfaces fall off much more slowly with distance than those for individual atoms or molecules, the retardation effect mentioned above becomes more significant. However, while the retardation effect is important in quantitative theoretical discussions of surface interactions, from a practical standpoint, it is still relatively insignificant compared to other factors.

4.5.1. Surface Interactions in Nonvacuum Media

The equations for surface interactions given above were derived for the situation in which the interacting units were separated by a vacuum. Obviously, for practical purposes, that usually represents a rather unrealistic situation. "Real life" dictates that in all but a few situations, interacting units be separated by some medium that itself contains atoms or molecules that will impose their own effects on the system as a whole. How will the relevant equations be modified by the presence of the intervening medium?

Surfaces interacting through an intervening fluid medium will experience a reduced mutual attraction due to the presence of the units of the third component. The calculation of interactions through a vacuum involves certain simplifying assumptions, therefore it is not surprising to find that models for three component systems are even more theoretically complex. Although a number of elegant approaches to the problem have been developed over the years, for most purposes a simple approximation of a composite Hamaker constant is found to be sufficient. When two surfaces of component 1 are separated by a medium of component 2, the effective Hamaker constant (A_{Heff}) is approximated by

$$A_{\mathrm{Heff}} = [A_{10}^{1/2} - A_{20}^{1/2}]^2 \qquad (4.46)$$

where A_{10} is the Hamaker constant for component 1 in a vacuum, and A_{20} is that for component 2. A result of the relationship in this equation is that as the vacuum Hamaker constants for 1 and 2 become more alike, the effective Hamaker constant tends toward zero, and the free energy of attraction between the two surfaces of component 1 is also reduced to zero. As we will see in Chapter 9, such a reduction in the attractive forces due to an intervening medium gives one a handle on ways to successfully prevent the spontaneous joining of surfaces thereby imparting a certain added stability to the separated system. Since Equation (4.46) involves the square of the difference between the Hamaker constants for components 1 and 2, the same will be true for surfaces of component 2 separated by a medium of 1. The form of the interaction curve for the above situation will be the same as that for the vacuum case, although the exact values will differ because of the different value of the effective Hamaker constant.

Qualitatively, the preceding discussion of surface interactions tells us that free surfaces are inherently unstable and will usually experience a net attraction for similar surfaces in the vicinity. The practical repercussion is that if only the van der Waals forces were involved, systems involving the formation and maintenance of expanded interfaces would all be unstable and spontaneously revert to the condition of minimum interfacial area, thereby making impossible the preparation of paints, inks, cosmetics, many pharmaceuticals, many food products, emulsions of all kinds, foams, bilayer membranes, etc. It would be a decidedly different world we lived in. In fact, life as we know it (or can conceive of it) would not exist! Obviously, something is or can be involved at interfaces that alters the simple situation described above and makes things work. In the following chapters we will introduce other "actors" that allow nature (and humankind, when we're lucky) to manipulate surfaces and interfaces to suit our purposes.

4.5.2. Dipole, Induced Dipole, and Hydrogen Bonding (Acid–Base) Interactions

In the preceding sections the discussion centered on the source of attractive forces between atoms, molecules, and larger material units. The London-van der Waals forces were characterized as being universal and almost always attractive over relatively long distances. Many, if not most, practical systems, however, involve situations in which the substances and surfaces in question are composed of materials that can interact through forces other than London–van der Waals attractions. Such non-quantum-mechanical interactions can be classed generally as electrostatic in nature. That is, they involve to one degree or another the interaction of partial or complete electronic charges. Because electrostatic interactions can occur between like charges (repulsive) or unlike charges (attractive), they can have different and significant effects on the characteristics of atomic, molecular, and interfacial interactions.

4.6. LIFSHITZ THEORY: A CONTINUUM APPROACH

As shown above, there have been identified several mechanisms involved in the interactions between atoms and molecules, denominated collectively as the van der Waals forces. In atomic and completely nonpolar molecular systems (hydrocarbons, fluorocarbons, etc.) the London dispersion forces provide the major contribution to the total interaction potential. However, in many molecular systems containing atoms of very different electronegativities and polarizabilities the dipole–dipole (Keesom) and dipole-induced dipole (Debye) forces may also make significant contributions to the total interaction.

The basic derivations of the van der Waals forces is based on isolated atoms and molecules. However, in many particle calculations or in the condensed state major difficulties arise in calculating the net potential over all possible interactions. The Debye interaction, for example is non additive so that a simple integration of Equation (4.27) over all units will not provide the total dipole-induced dipole interaction. A similar problem is encountered with the dipole–dipole interactions which depend not only on the simple electrostatic interaction analysis, but must include the relative spatial orientation of each interacting pair of dipoles. Additionally, in the condensed state, the calculation must include an average of all rotational motion. In simple electrolyte solutions, the (approximately) symmetric point charge ionic interactions can be handled in terms of a dielectric. The problem of van der Waals forces can, in principle, be approached similarly, however, the mathematical complexity of a complete analysis makes the Keesom force, like the Debye interaction, effectively nonadditive.

The problem was eventually solved (in so far as a theory can be considered a solution) by Lifshitz and co-workers by employing a continuum electrodynamics approach in which each unit or medium is described by its frequency-dependent dielectric permittivity $\varepsilon_r(\omega)$. Because of the nature of the "beast," an extensive derivation of the Lifshitz theory lies well beyond the scope of this book. However, a brief discussion will aid the reader in seeing the differences and similarities between it and the Hamaker approach.

Even though the two theories of atomic and molecular interaction appear to arise from distinctly different initial premises, the end results of the two are perhaps surprisingly alike. However, the interactions involved are the same, but they are expressed in different ways. For example, the Lifshitz theory predicts the same distance dependence of the overall interactions as that of Hamaker.

The Lifshitz formulation of the Hamaker constant (or its equivalent) for two like bodies (1) interacting through a second medium (2) is given by

$$A_{H_{121}} = \tfrac{3}{4}kT\left\{\frac{\varepsilon_1(0) - \varepsilon_2(2)}{\varepsilon_1(0) + \varepsilon_2(0)}\right\}^2 + \frac{3h\nu}{4\pi}\int_{\frac{kT}{h}}^{\infty}\left\{\frac{\varepsilon_1(i\omega) - \varepsilon_2(i\omega)}{\varepsilon_1(i\omega) + \varepsilon_2(i\omega)}\right\}^2 dw \quad (4.47)$$

in which the first term on the right corresponds to the Keesom and Debye (i.e., electrostatic) interactions. The term kT defines the interactions as being primarily entropic in nature with an maximum value of $\frac{3}{4}(kT)$, since

$$\left\{\frac{(\varepsilon_1(0) - \varepsilon_2(2))}{(\varepsilon_1(0) + \varepsilon_2(0))}\right\} \leq 1$$

At $T = 300$ K, $kT \approx 3 \times 10^{-21}$J, which is an order of magnitude less that the dispersion contribution. The actual difference between the two terms (dispersion and electrostatic) will be reduced by mathematical cancellations in the second (dispersion) term in Equation (4.47), but only rarely will the electrostatic contribution constitute the dominant factor in the total interaction. The presence of imaginary frequencies in the second term may cause some problems in terms of physical concepts of the processes involved; however, their use is actually a result of mathematical manipulations (i.e., tricks) that disappear as one works through the complete calculation.

For atoms and molecules of low to moderate polarizability, one can estimate the relative permittivity ε_r based on polarizability a using the Clausius–Mossatti relationship

$$\frac{\varepsilon_r - 1}{\varepsilon_r + 2} = \frac{\alpha\rho}{3\varepsilon_0} \tag{4.48}$$

where ρ is the number density of units involved. For optical frequencies, one can estimate ε_r as the square of the refractive index, so that for most materials of interest $2 < \varepsilon_r < 3$ and

$$\frac{0.2 < \alpha\rho}{3\varepsilon_0} < 0.4$$

Rearranging this equation 4.48 gives

$$\varepsilon_r = \frac{1 + (2\alpha\rho)/(3\varepsilon_0)}{1 - (\alpha\rho)/(3\varepsilon_0)} \tag{4.49}$$

which can then be used in the integration of Equation (4.47).

If the intervening medium (2) is taken as vacuum, then $\varepsilon_2 = 1$ at all frequencies and the dispersion contribution in (4.47) reduces to

$$\frac{3h\nu}{4\pi} \int_{\frac{kT}{h}}^{\infty} \left\{\frac{(\alpha(i\omega)\rho)/\varepsilon_0}{2 + (\alpha(i\omega)\rho/3\varepsilon_0)}\right\}^2 d\omega \tag{4.50}$$

which can be further simplified by assuming that the second term in the denominator (<0.4; see above) can be neglected. The final expression for the dispersion term then becomes

$$H_{11} \approx \frac{3(\rho/\varepsilon_0)^2}{16\pi} \int\limits_{kT}^{\infty} \alpha^2(i\omega) \, d(h\omega) \tag{4.51}$$

With the appropriate expression for $\alpha^2(i\omega)$, one can (with a little effort) integrate and show that the dispersion contribution is essentially equivalent to that determined using the simpler Hamaker theory. Some representative values for Hamaker constants for typically encountered materials are shown in Table 4.5. But why go to all that trouble? you may ask. As pointed out at the beginning, the assumptions and approximations that lead to the result in Table 4.5 are based on a low to moderate polarizability for the interacting units. For strongly interacting and polarizable materials such as most metals, the Hamaker approach begins to break down significantly and one must make recourse to the more complex Lifshitz theory.

TABLE 4.5. Typical Values of the Hamaker Constant ($\times 10^{20}$ J^{-1}) for Commonly Encountered Materials (A and B) and Intervening Phases (2)

Material A	Material B	Intervening Phase (2)	A_H
Water	Water	Air	3.7
C_5 alkane	C_5 alkane	Air	3.8
C_5 alkane	C_5 alkane	Water	0.3
C_5 alkane	Water	Air	3.6
C_5 alkane	Air	Water	0.15
C_{10} alkane	C_{10} alkane	Air	4.8
C_{10} alkane	Water	Air	4.1
C_{10} alkane	Air	Water	−0.3
C_{16} alkane	C_{16} alkane	Air	5.2
C_{16} alkane	Water	Air	4.3
C_{16} alkane	Air	Water	−0.5
Fused quartz	Fused quartz	Air	6.5
Fused quartz	Fused quartz	Water	0.8
Fused quartz	Water	Air	4.8
Fused quartz	Air	Water	−1.0
Polystyrene	Polystyrene	Air	6.6
Polystyrene	Polystyrene	Water	1.0
Polystyrene	Water	Air	4.8
Polystyrene	Air	Water	−1.1
PTFE	PTFE	Air	3.8
PTFE	PTFE	Water	0.1
PTFE	Water	Air	3.7
PTFE	Air	Water	0.1

As mentioned in an earlier section, the dispersion interactions exhibit a quantum-mechanical retardation effect at large (on the atomic or molecular scale) distances. Such effects are brought out explicitly by the Lifshitz theory, so that, for example, long range interactions become proportional to r^{-3} instead of r^{-2}, where r is the distance of separation, as is observed experimentally. Luckily, however, the effects are seldom of concern in practical systems since their magnitude is extremely small relative to other factors.

4.6.1 Some Shortcomings of the Hamaker and Lifshitz Theories

While the Lifshitz model of atomic and molecular interactions performs nobly for the analysis of interactions of atoms, molecules, or particles in a continuous medium, when one reaches down to the atomic or molecular level, the concept of continuity must per force be abandoned: atoms and molecules exist as definite individual species with defined (if fuzzily, at times) shapes and sizes. While we can understand, or at least rationalize, observed effects in the context defined by the Lifshitz and Hamaker models, once separation distances begin to approach molecular dimensions, we enter into a world in which the concept of continuity has no real physical meaning. Modern experimental techniques now allow us to reach to that level and the observed results demand new explanations.

As an example, it has been pointed out that the Hamaker and Lifshitz theories assume (explicitly and implicitly, respectively) that intensive physical properties of the media involved such as density, and dielectric constant, remain unchanged throughout the phase—that is, right up to the interface between phases. We know, however, that at the atomic or molecular level solids and liquids (and gases under certain circumstances) exhibit short-range periodic fluctuations; they are damped oscillating functions. Conceptually, if one visualizes a liquid in contact with a flat solid surface (Fig. 4.8a), one can see that the molecules (assumed to be approximately spherical, in this case) "trapped" between the surface and the bulk of the liquid will have less translational freedom relative to the bulk and therefore be more structured. That "structure" will (or may) result in changes in effective intensive properties near the surface.

4.6.2. Hard Sphere Diameter Effects

Intermolecular repulsion forces will define a hard-sphere diameter for the molecules allowing the calculation and/or measurement of the density fluctuation as one moves away from the surface. As shown in Figure 4.8b, the result is a damped density curve (density distribution versus distance in molecular diameters from the surface). For a liquid between two surfaces at short separation distances, the two effects will overlap producing an interference pattern as illustrated in Figure 4.9. When the distance of separation is some integral multiple of the hard sphere diameter, $n\lambda_{hs}$, reinforcement occurs producing a local free energy minimum. For separations that correspond to fractional

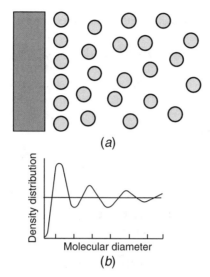

FIGURE 4.8. The structuring of liquid molecules near a surface (*a*) leads to an ordering of the "trapped" molecules to produce a regular, damped molecular density distribution as a function of molecular diameter as one moves away from the surface. The structuring near a surface may result in changes in the properties of the liquid near the surface (*b*).

FIGURE 4.9. A liquid trapped between two closely spaced surfaces can produce a highly ordered structure leading to possibly "interesting" changes in the characteristics of the liquid.

multiples, $(n + \frac{1}{2})\lambda_{hs}$, destructive interference occurs and a local energy maximum results. For molecules that diverge significantly from spherical the periodic effect decreases or disappears completely.

Although the impact of this molecular-level effect is quite small (negligible in most cases), it can, under some circumstances, produce an appreciable energetic effect—repulsive in the case of two approaching surfaces—that is, it will be difficult to displace the last molecular layers separating the surfaces—or attractive in the separation of two contacting (adhesive) surfaces. These topics will appear again in Chapters 10 (on colloids and colloidal stability), 15 on association colloids: micelles, vesicles and membranes), and 19 on adhesion).

4.7. HYDRODYNAMIC FLOW EFFECTS IN INTERFACIAL INTERACTIONS

The discussions so far have concentrated on equilibrium interactions between atoms, molecules, particles, and surfaces with little attention being paid to possible kinetic effects. In vacuum, one may assume that such effects will be negligible or absent until distances of separation approach molecular dimension. At that level effects may be seen that reflect any or all of the following processes:

1. The kinetics of ionic association or dissociation at the interface. The rate of ionization or ion pairing of surface charges will retard shifts to the "equilibrium" electrostatic interactions.
2. The slow movement or displacement of adsorbed species such as polymers. Changes in the configuration of adsorbed species (i.e., polymers or surfactants) may be slow relative to the velocity of approach of the two surfaces, or desorption rates (especially for polymers) will be much smaller.
3. The periodic packing effects discussed above come into play. The removal of the last intervening molecular layers of solvent will be more difficult due to the energy advantage gained by the periodic packing.
4. Other interesting effects that may occur to the perplexed investigator looking for an explanation for unexpected results.

An intervening gaseous phase would likewise be expected to have little effect. In an incompressible liquid, however, the potential role of kinetic (or hydrodynamic) effects at separations much larger than molecular dimensions can become significant. When a particle or large molecule moves through a liquid, molecules of the continuous phase must be moved from in front of the moving unit and replaced in the space vacated immediately behind. While the step-by-step process of such movement has a net energy change of zero (for an isolated particle), it does involve the application of a force, F, that will be transmitted through the surrounding liquid. The idea is represented

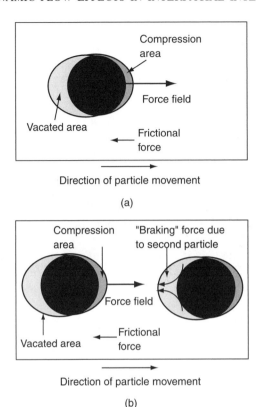

FIGURE 4.10. An isolated particle moving through a liquid medium (*a*) will exhibit a diffusion coefficient depending basically on the size of the particle and the viscosity of the liquid. In the presence of additional particles (*b*) the diffusion coefficient may be altered by a "braking" interaction due to interactions with adjacent particles.

schematically in Figure 4.10*a*. The force required for movement and the resulting velocity, *v*, are related by the friction coefficient, μ_f

$$\mu_f = \frac{F}{v} \tag{4.52}$$

The friction coefficient can be calculated from the Stokes equation

$$\mu_f = 6\pi R \eta \tag{4.53}$$

where R is the particle radius and η is the viscosity of the liquid. The diffusion coefficient for the particle (or molecule), D_0, is given by the Einstein equation as

$$D_0 = \frac{kT}{\mu_f} \tag{4.54}$$

Combining Equations (4.53) and (4.54) produces the equation for the diffusion (self-diffusion, in this case) of an isolated particle in a liquid medium

$$D_0 = \frac{kT}{6\pi\eta R} \tag{4.55}$$

If a second particle is present in the system, as will be the case in almost any imaginable situation, the force field produced by one moving particle will affect and be affected by the second (Fig. 4.10b). Because the second particle produces a "braking" effect on the movement of the first, the effective diffusion coefficient of particle 1 will decrease in a manner proportional to the distance separating the two. If it is assumed for purposes of simplification that particle 2 does not move, the relationship between the braking effect and the self-diffusion coefficient can be written as

$$D_r = D_0 G_r \tag{4.56}$$

where D_r is the retarded (or "braked") diffusion coefficient and G_r is a complex correction factor dependent on the relationship between separation distance, particle shape, and so on and takes the form

$$G_r = \frac{C_1 x + C_2 x^2 + x^3}{C_3 + C_4 x + C_5 x^2 + x^3} \tag{4.57}$$

The derivation of the expression for G_r involves numerical approximations that will not be presented here. The interested reader is referred to the original work for specifics. Values for sphere–sphere and sphere–plane interactions are presented in Table 4.6. The value x relates the distance of separation, r, to the particle radius R.

$$x = \frac{r - 2R}{R} \tag{4.58}$$

TABLE 4.6. Hydrodynamic Correction Factors, G_r, for Diffusion in Two-Particle or Particle-Surface Systems

	C_1	C_2	C_3	C_4	C_5
Sphere–sphere	0.154030	1.29993	0.0782416	1.10529	2.81955
Sphere–plane	2.04185	5.60414	2.06393	8.59190	6.72180

Source: D. Y. Chan and B. Halle, *J. Colloid Interface Sci.* **102,** 400 (1984).

At very small separation distances, the value of G_r becomes very small indicating that the velocity of approach of the moving particle to the second particle or surface becomes small as well. Such an effect may be expected to become significant in contexts such as particle flocculation, emulsion stability, and particle deposition onto surfaces. In summary, these hydrodynamic effects will be "repulsive" for approaching particles or surfaces and "attractive" for receding systems. It can be seen from Table 4.6 that the effect is much greater in the case of sphere–plane approach than that of sphere–sphere interaction.

It should be kept in mind that the hydrodynamic effects mentioned here are kinetic in nature. They do not affect the overall thermodynamic equilibrium of the final system configuration. However, in practical multiparticle systems the simple picture given here breaks down in terms of a quantitative evaluation, although the qualitative importance of the effects remains.

PROBLEMS

4.1. If two water molecules hydrogen bond in such a way that all six atoms lie in the same plane, calculate the electrostatic interaction between the molecules at an average oxygen–oxygen distance of 0.28 nm.

4.2. The Keesom interaction energy for freely rotating dipoles is obtained from an expansion series. That series becomes inaccurate as the interaction energy approaches a value of kT. Calculate the dipole moment at which this occurs at room temperature and a separation distance of 0.28 nm?

4.3. Estimate the dispersion interaction between two parallel plates of thickness d separated by a distance h using the Hamaker theory.

4.4. The Hamaker constant for a material composed of two different kinds of atoms can be calculated by replacing the quantity $n^2\alpha^2$ in Equation (4.44) by $(n_1^2\alpha_1^2 + n_2^2\alpha_2^2 + 2n_1n_1\alpha_1\alpha_1)$. Explain why this formulation should be correct and calculate A_H for ice, taking the polarizability of H and O to be 0.67×10^{-24} and 3.0×10^{-24} cm^3, respectively. Calculate the expected force between two thick flat slabs of ice 1 nm apart. Do the same for two spherical particles of ice 2000 nm in diameter and 1 nm apart.

4.5. The Hamaker constant can be calculated from the following expression

$$A_H = \pi^2 q^2 \lambda = \tfrac{3}{4}\pi^2 q^2 h \nu_0 \, \alpha_0^2$$

where n_0 is the characteristic frequency of the material and α_0 the electron polarizability near $n = 0$. α_0 and n_0 can be obtained from the Lorentz–Lorenz relationship

$$\frac{n^2-1}{n^2+2} \frac{M}{\rho} = 4/3 \ \pi N_A \alpha(\nu)$$

where n is the refractive index, M the molecular weight, ρ the density, and N_A Avagadro's number, if $\alpha(\nu)$ assumes the simplified form

$$\alpha(\nu) = \frac{e^2}{4\pi^2 m_e} \frac{\sigma_N}{\nu_0^2 \nu^2}$$

where e = electron charge, m_e = electron mass, σ_N = effective number of electrons per molecule. Using the above procedure, calculate A_H for polystyrene using the following data:

$\nu \times 10^{-14}$ (s^{-1})	$(n^2-1)/(n^2+2)$	$\nu \times 10^{-14}$ (s^{-1})	$(n^2-1)/(n^2+2)$
6.16	2.900	4.58	2.980
5.10	2.947	4.06	3.005

given that ρ = 1.08 g cm^{-3}, h = 6.63 $\times$ 10^{-34} J s^{-1}.

4.6. Calculate the contributions of the Keesom, Debye, and London forces separately for H_2O and HI given the following data:

	$\mu \times 10^{18}$(e.s.u.)	$\alpha \times 10^{24}$ (cm^3)	$h\nu_0$ (eV)
H_2O	0.38	5.4	12
HI	1.84	1.48	18

4.7. Estimate the free energy of attraction between two spheres of radius 500 nm at separation distances of 1, 10, and 100 nm for the following systems: (a) water–air–water; (b) pentane–water–pentane; (c) hexadecane–water–hexadecane; (d) quartz–air–quartz; (e) quartz–air–hexadecane; and (f) Teflon–water–Teflon.

4.8. Compare the two expressions for the Hamaker constant according to the van der Waals and Lifshitz theories presented in the chapter and discuss any significant differences one might expect from the application of each theory for the interactions between two spherical particles in a second liquid medium.

4.9. Given two spherical particles in a liquid medium, discuss qualitatively the different interaction energies to be expected if (a) the continuous medium is quiescent; (b) the continuous phase is undergoing laminar flow; and (c) the two particles are approaching one another "head on" in a situation in which two columns of liquid are flowing in opposite directions. Sketch each situation.

5 Electrostatic Forces and the Electrical Double Layer

Most solid surfaces in contact with water or an aqueous solution will be found to develop some type of electrical charge. The magnitude of that charge may be quite small or very large, but it will almost always exist. In macroscopic systems, the presence or absence of a surface charge may often be overlooked (except, perhaps, on a dry winter day when one walks across a rug and gets a rude shock when reaching for the door handle). However, in the microscopic world of colloids and interfaces, the presence or absence of even a small surface charge can have significant ramifications in terms of stability, sensitivity to environment, electrokinetic properties, and other factors. The previous chapter introduced the basic concepts governing coulombic interactions between atoms, molecules, and bulk materials. The following will build on that base and lay the groundwork for the application of electrostatic principles to the problems of stability in colloidal systems discussed in later chapters.

As a result of the presence of electrical charges, surfaces exhibit various properties that are not present in systems having no surface charge (e.g., nonpolar solids dispersed in a nonpolar medium). Of particular interest are the so-called electrokinetic properties of colloids, discussed later, and, perhaps more significantly, the important mechanism for the stabilization of many colloids, both natural and manmade, that they provide. As a result of their unique characteristics, charged surfaces provide a handle for manipulating a wide variety of multiphase systems that are of vital importance to life and technology. By the proper control of surface charges, we may be able to transform an interesting but useless system into one of great practical importance. Likewise, we may be able to take a system that is a nuisance colloid and, by changing its electrical environment, remove it completely or at least attenuate its effects.

5.1. SOURCES OF INTERFACIAL CHARGE

An interface may acquire an electrical charge by one or more of several mechanisms, the most common of which include (1) preferential (or differential) solution of surface ions, (2) direct ionization of surface groups, (3) substitution of surface ions, (4) specific ion adsorption, and (5) charges deriving from specific crystal structures. While other mechanisms can be invoked, these five represent the most common and most important encountered in colloidal systems. Each type of surface charge carries with it certain characteristics that define, partially at least, the electronic nature of the resulting colloid or surface. The five main classes are illustrated in Figure 5.1.

FIGURE 5.1. The principle sources of surface charge in solids include (*a*) differential ion solubility phenomena, (*b*) direct ionization of surface groups, (*c*) isomorphous substitution of ions from solution, and (*d*) specific-ion adsorption from the solution phase; (*e*) anisotropic crystal lattice structures.

5.1.1. Differential Ion Solubility

The preferential solubilization of ions from the surface of a sparingly soluble crystalline material represents one of the most common and earliest recognized mechanisms for the development of electrical charges on solid surfaces. A widely encountered example of such an effect is in the silver halide colloids (incorrectly, but commonly referred to in the business as "emulsions") used in photographic products.

The silver salts of chlorine, bromine, and iodine are of very limited solubility in water. When crystals of, for example, silver iodide are placed in water, ions dissolve from the surface until the product of the concentration of the two ions is equal to the solubility product of the material. For silver iodide (AgI), that would be $K_{SP} = [Ag^+][I^-] = 10^{-16}$ M. If the two ions were dissolved equally readily, their concentrations in solution would be equal (10^{-8} M), as would their occurrence on the surface of the crystal, leaving a net surface charge of zero. One finds, however, that under certain conditions silver ions may be more readily dissolved, so that their concentration in solution is greater than that of iodide, while the concentration of I^- on the surface of the crystal is enhanced, leading to the formation of a net negative charge on the surface. The situation is shown in Figure 5.1a.

For example, if a soluble silver salt (e.g., $AgNO_3$) is added to the solution, the dissolution of Ag^+ from the crystal surface will be suppressed (by the common ion effect) and the negative surface charge correspondingly reduced. At some characteristic silver ion concentration, the dissolution of silver ion from the crystal will be zero, and there will be no surface charge. That point is termed the *point of zero charge* (ZPC). If addition of Ag^+ is continued, a net positive surface charge will develop. If soluble iodide salts had been added to the solution, the solution of I^- would have been suppressed further, leading to an increase in the net negative surface charge on the crystal.

The control of the magnitude and sign of the surface charge on materials such as the silver halides by controlling the concentration of one of the two ions through the common ion effect gives one a useful handle for the manipulation of colloidal systems containing such materials. In fact, most of the original classical studies of inorganic colloids or "sols" were based on this stabilization mechanism.

5.1.2. Direct Ionization of Surface Groups

Materials containing surface groups that can be directly ionized, but in which one of the ions is permanently bound to the surface, illustrate a second important mechanism for the development of surface charge. This group of materials includes many metal oxides as well as many polymer latexes (Fig. 5.1b). Some metal oxides are amphoteric in that they can develop either negative or positive surfaces, depending on the pH of the solution. Such surfaces will obviously exhibit a characteristic point of zero charge such as that found for the silver halides, except that it will more likely be controlled by pH rather than the concentration of a common ion.

Typical examples of ionizable polymer surfaces include those containing carboxylic and sulfonic acids and their salts ($-COOH$, $-COO^-\ M^+$, $-SO_3H$, and $-SO_3^-\ M^+$), sulfuric acid esters and their salts ($-OSO_3H$ and $-OSO_3^-$ M^+), basic amino groups ($-NR_3$, $R = H$ or an organic group), and quaternary ammonium groups ($-NR_4^+X^-$). In some cases, the degree of ionization will be determined by the pH and the acid (or base) strength of the ionizable group. For weak acids and bases, such as carboxyl and amino groups, the surface charge will be specific, either negative or positive, respectively, or zero. In the case of the sulfonates and sulfuric acid esters, since they are strong acids, the complete suppression of their charge will require a very low pH (all $pK_a \approx 1$). Similarly, quaternary ammonium salts will be essentially independent of pH, giving a positively charged surface under all conditions, although the degree of ionization can be suppressed by high ionic strength in the solution. In the special case of surfaces containing amino acid groups (e.g., $-C(NH_2)COOH$) the surface may acquire a net positive or negative charge, depending on the pH.

5.1.3. Substitution of Surface Ions

Many minerals, clays, oxides, and other compounds can undergo isomorphous substitution, which means that structural ions are substituted by ions of valency one less than the original. For example, a silicon atom (valency $= 4^+$) in clay may be replaced by aluminum (3^+), producing a surface with a net negative charge (Fig. 5.1c). Such a surface can be brought to its ZPC by lowering the pH. A similar effect can be observed in the substitution of, for example, sulfate (SO_4^-) for chloride (Cl^-) in a crystal lattice.

5.1.4. Specific-Ion Adsorption

Some surfaces that do not possess a direct mechanism for acquiring a surface charge may do so by the adsorption of specific ions that impart a charge to the surface. Gold sols produced by the reduction of $HAuCl_4$ adsorb "free" chloride ions to produce a surface that apparently has a structure related to $AuCl_4^-$. Of particular practical importance is the adsorption of surfactant ions onto surfaces. In that case, the adsorption of an anionic surfactant produces a negatively charged surface while the adsorption of a cationic surfactant produces a positively charged surface (Figure 5.1d). The classification of adsorption phenomena and their consequences will be discussed in more detail in later chapters.

A variation on the theme of specific-ion adsorption is encountered when a surface charge arises from the dissociation of a salt, say a sodium carboxylate ($-COO^- + Na^+$), to produce a negatively charged surface. If di- or trivalent ions are present in the solution, they may adsorb onto the surface in such a manner that the net result is a charge inversion from a negative to a positive surface charge. Such an ion-exchange mechanism has been seen in many systems, including biologically important bilayer membrane structures.

5.1.5. Anisotropic Crystals

Some important materials, such as kaolinite clay used for making fine china, are composed of aluminosilicates and have crystal structures that, when cleaved, can result in the production of both positively and negatively charged surfaces. Depending on the crystal face exposed, there may be positive $AlOH_2^+$ groups or negatively charged basal groups, or (more likely) both. As a result, such materials may exhibit very special properties, including the formation of characteristic open structures as illustrated in Figure 5.1e. They may also show the apparent existence of more than one ZPC. Typically, an aluminosilicate clay may be found to increase its volume tenfold on addition of water due to the special properties of its anisotropic crystal structure.

The above five mechanisms for the formation of surface charges cover the great majority of examples encountered in colloidal system. In order to appreciate the significance of their formation and activity in the context of colloidal systems, it is necessary to return to some basic principles of electrostatics.

5.2. ELECTROSTATIC THEORY: COULOMB'S LAW

The fundamental law governing interactions between charged species was introduced in the preceding chapter. As a reminder, however, for the interaction of two charges q_1 and q_2 in a vacuum (F_{el}) separated by a distance r, the law takes the form

$$F_{el} = \frac{q_1 q_2}{4\pi\varepsilon_0 \varepsilon r^2} \tag{5.1}$$

where the symbols are as defined previously. The work necessary to bring the two charges together from infinity to the distance r is

$$w(r) = -\int_\infty^r F_{el} \, dr = \frac{q_1 q_2}{(4\pi\varepsilon_0 \varepsilon r)} \tag{5.2}$$

For charges of the same sign, $w(r)$ will be positive and the interaction will be repulsive; if of opposite charge, it will be attractive.

Consider for a moment that a charge q_1 is isolated in space. It will produce an electric field at a point r, such that the work necessary to bring a unit electrical charge from infinity to distance r from q_1 will be equal to $q_1/(4\pi\varepsilon_0 \varepsilon r)$. That quantity of work is defined as the electrical potential at r due to the charge q_1, and is given the symbol ψ. According to Equation (5.1), the force involved in bringing a charge q_2 to within a distance r of q_1 will be simply $q_2 \psi$.

5.2.1. Boltzmann's Distribution and the Electrical Double Layer

While Coulomb's law is simple enough in isolation, in order to make use of it in more realistic situations, the presence of all ions in the system must be taken into consideration. In practical applications, one is concerned with solutions containing many charges (i.e., dissolved ions), particles, and surfaces that may also contain charges of the same or different sign. In order to apply Coulomb's law to solutions of electrolytes and colloids, it is necessary to employ Boltzmann's distribution law, which relates the probability of a unit (atom, molecule, ion, particle, etc.) being at a certain point with a specified free energy (or potential energy), ΔG, relative to a specified reference state. The probability is generally expressed in terms of an average unit concentration, c, at the point r relative to a concentration, c_0, at some reference distance at which the energy is taken as zero. At a temperature, T, the Boltzmann distribution is given as

$$c = c_0 \exp \frac{-\Delta G}{kT} \tag{5.3}$$

When applied to the situation involving charged particles and Coulomb's law, Equation (5.3) predicts that if there exists a negative electrical potential ψ at some point in an electrolyte solution, then in the region of that point the concentration of positive charges, c_+, will be given by

$$c_+ = c_0 \exp \frac{-z^+ e\psi}{kT} \tag{5.4}$$

where z^+ is the valency of the positive ion and c_0 is the concentration of the positive ion in a region where $\psi = 0$. A similar expression can be written for the negative ions in the solution

$$c_- = c_0 \exp \frac{+ z^- e\psi}{kT} \tag{5.5}$$

Although the solution as a whole will be electrically neutral, in the vicinity of the electrical potential there will exist an imbalance of electrical charges. Thus, for a negative ψ, there will be more positive ions in the region than negative ions. For the case where $z^- = z^+ = 1$, the excess is given by

$$c_+ - c_- = c_0 \left[\exp \left(\frac{-e\psi}{kT} \right) - \exp \left(\frac{+e\psi}{kT} \right) \right] \tag{5.6}$$

The region of excess charge of the opposite sign around a potential is commonly referred to as the "ionic atmosphere" or "charge cloud" associated with that potential.

5.2.2. Double-Layer Thickness: The Debye Length

In the broad field of physical chemistry, the Boltzmann distribution law is fundamental to the derivation of the Debye–Hückel theory of electrolyte solutions. In the more narrow arena of interfacial and colloid science, it is applied to the determination of the ionic atmosphere around charged interfaces. In that context, the charge cloud is more commonly referred to as the electrical double layer (EDL). The concept is illustrated schematically (Fig. 5.2) for the situation in which a particle possesses an evenly distributed charge that is just balanced by the total opposite charge, the counterions in the electrical double layer.

The idea of the electrical double layer was first formally proposed by Helmholtz, who developed the concept of a system having charges arranged in two parallel planes as illustrated in Figure 5.2a. Such a situation describes, in essence, a molecular capacitor and is relatively easy to handle mathematically on that basis. In reality, of course, the thermal motion of ions in solution introduces a certain degree of chaos causing the ions to be spread out in the region of the charged surface, forming a "diffuse" double layer in which the local ion concentration is determined by Equation (5.6). In that case, the analysis of the electronic environment near the surface is more complex and requires more detailed analysis. Such analysis gave rise to the more accurate Gouy–Chapman model of the electrical double layer illustrated in Figure 5.2b.

An additional "reality" is that charges occupy a finite amount of space and therefore have certain steric requirements, leading to the postulation of the

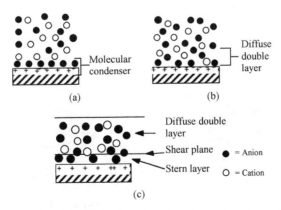

FIGURE 5.2. Classic models of the interface for charged surfaces include. (a) the early Helmholtz model of a molecular capacitor, (b) the Gouy–Chapman model of the diffuse double layer, and (c) the Stern model.

so-called Stern layer. The Stern layer is a small space separating the ionic atmosphere near an interface, the diffuse double layer, from the steric "wall" of the charged plane just adjacent to the interface (Fig. 5.2c). The thickness of the Stern layer, d in Figure 5.2c, is usually on the order of a few nanometers and reflects the finite size of charged groups and ions specifically associated with the surface.

While the Boltzmann distribution is relatively easy to evaluate for a single point charge, the situation can become quite complex in the case of a surface having many charges. For a detailed discussion of the finer points of double-layer theory, the reader is referred to the works of Adamson, Kruyt, or Hiemenez cited in the Bibliography. From a practical (and very simplistic) point of view, it is normally assumed that the electrical potential in the solution surrounding the surface in question falls off exponentially with distance from the surface (Fig. 5.3) according to the Debye–Hückel approximation

$$\psi = \psi_0 \exp(-\kappa z) \tag{5.7}$$

where κ is identified as the reciprocal of the thickness of the electrical double layer, also commonly referred to as the "Debye length." Thus at a distance of $1/\kappa$ from a charged surface, the potential has fallen off by a factor of $1/e$. For low surface potentials (< 25 mV), the theoretical equation for the double-layer thickness, $1/\kappa$, is

$$\frac{1}{\kappa} = \left(\frac{\varepsilon_0 \varepsilon \, kT}{e_i^2 \Sigma c_i z_i^2} \right)^{1/2} \tag{5.8}$$

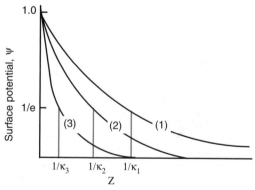

FIGURE 5.3. The decay of surface potential with distance from a surface in various electrolyte concentration ranges has a characteristic shape: (1) low electrolyte concentration—slower decay rate; (2) intermediate concentration—intermediate decay rate; (3) high concentration—rapid decay.

One can see that the thickness of the electrical double layer is inversely proportional to the concentration of electrolyte in the system and to the square of the valency of the ions involved. In terms of colloidal stability, this means that the distance of separation between two particles that can be maintained under a given set of circumstances will depend on, among other things, those two factors. And their important effect gives one a handle for manipulating the characteristics and stability of many colloidal systems.

In Table 5.1 are given the values of $1/\kappa$ for various concentrations of a selection of electrolytes in water at 25°C calculated according to Equation (5.8). One can see that the double-layer thickness drops off very rapidly as the concentration of electrolyte increases. This effect of electrolyte concentration has important ramifications in the world of practical colloids, as does the effect of the ionic charge (z) involved.

For ease of calculation of $1/\kappa$, Equation (5.8) can be simplified to the following relationships:

For 1:1 electrolytes:

$$1/\kappa = 0.304 \ [MX]^{-1/2} \tag{5.9}$$

For 2:1 and 1:2 electrolytes:

$$1/\kappa = 0.178 \ [MX^2]^{-1/2} \quad (\text{or} \quad [M_2X]^{-1/2}) \tag{5.10}$$

For 2:2, 3:1, or 1:3 electrolytes:

$$\frac{1}{\kappa} = 0.152 \ [M_2X_2]^{-1/2} \quad (\text{or} \quad [MX_3]^{-1/2} \quad \text{or} \quad [M_3X]^{-1/2}) \tag{5.11}$$

For 2:3 or 3:2 electrolytes:

$$\frac{1}{\kappa} = 0.136 \ [M_3X_2]^{-1/2} \quad (\text{or} \quad [M_2X_3]^{1/2}) \tag{5.12}$$

TABLE 5.1. Double-Layer Thickness ($1/\kappa$) for Various Electrolytes in Water

Electrolyte Concentration	$1/\kappa$ (nm)				
	1:1 (MX)	1:2 (MX2)	2:2 (MX)	1:3 (MX3)	2:3 (M2X3)
10^{-4}	30.4	17.6	15.2	15.2	13.6
10^{-3}	9.6	5.57	4.81	4.81	4.30
10^{-2}	3.0	1.76	1.52	1.52	1.36
10^{-1}	0.96	0.56	0.48	0.48	0.43
1	0.30	0.18	0.15	0.15	0.14

The picture of the electrical double layer around a colloidal particle that arises as a consequence of Equation (5.8) is that of a cloud of ions surrounding the particle dominated by charges opposite to that of the surface. The distance over which this ionic "sheath" extends is a function of the concentration and valency of the ions in the solution and the charge on the surface (Fig. 5.3). According to (5.8), the local concentration of ions near the surface varies as shown in Figure 5.4a, while the local charge density ($c_+ - c_-$) varies as shown in Figure 5.4b. For the situation in which the surface charge density is constant, that is, progressive adsorption of ions with increased c_0 does not reduce the surface charge (see discussion on charge regulation, below), the area under curve b will be equal to the charge on the surface.

In practical situations, the stipulation of constant surface charge is often found to be invalid, especially in concentrated colloidal systems where the distance between interacting surfaces is relatively small. In those cases, a number of events can occur that will result in changes in the net surface charge, and therefore the overall electrical characteristics of the system. The most important of these processes is the specific absorption of ions at the interface.

5.2.3. Specific Ion Adsorption and the Stern Layer

The derivation of equations related to the charges at interfaces has, to this point, been made with the aid of several important assumptions, most of which are valid only up to a certain point. The most significant of those included the following:

1. The ions involved, both on the surface and in solution, are point charges—that is, they have no finite volume. In fact, of course, ions possess characteristic radii which vary significantly with the ion involved.

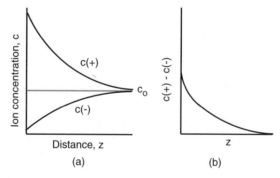

FIGURE 5.4. The distribution of charges in the diffuse double layer around a negatively charged surface will have a curve shape similar to those in figure 5.3: (a) the total concentration of ions near the surface; (b) the net local charge density.

In addition, the effective radius of an ion will depend on the number of solvent ions associated closely with it (its hydration number).

2. Charges on a surface are "smeared out" over the surface so that a uniform charge exists, when in fact charges arise from discrete points on the surface.

3. Charges of opposite sign can approach infinitely closely, in obvious contradiction to (1).

4. The solvent influences the double layer only through its dielectric constant and that the dielectric constant remains unchanged throughout the double layer. The latter is known not to be the case for water, in which the dielectric constant may be reduced by an order of magnitude near a surface due to orientation of the water molecules by charges in the area. In addition, as an ion is specifically adsorbed onto the surface, it is probable that it becomes dehydrated (desolvated), to some extent.

Luckily, most of those erroneous assumptions introduce only minor difficulties in the interpretation of electrostatic phenomena at interfaces, especially for practical purposes. Of perhaps more significance is the assumption that the charge density on a surface remains constant under all conditions. In fact, as the environment of an interface is altered (e.g., electrolyte concentration increases or two surfaces are brought into very close proximity—say, 1–2 nm) the number of "free" charges on the surface may be reduced as a result of the specific adsorption of ions of opposite charge. Thus, as two charged surfaces are brought together, the surface charge density becomes a function of the distance of separation and tends toward zero at contact. That phenomenon is known as "charge regulation." The end effect of charge regulation is to reduce the electrical potential of the surface, reduce the thickness of the electrical double layer, and reduce the effective repulsion between surfaces relative to that expected based on the "theoretical" potential, ψ_0.

Normally, a certain number of counterions will be strongly adsorbed in an area close to the actual interface in the Stern layer. Such ions will be adsorbed strongly enough that they will not be displaced (for a relatively long time, at least) by thermal Brownian motion. Because ions in the Stern layer are "fixed" relative to the ions further from the surface, in the so-called diffuse double layer, they effectively screen or neutralize a portion of the inherent surface charge. In that case, the surface potential ψ_0 is replaced in the Gouy–Chapman treatment by ψ_s which is the Stern potential (Fig. 5.5a). The curve shown is typical of a system with no specific adsorption; the electrical potential decreases rapidly from ψ_0 to ψ_s within the Stern layer followed by the "normal" decay of ψ_s to zero in the diffuse double layer. The existence of the Stern layer should be considered as a general phenomenon in charged systems, as distinct from specific adsorption which can have special consequences not directly related to the Stern layer.

Figure 5.5b illustrates schematically the effects of two examples of specific adsorption on the electrical nature of the double layer. Curve 1 shows the

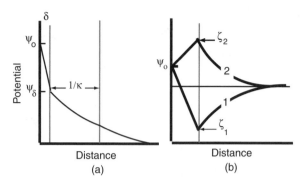

FIGURE 5.5. (*a*) Because of some strong ion adsorption at the surface, the actual electrostatic potential-energy curve may show a break at the Stern layer; (*b*) additional specific adsorption processes may further alter the curve shape to produce a more rapid dropoff in charge density (curve 1) or even charge reversal (curve 2).

effect of adsorption of a polyvalent counter ion or a surface active ion of opposite charge, leading to charge reversal (i.e., ψ_s has the opposite sign to ψ_0). Curve 2 represents the situation in which the adsorption of ions or surface-active species of like charge causes the Stern potential to increase relative to that of the surface ($\psi_s > \psi_0$).

The complete mathematical expression for the double layer incorporating the Stern layer is quite complex and will not be given here. However, its existence and related effects are quite significant for practical studies of electro-kinetic phenomena discussed below because it is ψ_s that is actually being estimated in such procedures. When a charged particle moves relative to an electrolyte solution, or a solution moves relative to a charged surface, viscosity effects dictate that only that portion of the electrical double layer up to (approximately) the Stern layer will move. The ions in the Stern layer will remain with the surface. The dividing line between movement with the solution and that with the surface is referred to as the shear plane (Fig. 5.6). The exact

FIGURE 5.6. The plane of shear is that distance from the surface at which ions become bound (essentially immobile) with respect to changes in the surface potential, ion mobility, and solvent or particle movement in electrophoretic phenomena.

location of the shear plane, which is actually a very thin region in which viscosity effects change rapidly, is difficult to determine, but it is usually assumed to be just outside the Stern layer, which implies that the potential at that point will be slightly less than ψ_s.

The potential at the shear plane is termed the electrokinetic or ζ (zeta) potential and represents the actual value determined in the procedures discussed in the next section. It is generally assumed in tests of double-layer theory that the ζ potential and ψ_s are the same, since any error introduced will be small under ordinary circumstances. More significant errors may be introduced at high potentials, high electrolyte concentrations, or in the presence of adsorbed bulky nonionic species that force the shear plane further away from the surface, reducing the ζ potential relative to ψ_s.

Considering all the assumptions and approximations involved in the derivation of the Gouy–Chapman model of the double layer, it should be obvious that a "real" situation is likely to be much more complex. Nevertheless, results obtained based on that model have served (and continue to serve) well in furthering our understanding of electrical phenomena in colloidal systems. Further refinements of double layer theory have succeeded in explaining a number of bothersome observations in specific situations, especially very high surface potentials. However, the complications involved in their application, and the benefits derived, do not generally warrant such effort in most practical situations.

5.3. ELECTROKINETIC PHENOMENA

An important consequence of the existence of electrical charges at interfaces, whether they are colloids, porous materials, or some other system, is that they will exhibit certain phenomena under the influence of an applied electric field related to movement of some part of their electrical double layer. Those phenomena (illustrated schematically in Fig. 5.7) are collectively defined as electrokinetic phenomena and include four main classes:

1. *Electroosmosis.* The movement of a liquid relative to a stationary charged interface under the influence of an electric field. The fixed surface will typically be a capillary tube or porous plug.

2. *Electrophoresis.* The movement of a charged interface (usually colloidal particles or macromolecules) plus its electrical double layer relative to a stationary liquid, under the influence of an applied field. Electrophoresis is, of course, the complement of electroosmosis.

3. *Streaming Potential.* The electric field generated when a liquid is forced to flow past a stationary charged interface.

4. *Sedimentation Potential.* The electric field produced when charged particles move relative to a stationary liquid.

FIGURE 5.7. The four principal classes of electrophoretic phenomena most often studied include (*a*) electrophoresis, (*b*) electroosmosis, (*c*) streaming potential, and (*d*) sedimentation potential.

Of these four, the phenomenon of greatest practical interest is electrophoresis. Over the years, several relatively easy techniques for the study and application of electrophoresis have been developed and today these are important tools in many areas of science and technology, including colloid science, polymer science, biology, and medicine. Of lesser practical importance, and less intensely studied, are electroosmosis and streaming potential. Sedimentation potential has received relatively little attention because of experimental difficulties.

While a thorough discussion of the details of those techniques is not possible here, a brief conceptual description of the more important phenomena and their practical applications will be useful in guiding the interested reader to the method of choice for a specific application. Theoretical and experimental details can be found in the comprehensive colloid and surface chemistry works cited in the Bibliography.

5.3.1. Particle Electrophoresis

Particle electrophoresis, also sometimes known as microscope electrophoresis or microelectrophoresis, is one of the easiest and most useful techniques for investigating the electrical properties of colloidal particles. If the system of interest is in the form of a reasonably stable dispersion of particle size observable by light microscopy (say, larger than 200 nm for practical application), the electrokinetic behavior of the system can be observed and measured directly. Several commercial instruments are available for the purpose. For smaller particles, laser scattering instruments are now readily available.

The measurement of the electrophoretic characteristics of a system can be very useful in evaluating the electrical nature of a surface—its charge sign and, indirectly, its charge density—and the effects of changes of pH, electrolyte content, electrolyte valency, and other factors on the charge. It can also be used to evaluate the electrical nature of soluble species (e.g., surfactants or polymers) that may be adsorbed onto the surface of a neutral particle. For example, if a colloid is prepared that has little or no inherent charge and then placed in a solution of a polymer such as a protein, adsorbed protein will impart an electrical charge to the surface and measurably change the mobility of the particles in an electric field. From that information the electrical nature of the protein can be deduced, information which would be much more difficult to obtain otherwise.

Particle electrophoresis has proved to be very useful in many areas of theoretical and practical interface and colloid science, including "model" polymer latex and silver halide systems, and more practical problems related to water purification, detergency, emulsion science, the characterization of bacterial surfaces, blood cells, viruses, and so on. With the advent of more sophisticated computer data analysis and laser light sources, the limits of resolution for particle sizes that can be analyzed has been, and is being, steadily reduced, so that with proper (and more expensive) instrumentation, the electrophoretic nature of particles in the size range of a few nanometers can be readily determined.

5.3.2. Moving-Boundary Electrophoresis

An alternative to particle electrophoresis is moving-boundary electrophoresis. The technique is used to study the movement of a boundary formed between a colloidal sol or solution and the pure dispersion medium under the influence of the electric field. The technique has found some application for determining not only electrophoretic mobility, but also for small-scale separation of species from a mixture for further identification. It found early application in the study of proteins and other dissolved marcomolecules.

If a protein solution contains a number of different species of different charge characteristics (and therefore different mobilities in a given charge field), the technique may be able to separate the fractions, or at least the peaks, sufficiently to indicate the number of distinct species present. The technique has largely been displaced by more sensitive, and experimentally easy, techniques; however, the inexpensive nature of the process still carries some weight in choosing an approach for some applications.

5.3.3. Gel (or Zone) Electrophoresis

Another alternative technique commonly used to investigate the electrophoretic properties of a material (especially soluble macromolecules) or mixtures thereof is gel or zone electrophoresis. The technique involves the use of a relatively inert solid or gel support for the solution of interest, which minimizes many of the experimental difficulties encountered in the moving-boundary technique, especially convection and vibrational disturbances. It is also much simpler, since there now exist a number of commercially available setups that require little manipulation and can be handled very easily by technicians with little advanced training or experience.

Gel electrophoresis requires very small sample sizes and can, in theory at least, give complete separation of a mixture of substances. While it cannot be used to determine electrophoretic mobilities, it allows for the separation and identification of components that would be extremely difficult or impossible to separate using other techniques. It is especially applicable to biological systems where sample availability may be a problem. It may even be used as a small-scale preparative procedure,

Zone or gel electrophoresis is limited in quantitative terms because it separates components according to two criteria: charge and molecular weight or size. It may happen that two components coincidentally have the right combination of charge and size so that they move together under a given field. To overcome that problem, techniques have been developed that may be called "two-dimensional" electrophoresis in which an electric field is applied in one direction for a given period of time, followed by another field of different strength in a perpendicular direction. The net result is, hopefully, that components that fortuitously move together in the first field will be separated by the second, since only molecules of the same size and same charge would be expected to move together under two different electric fields.

Since the other electrokinetic phenomena are of significantly lesser practical importance, they will not be discussed here. For further theoretical and experimental details, the reader is referred to the works cited in the Bibliography.

5.3.4. Some Practical Comments on Electrokinetic Characteristics

Although often slighted or completely ignored, the nature of the electrical double layer can have an important influence in the characteristics and function of practical colloidal systems. The ζ potential, for example is a number characterizing the EDL that is often suspect because of doubts related to the technique used in its determination. Even approximate numbers, however, can be useful in practice since they may help avoid situations that would obviously be detrimental to the proper functioning of the system, if known and understood. Some practical examples of the importance of information about surface charge and ζ potential include the following areas.

In mineral foam or froth flotation the sign and magnitude of the surface charge will influence the adsorption of additives (collectors) onto the mineral surface that will determine whether a specific mineral will float or sink, and therefore the efficiency of its separation. The process of flotation is based on the interactions at the solid–liquid and solid–liquid–air interfaces. The addition of the proper collector determines what mineral fraction will become attached to foam bubbles and therefore float (and be separated), and which will be wetted by the solution and sink. The process is illustrated schematically in Figure 5.8.

For the clarification and filtration of many dilute colloidal systems such as industrial wastes and municipal water supplies it is generally necessary to add a flocculating agent such as aluminum sulfate that markedly alters the ζ potential of the colloids, causing flocculation and greater ease of separation.

Soil is composed of a variety of colloidal species including inorganic mineral particles and organic macromolecules. The structure and quality of the soil is determined, in part, by its porosity which is determined by the state of flocculation of the various colloidal species present. A loosely flocculated soil is optimum in terms factors such as air and moisture penetration and root growth. The presence of bivalent ions, especially calcium, promotes such loose flocculation. Soils in which the calcium content has been excessively depleted by flooding, exposure to sea water (ion exchange in which the calcium is replaced by sodium) or where calcium was never present tend to form hard, compacted masses after wetting and drying. The reintroduction of calcium in the form of gypsum, for example, can return the soil to its flocculated state and greatly increase its productivity.

Electrophoretic deposition is an important process for the undercoating and painting of metallic products, especially automobiles. It is obviously necessary for the processor to know and control the electrical properties of the pigments being used in order to optimize the process and produce the best finished surface possible.

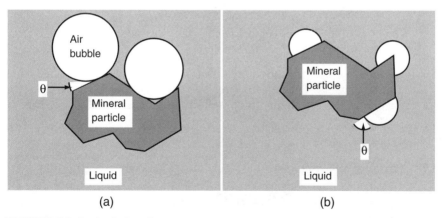

(a)   (b)

FIGURE 5.8. In froth flotation, the buoyant force of the foam bubble on the particle to be floated must be greater than that of gravity. In order to achieve that result it is necessary to produce a large finite contact angle between particle and liquid, usually accomplished by the additin of "collectors" that absorb on the particle surface and increase the contact angle.

A number of additional processes of practical importance involving colloids and surface interactions can be identified that require some knowledge of the electrical properties of the system involved. Investigators are always well advised to spend a little time assuring themselves that the proper electrokinetic information is available in order to avoid possible problems at some later date.

The true practical significance of the material presented in this chapter will become more apparent in the following chapters. Obviously, the more complex a colloidal system becomes, the more difficult it will be to pin down its behavior in terms of specific phenomena discussed so far. However, small pieces of information (such as electrophoretic mobility) can be important in helping to determine the performance of a given system.

PROBLEMS

5.1. Given the bond length in the water molecule as 0.096 nm and the bond angle 104°, calculate a value of the charge on the oxygen atom, q_0, consistent with a molecular dipole moment of 1.87 D, assuming that the molecule has no net charge and that the two hydrogen atoms are equivalent ($q_{H1} = q_{H2}$).

5.2. Calculate the Debye length κ^{-1} for the following solutions: (a) pure water ($K_w = 10^{-14}$; (b) 0.1 M NaCl; (c) 0.01 M NaCl; (d) 1mM NaCl; (e) 0.1 M CaSO$_4$; (f) 1.0 mM La(NO$_3$)$_3$.

5.3 Calculate the free energy of repulsion, V_R, between two double layers each having an area of 1.0 cm^2, a surface potential $\phi_0 = (4RT/F)$ in a 1:1 electrolyte producing $\kappa = 10^{-6}$ cm, and a separation distance of 10 nm. (Use the dielectric constant for water at 25°C).

5.4. How does double-layer repulsion change if the medium is changed from water to methanol when ϕ_0 and the ionic concentration remain the same?

5.5 If a monovalent electrolyte is replaced by a divalent one but κ remains the same, how must the electrolyte concentration change? How does the repulsion energy V_R change at constant H_0 if ϕ_0 is small? If $\phi_0 = 100$ mV?

5.6. What would be the probable effect on clay soils after being flooded by sea water?

5.7. In areas of highly compacted soils, porosity and agricultural productivity is sometimes improved by treatment with calcium salts. Propose a valid colloidal mechanism to explain such a result.

6 Capillarity

Although defined in various ways depending on the context, "capillarity" for current purposes will be defined as the macroscopic motion of a fluid system under the influence of its own surface and interfacial forces. Such flow is similar to other types of hydraulic flow in that it results from the presence of a pressure differential between two hydraulically connected regions of the liquid mass (Fig. 6.1). The direction of flow is such as to decrease the pressure difference. When the difference vanishes, or when there is no longer a mechanism to reduce the difference, flow ceases.

Capillary effects are encountered in many areas of interface and colloid science, with its importance relative to other processes (e.g., fluid dynamics) depending on the exact situation. For example, when two spherical drops of a liquid in an emulsion make contact and coalesce to form a larger drop (Fig. 6.2a), the extent and duration of flow due to the capillary phenomenon is limited and fluid dynamics is of little practical importance. When there is an extensive amount of flow, on the other hand, such as in capillary imbibition, wicking processes, or capillary displacement (Fig. 6.2b) fluid dynamics may become important.

6.1. FLUID PROPERTIES AND DYNAMICS

Although this chapter is concerned specifically with fluid flow resulting from surface and interfacial tension effects, it will be useful to introduce briefly some basic ideas about the nature and properties of fluids, especially liquids. In order to understand the concepts involved it is necessary to understand the "language" so that a number of terms must be introduced.

Mechanics is a field of science concerned with the movement of bodies and the specific conditions controlling that motion. It is commonly divided into two subsections: (1) kinematics, concerned with the geometry of motion with no concern for the forces involved; and (2) dynamics, which concerns itself with the forces acting to cause the motion.

In a physical sense a fluid may be defined as a state of matter in which the atoms or molecules composing the phase have a spacing and mobility greater than that found in the solid that allows a relatively free (compared to the solid) movement of the units when an external force is applied; that is, in a fluid the range of motion is greater than that in a solid. A useful definition

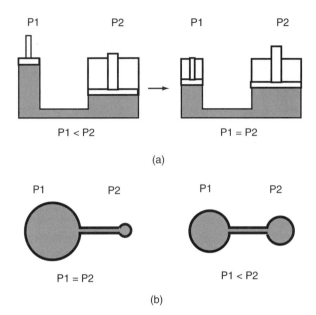

FIGURE 6.1. Capillary flow can be compared to hydraulic flow. When there exists an unbalanced pressure in a hydraulic system (*a*) flow occurs until the pressure is balanced. In a capillary system (*b*), where pressure difference is a result of different radii of curvature, surface tension, and related effects [Eq. (6.1)], flow also occurs until the pressure balance is achieved, or until no mechanism remains for reaching equilibrium.

of "fluid" is a substance that, when a tangential or shearing force is applied, cannot sustain its equilibrium position no matter how small the shearing force may be.

Fluids are divide into two subclasses: liquids and gases. A liquid occupies a definite volume, independent of the total volume available in the container. It can have a free surface, such as that of a bowl of water, with the resulting surface properties introduced in Chapter 2. A gas, on the other hand, can have no free surface and will, in general, completely fill the volume of its container, although in some circumstances there may exist a density gradient due to external forces. While the terms "gas" and "vapor" are routinely used interchangeably, the usual convention is that a vapor is relatively easily condensed to a liquid while a gas would require significant effort for such a process.

It is commonly assumed that gases (and vapors) are compressible while liquids are not. In fact, all fluids are compressible, although the much greater unit spacing in gases results in a more obvious compressibility (Fig. 6.3). Many common physical phenomena result from the compression of liquids, perhaps the most common being the transmission of pressure and sound waves through water or other liquids, which depends on the compressibility or elasticity of the liquid. The results of the greater mobility of atoms or molecules in liquids

Drop deformation and thinning
of external liquid film

(a)

Nonwetting liquid   Wetting liquid

(b)

FIGURE 6.2. Important functions of capillary forces in practical situations. (*a*) As two emulsion drops approach, the pressure at the nearest surfaces increases, deforming the drops and enlarging the radius of curvature in the immediate area. That deformation causes the capillary pressure in the regions outside that area to decrease in a relative sense, suctioning continuous phase from between the drops and increasing the likelihood of contact and film rupture or coalescence. (*b*) In capillary displacement, the liquid that preferentially wets the soild will displace the less wetting liquid.

relative to solids has already been pointed out in relation to the surface properties of solids versus those of liquids.

It is common practice to consider fluids to be continuous media, although they are in fact composed of individual atoms or molecules. We know, however, that fluid properties and related phenomena are controlled by the interactions among neighboring units and that considerations at the unit level cannot always be ignored. Many, if not most, practical problems related to fluid flow occur on a scale such that the range of movement (i.e., mean free path) of the individual units is extremely small relative to the scale of the phenomena being considered and reference to bulk properties is justified. That is also commonly assumed to be the case at surfaces and interfaces, although following chapters will show that as one enters this twilight zone, the movement and characteristics of individual units assume greater, if not determinant importance.

Most discussions of capillary action tend to concern themselves with the interfacial driving forces behind the phenomena, little attention is paid to the fluid dynamics aspect. In many important practical applications, however, fluid

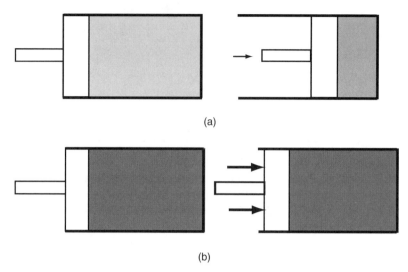

(a)

(b)

FIGURE 6.3. Although gases are easily recognized as being compressible, liquids may also be compressed, although the extent of such compression is much more limited by the reduced distance between molecules in the liquid phase.

characteristics (e.g., viscosity) play an important role and must be considered. For the most part, the following comments will not take such factors into consideration but will address the capillary phenomena in isolation. The discussion will be limited to a qualitative and descriptive presentation rather than a rigorous derivation of the principles involved. A number of secondary references are provided that go into detail on the more theoretical questions.

Capillary phenomena arise as a result of differences in pressure across a system containing at least one liquid phase and another liquid, vapor, and/or solid phase. As illustrated below, such pressure differences may result from differences in curvature in different regions of liquid–fluid phases in a system and/or due to the presence of an effective mechanical tension in the interface, the interfacial tension. The differences in curvature giving rise to the pressure differentials may result from various sources including the application of external forces, the contacting and coalescence of two masses of the liquid phase, or from the contact of the liquid phase with an second fluid phase and a solid surface.

6.2. A CAPILLARY MODEL

Capillary flow systems of most practical interest are those that involve a solid, a liquid, and a second fluid phase. In the absence of other external forces, the net driving force for capillary flow in such a system will be controlled by

FIGURE 6.4. When a liquid is placed on a solid surface, it may form an even duplex film or form a drop with a finite contact angle. The angle, measured through the liquid, may be small (*a*) or large (*b*), reflecting the extent of interaction between solid and liquid.

three basic quantities, including the various interfacial tensions, the geometry of the solid–liquid–fluid interface, and the geometry of the solid surface at the three-phase boundary line. When a liquid contacts a second fluid phase and a solid surface, there is produced a force imbalance orienting the liquid–fluid interface by causing it to assume a characteristic equilibrium orientation with respect to the solid surface, reflected in the so-called contact angle, θ (Fig. 6.4). At equilibrium, the contact angle may be considered to be an intensive material constant depending only on the natures of the three component phases, but independent of the quantities present (within limits to the molecular level). Being an experimentally accessible quantity, the contact angle can be, as will be seen later, a very useful tool for studying interfacial effects.

Except under special circumstances (such as zero-gravity environments), practical capillary systems will experience hydraulic pressure gradients in addition to those resulting from curvature, the most important of which is that due to gravity. Treatments of capillary flow, then, should theoretically take into consideration the effect of the gravity-induced hydrostatic head. The gravitational pressure gradient must be included algebraically in many calculations of capillary flow, especially those containing a significant vertical component. For primarily horizontal systems, however, the pure capillary pressure contribution is so much greater than any gravitational effect that the latter may be neglected. In the discussion to follow, the effects of gravity have been neglected unless specifically indicated.

6.3. CAPILLARY DRIVING FORCES IN LIQUID–FLUID SYSTEMS

Of the driving forces for capillary action mentioned above, the most fundamental are those of interfacial tension and related effects (e.g., contact angle). As pointed out in Chapter 2, a liquid–fluid interface behaves as if it is an elastic film stretched over (or between) the two phases and resisting any more stretching to produce greater interfacial area. The tension results fundamentally from the imbalance in the forces acting on the molecules at the interface, which tend to pull the molecules back into the bulk phases. At equilibrium, the surface tension is a material constant, the value of which can be determined by many convenient methods and then be applied as needed in other situations.

Mathematically, an interface is a two-dimensional region. In reality, it will be three-dimensional, although the third dimension may have the thickness of only one or two molecules. Because it is three-dimensional, the interfacial region may be treated in the context of hydrostatics, or in terms of molecular forces and distribution functions. Alternatively, a thermodynamic approach may be taken to arrive at the same conclusions.

The fact that a tension exists at a liquid–fluid interface implies that, if it is curved, there will be a difference in hydrostatic pressure across the interface. Laplace derived an expression for the pressure difference across a curved interface in terms of surface tension and curvature. The equation, referred to as the Laplace equation, is

$$P_1 - P_2 = \Delta P = \sigma \left(\frac{1}{r_1} + \frac{1}{r_2} \right) = P_{\text{cap}} \tag{6.1}$$

In which P_1 and P_2 are the pressures in the two phases forming the interface, and r_1 and r_2 are the principal radii of curvature of the interface at the point in question. For a spherical surface, $r_1 = r_2 = r$, and Equation (6.1) simplifies to

$$\Delta P = \frac{2\sigma}{r} \tag{6.2}$$

It is useful to work through the derivation of this equation for a spherical surface to be certain of the relationship between surface tension and pressure, since pressure is the driving force for capillary action. If one takes a spherical drop of liquid of radius r and adds more liquid so that the radius increases by a factor dr, the surface area of the drop will increase by a factor $8\pi r\, dr$ (Fig. 6.5). As seen in Chapter 2, the amount of work that must be done to expand a surface or interface is given by

$$W = \sigma\, dA \tag{6.3}$$

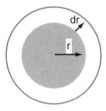

FIGURE 6.5. As liquid is added to a spherical drop, the volume and surface area will increase, requiring the addition of work as the surface area increases.

so that for the drop in question, the work will be $\sigma \times 8\pi r \, dr$. Under conditions of mechanical equilibrium, that work will come from the pressure difference across the interface $(P_1 - P_2) \times 4\pi r^2$ acting through the distance dr, where by convention P_1 is the pressure on the inside of the drop and P_2 is the pressure outside. If the two terms are equated, the result is

$$(P_1 - P_2) = \frac{2\sigma}{r} \tag{6.2a}$$

which is, of course, Equation (6.2). A similar analysis of nonspherical systems produces the same result, but in the more general form of Equation (6.1).

In a system involving purely capillary phenomena, the sole driving force is the pressure differential in various areas of the system. The sign of the pressure term in (6.1) will depend on the assignment of P_1 and P_2; however, as already pointed out, convention is to assign P_1 to the more highly curved (i.e., internal) phase of interest and P_2 to the adjoining fluid phase. In a flat surface where $r_1 = r_2 = \infty$, $\Delta P = 0$. Alternatively, one can say that if the surface is concave inward (toward the liquid), $\Delta P > 0$; if it is convex inward, $\Delta P < 0$. Where a pressure differential exists, the liquid will flow from high to low pressure until the differential is decreased and ultimately eliminated. In situations where the interfacial tension is uniform from point to point, the capillary pressure will depend only on the curvature of the interface.

In most circumstances it is common to ignore gravitational effects for most calculations of capillary forces. The validity of such a procedure can be seen with a simple example. Suppose that a spherical drop of liquid of unit density and surface tension of 50 mN m^{-1} has a diameter of 0.1 cm. The hydrodynamic pressure difference due to gravity between the top and the bottom of the drop will be 98 mJ m^{-2}. The capillary pressure difference will be, according to Equation (6.2), 2000 mJ m^{-2}. As the drop diameter and the radius of curvature decrease, the capillary pressure increases to approximately 1.01×10^5 mJ m^{-2} at $r = 10^{-4}$ cm and 1×10^7 mJ m^{-2} at $r = 10$ nm. As the drop diameter approaches molecular dimensions, Equations (6.1) and (6.2) should no longer be considered valid. Therefore the "calculation" of extremely high pressure in capillary systems of very small radius must be approached with great caution.

6.3.1. Solid–Liquid–Fluid Systems: The Effect of Contact Angle

The preceding discussion was concerned with capillary forces in a system containing only the liquid of interest and a second fluid phase. It has been stated, however, that the systems of most practical interest involve a third phase (usually a solid) resulting in a three-phase boundary line. The situation can be represented as a drop of liquid resting on a flat solid surface and contacting the second liquid or vapor phase as illustrated in Fig. 6.6. If the drop is allowed to spread over the surface a small distance from point a to point b, the new liquid–vapor interface will remain essentially parallel to the old, but increase its area by the amount $(bc) = (ab) \cos \theta$. In the process, the solid–liquid interface is increased by (ab) and the solid–vapor interface decreased by that amount. At equilibrium, the change in free energy will be zero, so that

$$\Delta G = \sigma_{SV} \, \Delta A_{SV} + \sigma_{SL} \Delta A_{SL} + \sigma_{LV} \Delta A_{LV} = 0 \tag{6.4}$$

leading to the result

$$\sigma_{SV} = \sigma_{SL} + \sigma_{LV} \cos \theta \tag{6.5}$$

which is generally known as Young's equation. While the exact physical interpretation of Young's equation is not clear for all situations, it is one of the fundamental equations of surface chemistry and will be encountered repeatedly.

6.3.2. Capillary Flow and Spreading Processes

Before continuing with the specific topic of capillary flow, it may be useful to divert our attention to the broader topic of the spreading of a liquid on a surface. The spontaneous spreading of liquids is a topic of great practical interest because of its importance in many applications. For that reason, such phenomena will be discussed in more detail in later chapters. However, it will be useful to introduce the basic concepts at this point in order to clarify some concepts related to the current topic.

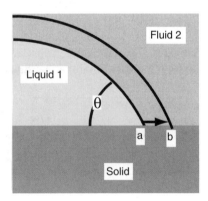

FIGURE 6.6. The wetting of a solid by a liquid in the presence of a second fluid phase will be controlled by the relative interfacial interactions among the three phases.

If a quantity of liquid is placed on a surface, which may be a solid or another liquid, one of two things may occur: (1) the liquid may spread across the surface to form a uniform duplex film; or (2) the liquid may form a drop (on a solid) or lens (on a liquid) with a finite, nonzero contact angle (Fig. 6.7). Thermodynamically, at constant temperature and pressure the change in the free energy of the system is given by

$$\Delta G = \frac{\delta G}{\delta A_A} dA_A + \frac{\delta G}{\delta A_{AB}} dA_{AB} + \frac{\delta G}{\delta A_B} dA_B \tag{6.6}$$

where subscript A designates the substrate and B, the liquid. Obviously, $dA_A = -dA_B = dA_{AB}$, so that

$$\frac{\delta G}{\delta A_A} = \sigma_A, \qquad \frac{\delta G}{\delta A_B} = \sigma_B, \qquad \frac{\delta G}{\delta A_{AB}} = \sigma_{AB}$$

The term $(\delta G/\delta A_B)$ gives the free energy change for the spreading of liquid B over the surface A and is called the "spreading coefficient" of B on A, $S_{B/A}$, given by

$$S_{B/A} = \sigma_A - \sigma_B - \sigma_{AB} \tag{6.7}$$

From the definitions of the works of cohesion and adhesion given in Chapter 2, it can be seen that $S_{B/A}$ is the difference between the work of adhesion of B to A and the work of cohesion of B:

$$S_{B/A} = W_{AB} - W_{BB} \tag{6.8}$$

From this analysis, the spreading coefficient will be positive if there is a decrease in free energy on spreading (i.e., adhesive forces dominate); that is, the spreading process will be spontaneous. If $S_{B/A}$ is negative, then cohesive forces will dominate and a drop or lens will result.

(a)   (b)

FIGURE 6.7. A liquid drop placed on a second, immiscible liquid may spread to form a duplex film (*a*) or form a lens (*b*).

What Equation (6.7) indicates, in general, is that when a liquid of low surface tension such as a hydrocarbon is placed on a liquid or solid of high surface energy such as clean glass or mercury, spontaneous spreading occurs. Conversely, if a liquid of high surface tension such as water is placed on a surface of lower surface energy such as teflon or paraffin wax, drop or lens formation results.

Unfortunately, complications arise in spreading phenomena because liquids, solids, and gases tend to interact in bulk processes as well as at interfaces, and those bulk-phase interactions may have significant effects on interfacial phenomena. In particular, gases tend to adsorb at solid interfaces and change the free energy of those surfaces, σ_{SV}; they may also become dissolved in liquid phases and thereby alter the liquid surface tension. More importantly, liquids in contact with other liquids tend to become mutually saturated, meaning that the composition of the two phases may not remain "pure" and no longer have the surface characteristics of the original materials. Finally, liquids and solutes, like gases, can adsorb at solid interfaces to alter the surface characteristics of the solid and thereby change the thermodynamics of the spreading process. Most of these situations will be addressed in the context of specific areas of interest in later chapters. However, the classic example of benzene–water systems will serve as a useful illustration.

For a drop of pure benzene ($\sigma_B = 28.9$ mN m^{-1}) placed on a surface of pure water ($\sigma_A = 72.8$ mN m^{-1}) with an interfacial tension, σ_{AB} of 35.0 mN m^{-1}, Equation (6.7) predicts a spreading coefficient of

$$S_{B/A} = 72.8 - 28.9 - 35.0 = 8.9 \text{ mN m}^{-1}$$

The positive spreading coefficient indicates that benzene should spread spontaneously on water. When the experiment is carried out, it is found that after an initial rapid spreading, the benzene layer will retract and form a lens on the water. How can this seemingly anomalous result be explained?

In this and many similar cases, it must be remembered that benzene and many other such water-"immiscible" liquids have, in fact, a small but finite solubility and the water will rapidly become saturated with benzene. Benzene, having a lower surface tension than water, will adsorb at the water–air interface so that the surface will no longer be that of pure water but that of water with a surface excess of benzene. The surface tension of benzene-saturated water can be measured and is found to be 62.2 mN m^{-1}, which is now the value that must be used in eq. 6.7 instead of that for pure water, so that

$$S_{B/A(B)} = 62.2 - 28.9 - 35.0 = -1.7$$

where the subscript A(B) indicates phase A saturated with phase B. The negative spreading coefficient indicates that lens formation should occur, as is observed. The saturation process occurs, of course, in both phases. However, since water is a material of relatively high surface tension, it will have little tendency to adsorb at the benzene–air interface and will therefore cause little change in the surface tension of the benzene. In this case $\sigma_{B(A)} = 28.8$ mN m^{-1} so that

$$S_{B(A)/A} = 72.8 - 28.8 - 35.0 = 9.0$$

If only the benzene layer were affected by the saturation process, spreading would still occur. Combining the two effects one obtains

$$S_{B(A)/A(B)} = 62.2 - 28.8 - 35.0 = -1.6$$

indicating that it is the effect of benzene in water that controls the spreading (or nonspreading) in this system. The interfacial tension of water–benzene is unchanged throughout because it inherently includes the mutual saturation process.

Situations like that for benzene are very general for low surface tension liquids on water. There may be initial spreading followed by retraction and lens formation. A similar effect can in principle be achieved if a third component (e.g., a surfactant) that strongly absorbs at the water–air interface, but not the oil–water interface, is added to the system. Conversely, if the material is strongly adsorbed at the oil–water interface, lowering the interfacial tension, spreading may be achieved where it did not occur otherwise. This is, of course, a technologically very important process and will be discussed in more detail in later chapters.

6.3.3. Geometric Considerations in Capillary Flow

When considering capillary flow problems there are several external factors that must be kept in mind, in addition to the question of pressure differentials due to surface curvature and surface tension effects. For a liquid–fluid system, if one assumes that σ_{LV} is constant and that there are no external factors inducing pressure differences in the system, then the capillary pressure, P_{cap}, is a function entirely of the curvature of the liquid–vapor (LV) interface. Put another way, when the system is at mechanical equilibrium, P_{cap} will be constant and at its minimum value, and curvature will be constant in all parts of the system. If some external force induces a change in curvature at some point, the resulting increase in P_{cap} returns the system to its original state, provided the disturbance has not been to great. If the perturbing force is sufficient to remove the system significantly from equilibrium, the resulting P_{cap} may cause the liquid mass to be divided rather than returning to the original configuration. This effect is often referred to as a "yielding of the surface." Such an effect can be either advantageous or disadvantageous, depending on the situation.

A commonly encountered example of the yielding effect can be seen in the breaking of a foam formed in a pure liquid. It is generally observed that pure liquids do not foam; or that any transitory foam formed on agitation will dissipate very rapidly once agitation ceases. A schematic representation of an unstable foam system is given in Figure 6.8, where the liquid phase lies in the thin lamellae between the vapor cells. Because of the large curvature differences in the plateau regions (P) relative to the lamellae there will exist a large pressure differential in the system. The P regions have a small convex radius of curvature, which leads to a large negative P_{cap}, while the lamellar regions have a much larger radius of curvature and a correspondingly smaller (and less negative) P_{cap}. As a result of the pressure differential, liquid will flow from the lamellar region to region P, thinning the lamellae until the cell ruptures and the foam breaks. In unstable foams, this process occurs very rapidly. In systems containing various additives such as surfactants or polymers, the process can be slowed sufficiently to produce "stable" foams. That topic will be discussed in Chapter 12.

An important and potentially costly disadvantage of surface yielding can be found in the process of secondary oil recovery. In many such operations, the oil to be recovered is forced out of the porous rock formation by an aqueous "plug" which displaces the oil in the rock capillaries (Fig. 6.9). If the moving aqueous front is not properly formulated and/or the process not properly implemented, the action of the aqueous plug may cause the crude oil phase to undergo yielding processes, leading to its breakup into small particles and isolation in pores from which recovery will be difficult or impossible.

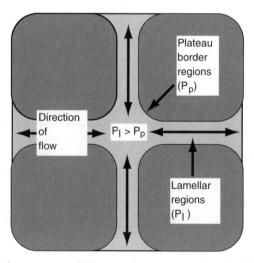

FIGURE 6.8. In a foam system, differences in curvature along the liquid–gas interface due to distortion give rise to pressure differences and subsequent capillary flow.

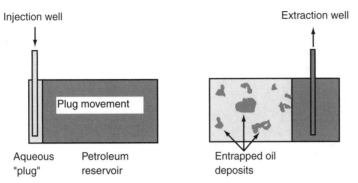

FIGURE 6.9. In the petroleum industry, crude oil is often recovered by the use of an aqueous "plug" in which a solution of surfactants, muds, polymers, and other compounds is forced into a reservoir to push the entrapped oil toward a central extraction well. If the proper solution balance and conditions of use are not employed, the crude oil may undergo surface yielding leading to the deposition of significant amounts of crude in the rock formation that cannot be recovered.

6.3.4. Measurement of Capillary Driving Forces

Classically, the approach used to calculate capillary flow has been to determine the curvature of liquid interfaces in the system and calculate P_{cap} from Equation (6.1). Those values could then be used to calculate the direction and magnitude of the driving forces. In systems of simple geometry such as liquids which form spherical interfaces and smooth cylindrical solid surfaces, the technique works out very well. Perhaps the best known example of such a system is the capillary rise method for determining the surface tension of a liquid, illustrated in Figure 6.10. In this system, capillary forces cause the liquid to rise in the tube due to differences in curvature of the liquid–air interface within the tube (a small radius of curvature) and that in the reservoir

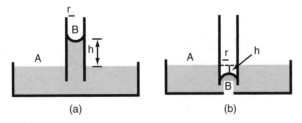

FIGURE 6.10. In the capillary rise method of surface tension measurement, surface tension effects cause the wetting liquid to rise in the small capillary to a height that just balances the hydrodynamic force due to gravity (*a*). For non-wetting liquids such as mercury, a depressing effect is observed (*b*).

$(R \approx \infty)$. Here there is also a constant opposing force due to gravity that must be included.

If the contact angle of the liquid on a capillary surface is θ, the radius of the tube is r, and it is assumed that r is sufficiently small that the liquid surface in the tube is spherical, then the radius of curvature of the liquid–air interface, R, $= r/\cos \theta$. According to Equation (6.1), the capillary pressure at point B will be $(-2\sigma \cos \theta/r)$ and the net driving force for capillary rise will be $P_{cap(A)} - P_{cap(B)}$, since at A, $P_{cap} = 0$, the total driving force comes from the curvature of the interface in the tube. The liquid movement will continue until the hydrostatic head of liquid in the tube, $\Delta \rho g h$, is equal to $P_{cap(B)}$, where $\Delta \rho$ is the difference in density between the liquid and the vapor, g is the acceleration due to gravity, and h is the height of the meniscus in the tube above the liquid level at A. Then

$$\Delta \rho g h = \frac{2\sigma \cos \theta}{r} \tag{6.9}$$

Equation 6.9 is the classic equation for determining the surface tension of a liquid by the capillary rise method. When $\theta = 0°$, the equation simplifies to

$$\sigma = \frac{\Delta \rho \, g h r}{2} \tag{6.10}$$

In the absence of gravity, a similar effect can be seen for two capillary tubes of unequal radius connected end-to-end as shown in Figure 6.11. In this case the net driving force for liquid flow arises as a result of the differences in curvature of the ends of the liquid mass in the two tubes at points A and B. The net capillary pressure is given by

$$P_{cap} = 2\sigma \cos \theta \left(\frac{1}{r} - \frac{1}{r'} \right) \tag{6.11}$$

where r is the radius of the small and r' that of the large capillary. Beginning with the situation as illustrated in Figure 6.11a, the liquid will flow into the smaller capillary until it attains configuration (b), at which time all pressures will be equal and flow will cease.

The situation illustrated by Figure 6.11 is potentially important in many practical areas because it represents an idealized system for many wicking, blotting, and absorption processes. Unfortunately, most real systems, such as textiles or paper products, have such complex geometries that it is not possible to determine simple values for r and r'. In such cases it is convenient to employ a thermodynamic approach rather than use pressure differentials.

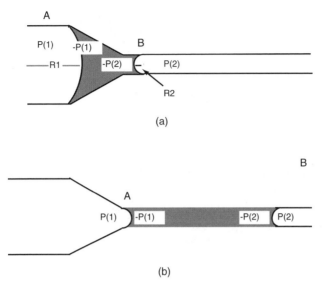

FIGURE 6.11. In a zero-gravity situation, or in horizontal capillary tubes where gravity can be ignored, capillary flow will occur spontaneously to take the system to equilibrium. In (a) the two radii or curvature are unequal causing a pressure differential between A and B. Capillary flow will occur until the pressure has equalized (b) or until other forces inhibit further movement.

In such a thermodynamic analysis, it is necessary to use Equations (6.4) and (6.5), and the fact that in the system the sum of the areas of solid–vapor (A_{SV}) and solid–liquid (A_{SL}) interfaces remains constant. Using the equations, it can be shown that the change in free energy δG caused by a change in position of the three-phase boundary by a distance δs can be represented as

$$\frac{\delta G}{\delta s} = \frac{\sigma_{LV}\,\delta A_{LV}}{\delta s} - \sigma_{LV}\frac{\cos\theta\,\delta A_{SL}}{\delta s} \tag{6.12}$$

The quantities σ_{LV} and θ are experimentally accessible and the area changes can be determined from the geometry of the system. According to eq. 6.12, the liquid will move in a capillary system if $\delta G/\delta s < 0$, where $\delta G/\delta s$ is numerically equal to the net driving force P_{cap}. The thermodynamic approach is in principle very general and has been successfully applied with a number of models including the pull of a liquid in a partially immersed rod (Fig. 6.12a), the movement of a liquid in a notch, and the movement of a liquid on two closely spaced rods. Such models have been found useful for studying various types of woven or interlacing systems in the textile and paper industries (Fig. 6.12b).

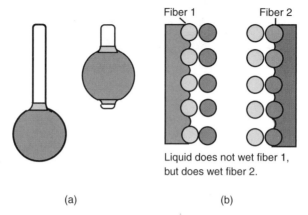

FIGURE 6.12. Some special examples of capillary flow: (*a*) a liquid climbing a partially immersed rod; (*b*) wicking—the spontaneous movement of a liquid from a nonwetting to a wetting situation.

6.3.5. Complications to Capillary Flow Analysis

In the discussions of capillary flow so far, it has been assumed that the values of surface tension and contact angle are constant and that other noncapillary factors such as external forces (except gravity in some cases), liquid viscosity, and electrical effects can be ignored. In practice, such is not always the case. The application of the principles of capillarity often involves the handling of variations in the surface tensions involved with time and location, changes in contact angles due to those surface tension gradients, and variations in contact angle due to compositional and geometric changes in the solid surface, hysteresis, and similar effects not to mention the fluid dynamic and other external factors. In practice, it is important to keep such variables in mind in order to better understand the capillary characteristics of a particular system.

Surface Tension Gradients and Related Effects. In many practical systems, one of the most commonly encountered complications to the analysis of capillary flow is that arising from variations in the solid–liquid and/or liquid–vapor interfacial tensions in the system. In particular, the value of σ_{LV} may vary significantly from point to point, leading to liquid flow unrelated to capillary phenomena. In a continuous liquid system in which surface tension gradients arise, liquid will spontaneously flow from regions of low to those of high surface tension. This flow is strictly a surface effect and is independent of curvature, which controls capillary flow. The rate of such flow will depend on the magnitude of the difference in σ_{LV} and the hydrodynamic characteristics of the liquid. Liquid flow at surfaces arising from surface tension gradients, commonly referred to as "Marangoni flow," can be important in many systems other than capillary phenomena. Some of those systems will be encountered in later chapters.

Marangoni effects can be encountered in both single- and multicomponent liquid systems. In a pure liquid, surface tension gradients result from differences in temperature (or evaporation rate) from one point to another in the system. It is generally found that an increase in temperature lowers σ_{LV} so that where "hot spots" occur, liquid flows away to cooler regions of the liquid (Fig. 6.13a). The result of such a phenomenon can be the formation of "dimples" in a surface that dries or solidifies under uneven temperature conditions.

In multicomponent systems (e.g., surfactant solutions), surface tension gradients usually are due to adsorption-related phenomena or, where possible, to different rates of evaporation from the system (although simple temperature variations can also be important). If the system contains two liquid components of differing volatility, the more volatile liquid may evaporate more quickly from the LV interface, resulting in localized compositional—and therefore surface tension—differences. It is also commonly found that when two or more components are present, one will be preferentially adsorbed at the LV interface and lower σ_{LV} of the system. If a "surface-active" component

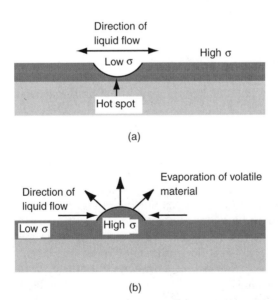

(a)

(b)

FIGURE 6.13. The Marangoni effect results from the presence of surface tension gradients in a liquid surface: (a) the presence of a "hot spot" will lower the surface tension near the heated area, causing flow in the direction of the cooler (higher surface tension) areas; (b) if a volatile surface-active material evaporates from a liquid surface, the "local" surface tension will increase resulting in flow toward the depleted area and the formation of a "bump" or drop as in "wine tears."

evaporates, the local surface tension of the liquid will rise and Marangoni flow toward the evaporation site will occur (Fig. 6.13b).

Multicomponent systems may also involve the selective adsorption of one component at the SL interface. Since the component that lowers the interfacial tension will be preferentially adsorbed, the rate of the adsorption process can affect the "local" tension and the contact angle. In many systems, the rate of adsorption at the solid surface is found to be quite slow compared to the rate of movement of the SLV contact line. As a result, the system does not have time for the various interfacial tensions to achieve their equilibrium values. Most surfactants, for example, require several seconds to attain adsorption equilibrium at a LV interface, and longer times at the SL interface. Therefore, if the liquid is flowing across fresh solid surface, or over any surface at a rate faster than the SL adsorption rate, the effective values of σ_{LV} and σ_{SL} (and therefore θ) will not be the equilibrium values one might obtain from more "static" measurements. More will be said about dynamic contact angles in later chapters.

An effect closely related to that of varying adsorption rates is that resulting from changes in σ_{SL} due to direct chemical or physical interaction between the liquid and the solid surface. Particularly important in the textile and paper industries would be the swelling of fibers on contact with water or other solvent liquids. As swelling occurs, the value of σ_{SL} will be continuously changing (usually decreasing) at an undetermined rate. Accurate analysis of the flow then becomes difficult or impossible. In addition, some polymeric surfaces that do not swell on contact with aqueous systems have specific monomer units that do interact with water (e.g., acid groups). Such surfaces may experience changes in σ_{SL} or θ as a result of specific interactions at the molecular level (e.g., ionization), again affecting the observed capillary flow.

Contact Angle Effects. In practical capillary systems such as textiles, paper products, and oil-bearing rock formations, anomalous effects due to contact angle variations can almost be considered a certainty. The effects may be due to heterogeneities in the solid surface (compositional effects), geometry (surface roughness), or other dynamic or molecular factors. It was previously stated that the equilibrium contact angle of a given solid–liquid–fluid system could be considered a material constant for the system. However, it is commonly found experimentally that the equilibrium angle measured as a drop advances across a fresh solid surface may differ significantly from an angle measured as a drop moves across a previously contacted area. Conventionally, the former situation is referred to as the advancing contact angle, θ_A, and the latter as the receding contact angle, θ_R. The difference between the two angles is the contact angle hysteresis (see also Chapter 17).

A commonly encountered illustration of contact angle hysteresis is that of a raindrop moving down a slanted car windshield (Fig. 6.14). In that case, gravity is causing the front of the drop to advance across the fresh glass surface with a relatively large θ_A while the trailing end moves with a much smaller

FIGURE 6.14. A common illustration of contact angle hysteresis is the movement of a liquid drop on an inclined surface such as a car windshield.

angle. There is, of course, some distortion of the drop due to gravity, but that only exaggerates the effect somewhat.

Contact angle hysteresis measured experimentally may be due to heterogeneities in the composition of the solid surface, surface irregularities, or dynamic effects due to adsorption or desorption phenomena, molecular reorientations, or similar.

The cause of contact angle hysteresis on a chemically uniform but rough surface is illustrated schematically in Figure 6.15a. From the illustration it can be inferred that the extent of hysteresis observed on a macroscopic scale may depend on the scale of observation of the angle and the size of the liquid sample employed. Obviously, hysteresis can occur from point to point along the SLV contact line. The extent to which the curvature of the drop exhibits localized hysteresis will depend on the length of the moving boundary line and the area of the moving front. It can be shown that if the "true" contact angle of a liquid on a smooth solid surface is $< 90°$, its apparent contact angle on a rough surface of the same material will be smaller than the actual value. Conversely, if the true angle is $> 90°$, the apparent angle will be greater than

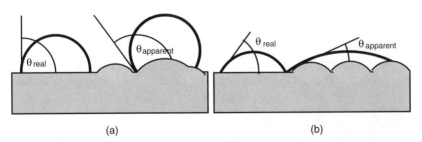

(a)   (b)

FIGURE 6.15. On a rough surface, the apparent contact angle can differ significantly from the "real" angle that would be observed on a molecularly smooth surface of the same material: (a) if the real contact angle is greater than 90°, the apparent angle will be even larger; (b) if the real angle is less than 90°, the apparent angle will be smaller.

the true value. A similar analysis of hysteresis on a nonuniform or composite surface has been employed to develop empirical relationships to quantitatively adjust observed contact angles for those complications. Those relationships and others concerning contact angles are discussed in more detail in Chapter 17.

Although hysteresis in many systems may be sufficiently small that it can be neglected, many important capillary systems are found to exhibit a hysteresis of 50°–60°. Prime examples of such systems would be many textile and paper products that have θ_A quite large but θ_R at or near zero. The same can be expected in some rock and soil formations in which contact with water alters the hydration state or electrical characteristics of the surface and thereby θ. In such capillary systems simple empirical adjustments such as those available for planar systems do not suffice. In capillaries where the driving force for liquid flow must be calculated, it is most convenient to employ a modified form of Equation (6.12), dividing the final term on the right into terms for θ_A and θ_R so that

$$\frac{\delta G}{\delta \sigma} = \sigma_{LV} \frac{\delta A_{LV}}{\delta \sigma} - \sigma_{LV}\cos \theta_A \frac{\delta A_{SL(A)}}{\delta \sigma} - \sigma_{LV} \cos \theta_R \frac{\delta \sigma A_{SL(R)}}{\delta \sigma} \quad (6.13)$$

where the subscripts SL(A) and SL(R) refer to the area of the advancing and receding liquid fronts, respectively. If one applies this equation to a fused capillary system such as that illustrated in Figure 6.11, where θ_A is significantly greater than θ_R, one will see that the liquid will remain stationary rather than move into the smaller capillary because the curvatures of the two ends of the liquid mass will be equal and P_{cap} will be zero.

Dynamic Contact Angle Effects. The comments on contact angle hysteresis effects above were made assuming that liquid movement was sufficiently slow for "equilibrium" or static values of θ_A and θ_R to be involved. In capillary systems in which liquid flow is relatively fast, the effects of a dynamic advancing contact angle, θ_{AD}, may become apparent. In such situations the advancing contact angle measured will be greater than θ_A. The difference between θ_A and θ_{AD} will generally be found to increase with the speed of liquid flow. In systems of large-bore, short-path capillaries or those with high static θ_A, the effects of the dynamic contact angle on liquid movement may appear at relatively slow flow rates, even becoming self-limiting. Where relatively long capillaries are involved, so that the area SLV_A is small relative to the area of the SL and SLV_R interfaces, fluid dynamic factors such as viscosity become more important than contact angle.

Phenomena related to dynamic contact angles are important in other, non-capillary applications, such as high-speed liquid coating operations. Some examples will be addressed in Chapter 17.

6.3.6. Rates and Patterns of Capillary Flow

Although it is important to analyze a capillary system with respect to the various factors affecting P_{cap} in order to understand liquid flow patterns, the effects of various fluid dynamic effects must also be kept in mind. While a thorough discussion of the topic is beyond the scope of this work, it may be useful to introduce a few simple ideas that can be applied in many situations.

When one considers the smooth, uniform (laminar) flow of a fluid in a narrow cylindrical tube, which is the classical model of a capillary, one can employ Poiseuille's equation to relate the volume rate of flow to various characteristics of the fluid and capillary system. The volume rate of flow, dv/dt (mL s^{-1}) is given by

$$\frac{dv}{dt} = \frac{\pi r^4 P}{8\eta l} \tag{6.14}$$

where r is the radius of the tube, η the viscosity of the fluid, l is the distance of fluid movement in the tube in time, t, and P is the pressure drop across the distance l. For linear rates of flow, Equation (6.14) becomes

$$\frac{dl}{dt} = \frac{r^2 P}{8\eta l} \tag{6.15}$$

In capillary systems, P is replaced by P_{cap}.

If Equation (6.12) is employed to calculate the driving force $dG/d\sigma$, it must be converted into an equivalent P_{cap} for use in (6.14) or (6.15). For a relatively simple system such as that illustrated in Figure 6.11, the conversion is straightforward. For example, if the normal projected cross-sectional area of the LV interface at point A (in cm^2) in the large end of the capillary system is significantly greater than that at point B at the small end, it can be assumed that P_{cap} at A will be negligible and only the contribution from P_{cap} at B need be considered. If $d\sigma$ is measured at B, the net driving force $-dG/d\sigma$ (in mJ) divided by the cross-sectional area at B will give the pressure drop from A to B (mJ cm^{-2}). In that case, Equation (6.15) becomes

$$\frac{dl}{dt} = \frac{2\sigma_{LV} \, r \cos \theta}{8\eta l} \tag{6.16}$$

In practice, unfortunately, most capillary systems do not involve nicely uniform cylindrical tubes, so the preceding analysis will not apply in a quantitative way. In not overly complicated systems, reasonable, and useful results can be obtained if r is replaced by a so-called hydraulic radius given by the volume of the liquid in the capillary section being considered, V, divided by the solid–liquid interfacial area, A, in the same section. In systems of still higher complexity calculation of a hypothetical value for r, or of a resistance factor, r/η, becomes quite difficult and only qualitative or semi-quantitative relationships can be expected.

Because many practical systems involving capillary flow involve irregular and ill-defined geometries, and therefore variations in curvature and P_{cap} from point to point in the system, there has developed an area of investigation into hydraulics that combines aspects of capillary flow and the associated field of fluid dynamics. Fortunately, many complex systems such as filtration and wetting of woven textile and paper products, can be approximated using models with the liquid mass in a given starting location and configuration and using Equation (6.12) to calculate the driving force for movement and the final configuration. If reasonable values of P_{cap} are assumed, Equation (6.14), (6.15), or (6.16) can be used to estimate initial rates of flow. From a comparison of predicted and measured initial flow rates, the validity of P_{cap} from the original assumptions can be evaluated and the model changed if necessary to improve the fit. By a series of such iterations, one may arrive at a workable practical model for the system in question.

6.4. SOME PRACTICAL CAPILLARY SYSTEMS

While many practical capillary systems do not lend themselves to direct analysis according to the simple concepts presented above, there are areas in which those principles, along with some intuition and a dollop of luck, can be constructively employed to analyze and predict capillary phenomena. Several areas in which various degrees of success have achieved are wetting and repellency of woven fibers, paper products, porous solids, wicking, and cleaning action in detergent baths. Approaches to some of those problems will be briefly discussed below.

6.4.1. Wetting in Woven Fibers and Papers

If one examines a woven fiber or paper system under a microscope, one sees an open capillary system that can, qualitatively at least, be approximated as shown in Figure 6.16. A typical fabric yarn will be composed of 100–200 approximately parallel fibers, each 15–20 μm in diameter and 3–8 cm in length. The yarn is then woven to produce the fabric. The spaces between the woven yarn will be several orders of magnitude larger than those between the individual fibers so that, in terms of capillary phenomena, the fibers act as the solid surface whose interaction at the SLV interface will determine the ultimate capillary driving force in the system.

In the treatment of woven fabrics for dyeing, waterproofing, sizing, and other procedures, important criteria for evaluation are the speed and completeness of the wetting process. In the textile industry, "wetting" is taken to mean the submersion of the fabric in an aqueous treating solution, replacing all the air in the cloth structure with the solution. Obviously, if the treatment

(a)   (b)

FIGURE 6.16. For paper or woven fabrics, the spaces between fibers represent a complex capillary system that can affect several aspects of the preparation (e.g., dying and use (e.g., waterproofing) of such materials: (*a*) if good wetting is not achieved by a dying solution, trapped air bubbles may produce an irregular finish; (*b*) for good waterproofing, a large contact angle is needed, which, if larger than 90°, will be aided by the roughness of the surface.

is to produce a product with uniform characteristics, all portions of the cloth must contact the solution for an optimum period of time to allow completion of the desired process. If air bubbles remain in the cloth structure during the treatment, areas of fiber that were not wetted will not receive the desired treatment and will have characteristics distinct from those of treated areas. Of course, given time, all the air may be removed from a system to produce the desired result. However, in industry, time is money, and it is usually desirable to have the wetting process occur evenly and rapidly. Since the interstitial spaces are the ones that dominate the capillary forces in the system, they will be the rate-determining factor in the overall process.

Neglecting the effect of gravity (i.e., buoyancy) as an external force helping to remove air from the fiber, the remaining effect will be almost entirely due to capillary forces. Analysis of the pertinent equation for capillary flow rates [Eq. (6.16)] indicates that the desired conditions for rapid wetting include a small contact angle and high surface tension. Unfortunately, due to the nature of the relationship between those two quantities, the most desirable conditions are normally contradictory. High-surface-tension aqueous solutions tend to have large contact angles on most fibers. Contact angles, however, can be reduced significantly by the addition of wetting agents or surfactants (Chapter 17). Unfortunately, surfactants by nature also lower surface tensions. All is not lost, however.

In aqueous systems, the range of surface tensions that will be encountered using common surfactants and practical concentrations is somewhat limited. In most practical cases, one might expect σ_{LV} to be reduced to the range of 40 mN m^{-1}, a change of a factor of ~ 2 from that of pure water (72.8). Lower values are normally attained only with very high surfactant concentrations (which can introduce foaming problems), with purer, more expensive surfactants, or with special materials such as fluorocarbon or silicone surfactants.

Significant contact angle reductions, on the other hand, can be achieved using much lower concentrations of less surface-active wetting agents. Table 6.1 presents hypothetical data illustrating the relative effects of changes in σ_{LV} and θ on potential wetting rates. One can see from the data that accessible changes in contact angle have a much greater relative (and positive) effect on the wetting process than the potentially detrimental effects of large reductions in surface tension, which are in most cases difficult to achieve.

The presence of surfactants or wetting agents in textile treatment solutions can also introduce other complications in the understanding of the dynamics of the wetting process. Because surfactants adsorb at the SL interface as well as the LV interface, as the liquid front moves across fresh solid surface, adsorption processes will tend to deplete the concentration of available surfactant and may cause localized changes in both σ_{LV} and θ. In many cases, however, adsorption rates at the SL interface is much slower than that at LV interfaces, so that such effects can be taken into consideration without too much difficulty.

Of more direct practical importance to textile processing is the fact that it becomes increasingly more difficult to remove the last traces of air from the fiber system. Fiber bundles in yarn can produce quite complex capillary systems that provide ample opportunity for air entrapment in very inaccessible nooks and crannies. It is found, for example, that yarns made of smooth essentially cylindrical fibers are much easier to wet completely than those composed of

TABLE 6.1. Effects of Change in Surface Tension σ_{LV} (mN m^{-1}) and Contact Angle on Linear Rate of Flow in a Hypothetical Capillary System[a]

Situation	θ (°)	dl/dt (cm s^{-1})	$\Delta(dl/dt)$ (x)
$\sigma_{LV} = 72$			
1	89	0.0016	—
2	75	0.023	15
3	50	0.058	36
4	25	0.082	51
5	0	0.091	56
6	89	0.0012	—
7	75	0.018	15
8	50	0.044	37
9	25	0.062	52
10	0	0.069	57
$\sigma_{LV} = 40$			
11	89	0.001	—
12	75	0.013	13
8	50	0.032	32
9	25	0.045	45
10	0	0.05	50

[a] Using Equation (6.16), where $r = 0.05$ cm, $\eta = 2.0$ cp, and $l = 5$ cm.

rough, irregular fibers. In any case, the last vestiges of air are probably not displaced by capillary process at all, but by direct solution of the small air bubbles in the aqueous solution. In that case, the dissolution process is probably "driven" by the high pressure in the bubble due to it small radius of curvature.

6.4.2. Waterproofing or Repellency Control

In dyeing and other processing of textiles, complete wetting is important. However, once the final product has been obtained it may be desirable to have a system that is no longer wetted by water or other liquid systems. Waterproofing or repellency, then, is the opposite of wetting and must be addressed with an essentially opposite approach. In this case, since the manufacturer will have no control over the surface tension of any contacting liquid, it is necessary to control wetting by attacking the problem from the aspect of the contact angle. Ideally, to have a completely nonwetting system for all possible liquids (water or oil) it is necessary to produce a fiber surface that exhibits a large θ for all liquids. Achieving that end for water solutions is not all that difficult because water will almost always have a higher contact angle (in air) than oils on organically treated fiber surfaces, and even untreated ones in many cases. For the case of oily liquids, it is usually necessary to treat the fibers with substances that produce high oil–solid θ values. Those materials, unfortunately, are quite limited and usually involve relatively expensive fluorocarbons or silicones.

Complete analysis of the effect of contact angle on a given system is complicated (naturally) by the possible existence of hysteresis, roughness effects, and surface compositional variations within the fibers. If $\theta_A > 90°$ for a given liquid, it will not spontaneously enter into the fiber network under the drive of capillary forces. Likewise, if θ_R is also $> 90°$, any liquid forced into the network by external physical forces will be spontaneously expelled from the system. Such a situation represents the "ideal" for most effective repellency control. Even if those conditions cannot be achieved, however, large values of $\theta_A < 90°$ will at least retard entrance of liquid into the fiber matrix or maintain a slow rate of movement.

If a liquid having $\theta < 90°$ is forced into a fabric, it can, in principle, be removed relatively easily by contacting the fabric with another matrix of essentially the same texture and weave (i.e., capillary characteristics), but having a lower contact angle with the liquid. In that case, the liquid will flow or "wick" into the more easily wetted fabric, leaving the first essentially dry (see Fig. 6.12b). For example, if one were interested in producing a material which would allow moisture to pass from the area contacting skin out to the air, but not allow movement in the opposite direction, a two-layer system composed of an inner layer of polyester or nylon (high θ) contacting a layer of cotton (low θ) would, in principle, provide just such a wicking/repellent fabric. For simpler repellency control, surface treatment of the fibers to produce large θ values is normally used.

Many other processes related to textile and paper treatments, impregnation of capillary systems, and wicking are of great industrial importance. The analysis of such problems begins with the same basic ideas as those employed for understanding and controlling wetting and repellency. However, they can be further complicated by such factors as irregularities in the capillary system (producing unsymmetric flow patterns), Marangoni effects due to variations in solvent evaporation rates and temperatures or swelling of fibers in contact with liquid. Such problems represent fruitful, but theoretically difficult areas for research and development.

6.4.3. Capillary Action in Detergency Processes

As a final topic for discussion, we will briefly address the problem of capillary action in the context of cleaning and detergency. In its simplest form (Fig. 6.17), detergency can be viewed as the process of separating a liquid or solid from contact with a solid by the action of a second liquid (L_2). The same principles can be applied to the case of separating two solid surfaces by the action of a liquid, but for present purposes the discussion will center on SL_1L_2 systems. Probably the most familiar detergency system is that involving the removal of an oily soil from a fabric through the action of an aqueous wash. However, the removal of aqueous liquid from a solid surface (usually metal or ceramic) by the action of an organic liquid is also of great industrial importance.

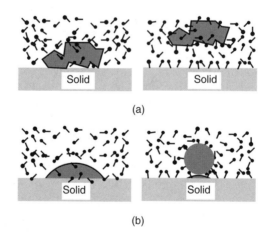

FIGURE 6.17. In detergency and cleaning, a major mechanism is the capillary penetration of the cleaning liquid, L(2), between the surface to be cleaned and the deposited solid or oily dirt, usually aided by the presence of surfactant and mechanical agitation: (*a*) adhering dirt particles are "lifted off" by detergent solution. Redeposition is retarded by the presence of adsorbed detergent molecules; (*b*) an oily dirt is "rolled up" by capillary action and removed by detergent as with solid dirt.

Although mechanical agitation is important in detergency processes, the fundamental physicochemical process for removal of the soil is the capillary displacement of one liquid on the solid surface by a second liquid. In such a process, the contact angles of the two liquids on the solid surface are the primary factors controlling the rate of capillary displacement, although viscosity (lower is better) and σ_{LL} (higher is better) are also important. For an aqueous detergency system, the optimum condition will be where $\theta_{A/water}$ is as small as possible relative to $\theta_{R/oil}$. If $\theta_{A/w}$ is greater than 90°, water will not penetrate the fabric and no capillary-driven cleaning action will result. Conventionally, such situations are avoided by the addition of surfactants that lower $\theta_{A/w}$. Since most natural fabrics are slightly swollen by water but not by oils, soaking can allow time for the aqueous solution to penetrate not only into the capillary system but also into the basic fiber network, swelling the fiber and further improving the contact angle situation in favor of oil removal.

Obviously, such a common process as fabric cleaning represents a very complex capillary system, and a quantitative understanding of the many factors involved in "detergent power" is still a goal to be pursued. The historic process of trial-and-error detergent formulation has slowly given way to scientific understanding which translates (hopefully) in more "bang for the buck" in cleaning products. In cleaning and many other important applications, capillary action cannot be isolated from other important surface, hydrodynamic, or mechanical processes. Reference to the ideas presented above, therefore, will be encountered in many sections of following chapters.

PROBLEMS

6.1. Two soap bubbles, A and B, are blown from the same solution and carefully brought into contact. Predict the resulting configuration for the following circumstances: (a) $R_A > R_B$; (b) $R_A = R_B$; (c) $R_A < R_B$. Illustrate your predicted results.

6.2. Two spherical particles 1000 nm in diameter are in contact and are held together by a liquid bridge. Assuming that (1) the size of the bridge is small relative to the size of the particles, (2) that the contact angle at the liquid–particle interface is zero, and (3) that the surface tension is 30 mN m^{-1}, calculate the force holding the spheres together.

6.3. Calculate the rate of shear necessary to separate the two spherical particles bound by van der Waals forces as given in Problem 6.2. The viscosity of the medium is 10^{-2} P (poise).

6.4. What pressure in atmospheres would be needed to force water at 25°C through a 1-cm-thick sintered-glass filter with a uniform pore size of 100 nm? What would be the force for mercury under the same conditions?

6.5. Two vertical cylindrical glass rods 1 cm in diameter with well polished flat ends are held togethet by a 150-nm-thick layer of water between the two ends. If the water meniscus is semicircular all around the cylinder circumference, what length can the lower cylinder have before it is detached by gravity from the upper one? Assume the density of glass to be 2.5 g cm^{-3}.

6.6. The capillary rise method was used to measure the surface tension of a liquid that wetted glass. A glass capillary tube of 0.0200 cm diameter was used and a capillary rise of 10.47 cm was measured. The liquid density was 1.000 g cm^{-3}. What is the surface tension of the liquid? What influence do corrections for the weight and nonsphericity of the meniscus have on the surface tension?

6.7. The surface tension of a liquid that wets glass is measured by determining the vertical distance, Δh, between levels of the two menisci in a U-tube having a small bore r_1 on one side and a larger bore r_2 on the other. The following data are known: $\Delta h = 1.90$ cm, $r_1 = 0.1$ cm, $r_2 = 1.00$ cm, and $\rho_L = 0.95$ g cm^{-3} Calculate the surface tension of the liquid using the simple capillary rise treatment.

6.8. A fabric made of fibers of diameter 2.0×10^{-4} cm and density 1.3 g cm^{-3} has a single fiber advancing contact angle for water of 120°. Calculate (a) the contact angle on a fabric so woven that its bulk density is 0.9 g cm^{-3} and (b) the depth of a water layer that the fabric could support before passage begins. Explain any necessary simplifications or assumptions.

6.9. The ability of soil to "lift" water from the water table level depends on capillary action. It has been calculated that the following soil types can lift water to the indicated heights: (a) fine gravel − 0.1 m; (b) coarse sand − 0.5 m; (c) fine sand − 2; (d) silt − 10 m; (e) clay − 50 m. Estimate the average pore size in each soil type.

6.10. The edges of two large glass plates are placed in water a distance of 0.5 mm apart. To what height will the water rise between them?

7 Solid Surfaces

A solid is by definition a phase of matter that is rigid and resists stress. Like the liquid surfaces discussed previously, a solid surface must be characterized by some surface free energy and total free energy terms. It should be evident, however, that such free energies for solids cannot be characterized using the capillary and related methods so useful in the study of fluid surfaces. While liquid surfaces can, in general, be assumed to be in equilibrium and equipotential, a solid surface is generally of such a nature that those two assumptions will not be valid.

On a normal time scale, a liquid surface under stress will undergo plastic flow—that is, as surface area is increased, molecules of the liquid phase will flow into the surface region from the bulk to maintain an equilibrium surface density (Fig. 7.1a,b). A solid surface, on the other hand, will normally undergo elastic flow, in that as a stress is applied and the surface area is increased, there will not be a significant flow of molecules from bulk to surface. Instead, the distance between surface molecules will increase to produce a lower surface density (Fig. 7.1c,d). As a result, since the distance between molecules increases, their lateral interactions decrease, producing a change in the energy of the stressed surface. Ultimately, events may occur that bring the surface back to equilibrium; however, for many solids, the time scale of those events may be years. Therefore the nature of a solid surface, with some exceptions depending on the exact nature of the surface and environmental conditions, will be determined at any given moment as much by its history as by equilibrium thermodynamics or surface tension forces. This "historical" effect on solid surfaces will be discussed further below.

7.1. SURFACE MOBILITY IN SOLIDS

For the purpose of placing events in the proper frame of reference, it is useful to do a simple calculation to estimate the mobility of atoms and molecules in a solid surface. To obtain such an approximation, one can view the surface as being in a dynamic state where there is a constant interchange of molecules between the surface, the bulk, and the vapor phase.

The number of moles of vapor hitting one cm^2 of surface per second, Z, is given by

$$Z = 0.23P \left(\frac{3}{MRT} \right)^{1/2} \tag{7.1}$$

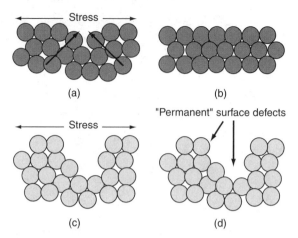

FIGURE 7.1. Schematic illustration of the responses of liquid and solid surfaces to stress: (*a*) a stressed liquid surface with "vacancies"; (*b*) molecules below the surface rapidly flow into the stressed area to "heal" the imbalance of forces; (*c*) a stressed solid surface with various defects; (*d*) due to lack of mobility, molecules cannot readily move into the stressed areas and the defects remain to produce a higher surface energy.

where P is the vapor pressure of the material, M its molecular weight, and R and T have their usual meanings. If the equation is applied to a solid such as tungsten with a room-temperature vapor pressure of $\approx 10^{-37}$ mmHg, then Z is equal to about 10–20 atoms cm^{-2} s^{-1}, and the average lifetime of a surface atom becomes 10^{37} seconds! More volatile but still refractory solids such as copper will also have very long surface residence times at room temperature.

For solids, there is defined a temperature, the so-called Tamman temperature, at which the atoms or molecules of the solid have acquired sufficient energy for their bulk mobility and reactivity (including sintering) to become appreciable. In general the Tamman temperature of a material will be approximately one-half its melting temperature (K). As the temperature at which noticeable sintering occurs is approached, one begins to see dramatic changes in the average surface residence time of the solid units. For example, at 725°C copper has a vapor pressure of about 10^{-8} mmHg. Equation (7.1) gives $Z \approx 10^{15}$ atoms cm^{-2} s^{-1} for a surface residence time of about 1 s for copper atoms. From the point of view of bulk diffusion processes under similar conditions, an average copper atom will move approximately 10 nm in 0.1 s. At room temperature, that diffusion rate falls to the range of 10^{27} s for a distance of 10 nm.

Such calculations serve to illustrate the relatively low mobility of atoms or molecules in solids and help dramatize the differences that must be considered when discussing and comparing solid and liquid surfaces. They do not, however, exactly describe events when an atom or molecule moves within the surface region. For example, in bulk diffusion, one considers the ease with which a unit moves from one position in the bulk phase to another, but with a net result (to a first approximation, at least) of zero change in the total free energy of the system. Likewise, the calculation of Z from Equation (7.1) relates to the movement of atoms or molecules from the surface to the vapor phase, again with no net change in free energy of the system. For units in the surface, however, there is a third option for movement that is of more importance in the temperature region below the Tamman temperature. That movement is surface diffusion.

An atom or molecule undergoing bulk diffusion will experience no net change in the extent of its interaction with its neighboring units (assuming no change in the bulk structure). However, in order to move from one site in the structure to another, it must move past other units, meaning that there must be a significant activation energy, which must be overcome. The process is illustrated (roughly!) in Figure 7.2a. For a unit in the surface (Fig. 7.2b), however, one would expect a much lower activation energy since the unit, by its location in the surface, will usually have one or more "empty" sites into which it can move without requiring the movement of neighboring units. The surface unit, therefore, will face a much lower barrier to diffusion and will have much greater mobility at a given temperature. Surface diffusion, then, because of its lower activation energy, might be expected to represent an important phenomenon in temperature regions below that where sintering becomes important, especially with respect to surface chemical effects.

In summary, when one considers the nature of solid surfaces, one must always take into consideration the history (especially thermal) of the material. Since atoms and molecules in solid surfaces at room temperature have a very

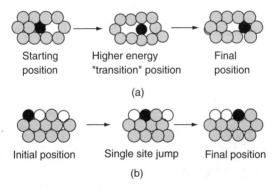

Starting position | Higher energy "transition" position | Final position

(a)

Initial position | Single site jump | Final position

(b)

FIGURE 7.2. Schematic illustration of the comparative energetics of diffusion in bulk and in a surface: (a) bulk diffusion, illustrated in cross section, involves the displacement of several nearest-neighbor units, representing a relatively large energy barrier; (b) surface diffusion, shown in a top view, involves the displacement of fewer neighboring units, and therefore a relatively lower activation energy for the process. As a result, systems that undergo bulk diffusion with difficulty may exhibit orders of magnitude more surface diffusion under the same conditions.

low mobility, they can be expected to retain the surface positions they acquired at the time of formation, even though they may be occupying positions of high relative energy. As a solid approaches its melting point, the surface units begin to acquire the properties of the bulk liquid phase, with overall greater mobility in terms of interchange with the vapor and bulk phases, and especially in lateral movement in the surface. In the intermediate temperature range, however, where bulk diffusion and evaporation–condensation are negligible, individual surface units can still move relatively rapidly toward more favorable energetic positions and thereby alter the surface chemical characteristics of the system. Such relatively subtle surface changes may be reflected in more dramatic alterations in characteristics such as adsorption, wetting, adhesion, friction, or lubrication.

7.1.1. Sintering

Under some conditions, especially close to the melting point, many materials usually considered to be solids will exhibit sufficient plastic flow in the surface that capillary forces will slowly, but within a reasonable time frame, come into play to move the surface toward equilibrium, or at least a lower energy situation. A prime practical example of such action is the sintering of solids. If a solid powder—metallic, crystalline, or amorphous—is heated to some temperature below its melting point, usually, but not always, with some applied pressure, sintering or fusion of adjacent particles will occur (Fig. 7.3).

It is assumed that the main driving force for sintering is surface tension rather than external pressure. However, because of the above mentioned "historical" nature of solid surfaces, various parts of the powder surface will experience different net driving forces, leading to a rather complex situation during the sintering process. For example, atoms or molecules located at sharp asperities on the surface will have higher local excess surface energies than others located in situations more similar to the bulk phase. As a result, they will experience a greater surface tension force and have a greater surface

Loose powder   Sintered mass

FIGURE 7.3. Schematic illustration of the stepwise process of particle sintering.

FIGURE 7.4. Schematic illustration of plastic deformation and flow under load at points of contact between asperities of a soft material and a relatively hard surface.

mobility. In addition, the normally rough surfaces of solids mean that, due to the presence of asperities, the actual area of contact between particles undergoing sintering will be small so that even at relatively low total pressures, the pressure applied at the points of contact will be multiplied so as to exceed the yield value of the material, producing plastic deformation and flow at the points of contact (Fig. 7.4).

In addition to local pressure effects, normal sintering temperatures will also usually allow significant amounts of diffusion, both in the bulk and in the surface. By such mechanisms, scratches on some metal surfaces will be "healed" if the sample is heated to a temperature well below the melting point.

7.2. "HISTORY" AND THE CHARACTERISTICS OF SOLID SURFACES

The relative immobility of atoms and molecules in the surface of a solid well below its melting point results in surface energy and related characteristics that depend to a great extent on the formative and environmental history of the sample. For example, a clean cleaved crystal surface will almost certainly have a different surface energy than a surface of the same material that has been ground or polished. In this case, the cleaved surface will probably be of lower energy (assuming the absence of surface contamination) since the cleaving process will tend to occur along the crystal face of lowest energy. Grinding and polishing, on the other hand, are rather indiscriminate in their action and tend to leave a significant number of small but energetically significant crystal defects that increase the surface energy.

Other factors that can affect the apparent energy of a surface include the action of friction (see Chapter 18), corrosive action (which actually changes the chemical nature of the surface such as oxidation in metals), and adsorption. All of these items are pointed out because they should always be taken into account when one attempts to determine the surface energy of a solid.

Of equal or greater importance to the nature of a surface is the question of the history of its formation (as opposed to the treatments noted above).

TABLE 7.1. Reported Surface Energies of Commonly Encountered Solids

Material	Surface Energy (mJ m^{-2})	Material	Surface Energy (mJ m^{-2})
Teflon	20	Lead iodide (PbI$_2$)	130
Paraffin wax	26	Silica	462
Polypropylene	28	Lead fluoride (PbF$_2$)	900
Polyethylene	36	Iron	1360
Polystyrene	44	Gold	1500
Ice	82	Mica	4500

For crystalline materials especially, it must be kept in mind that a crystal surface may contain any or all of the defects already mentioned, as well as missing layers, and screw and spiral dislocations. All such defects will alter the surface energy of the crystal and complicate the analysis of phenomena related to it. Obviously, solid surfaces are more difficult to analyze and understand not only because of their inherent anisotropic nature but also because of the potential role of history in determining the exact nature of the surface produced at formation.

Equally important to the energetic nature of a surface is the presence (or absence) of adsorbed species and surface contamination. For low-energy, amorphous solids such as most polymers, surface contamination due to adsorption is normally not a major problem and can be controlled relatively easily. Higher-energy crystalline, metallic, and inorganic glassy materials, however, pose significant experimental problems. "Clean" surfaces of such materials will routinely have high surface energies in the range of 10^2–10^3 mJ m^{-2} (Table 7.1). They will literally "do anything" in order to lower their surface energies, especially including the adsorption of almost any available molecule— nitrogen, oxygen, water, or any other material present in the environment. For that reason, it is difficult to prepare a truly clean surface of many solids without the use of exceedingly stringent environmental controls. Almost any low-energy material present in the environment will tend to be adsorbed leading to contamination of the surface and incorrect analysis of surface energy. As a result, most exact surface studies on solids are carried out in high vacuum and under ultraclean conditions. That is not to say, however, that meaningful and very useful results related to solid surface energies cannot be obtained by much easier techniques.

7.3. SOLID SURFACE FREE ENERGY VERSUS SURFACE TENSION

As might be guessed from the comments above, the surface tension or surface free energy of solids, unlike liquids, cannot be equated with the total energy in the surface layer—the "native" specific surface free energy plus "stress" energy due to factors mentioned above. By definition, the former is the work performed in forming a unit area of new surface while the latter pertains to the work spent in stretching the surface. In order to see the difference between the two concepts, it is useful to visualize new surface formation as a two-step process. First, the condensed phase is divided to produce the new surface, but the molecules in the new surface are held in the exact locations (relative to the remaining bulk phases) they occupied in the bulk, as illustrated schematically in Figure 7.5a. In the second step, the atoms in the newly formed surface are allowed to relocate into their most stable configuration. What this means, in effect, is that some of the units in the original (new) surface are "pulled" into the bulk by the unbalanced forces acting on them (Fig. 7.5b). In a liquid system these two steps will occur essentially simultaneously because of the mobility of the units.

In solids, on the other hand, greatly reduced atomic or molecular mobility means that the rearrangement will occur much more slowly, or perhaps not at all on a reasonable timescale. The density of units in the new solid surface will therefore be something other than the "equilibrium" value. The surface may be compressed or stretched with no coincident change in the surface unit density. What changes will be the distance between units, which, as we have seen from Chapters 4 and 5, means a change in their interaction forces and therefore their free energy.

A mechanical model is useful for understanding exactly what is meant by surface stress in a solid, as opposed to its surface tension. Suppose that a solid surface is cleaved in a direction perpendicular to the surface. As pointed out above, the solid units in the new surface will not be able to relocate to attain their equilibrium positions relative to the bulk. In order to "make" their positions into "equilibrium" positions, one can think of applying some external force or lateral pressure on the surface units to hold them in place. The force

FIGURE 7.5. Schematic illustration of a stepwise mechanism for the formation of new surface: (*a*) initial cleavage; (*b*) rearrangement of surface units due to "excess" attraction by bulk underlying units.

per unit length of new surface needed for this equilibrium situation is the surface stress. If one takes the average surface stress for two mutually perpendicular cuts, one will obtain the surface tension of the solid. For the cases of liquids or isotropic solids, the two stresses are equal and the surface stress and surface tension are equivalent. For an anisotropic solid, however, the two will not be equal (except perhaps by coincidence), so that the differentiation must be made.

7.4. THE FORMATION OF SOLID SURFACES

Because of the obvious importance of history to the nature of a solid surface, it is useful to understand some of the basic principles underlying the formation of such surfaces. The following section will briefly address the subject for two important classes of solids: crystals and amorphous solids. Because of the complexity of describing metallic surfaces, they will not be treated here.

7.4.1. Crystalline Surfaces

Crystallization is a process in which an ordered solid phase is precipitated from a gaseous, liquid, or solid phase. The liquid phase may be either a melt or a solution. For most, but certainly not all, of the most important crystallization processes, that from solution is most important and will be emphasized here. A solid phase is precipitated from a solution if the chemical potential of the solid phase is less than that of the material in solution. A solution in which the chemical potential of the dissolved component is the same as that of the solid phase is said to be in equilibrium under the given set of conditions and is termed a saturated solution. The equilibrium state is defined by the concentration of the saturated component at a given temperature and concentration of other components, that is, by its solubility under those conditions.

In order for crystallization to occur, the equilibrium concentration of the component of interest must be exceeded by some supersaturation method, including

1. Cooling a solution in which the solubility of the component increases with increasing temperature or heating a solution in which the solubility of the material decreases with increasing temperature
2. Evaporating the solvent under heating
3. Adiabatic evaporation of the solvent, where removal of the heat of vaporization of the solvent is reflected in a decrease in the temperature of the solution
4. Adding to the solution another solvent that is miscible with the primary solvent, but is a poorer solvent for the material being crystallized

5. Salting out by the addition of substances that may contain a common ion with the crystallized substance, thereby reducing its solubility, or by changing other factors affecting the ability of the solvent to solvate the material

6. Chemical reaction in the solution changing a soluble substance into an insoluble one.

7. Various other specialized processes.

In any of these ways, the supersaturated solution necessary for crystallization can be obtained. If the supersaturation is attained by cooling, then the difference in temperature between that exactly corresponding to saturation and the actual temperature of the solution is termed supercooling. If the supersaturation is not too great, the rate of formation of new crystals (nucleation) will be small and the solution will be in what is termed the "metastable region"; new crystals are formed to a limited extent and crystals already present grow. This, of course, corresponds to an ideal situation for the growth of a few very large (or single) crystals. If the degree of supersaturation is increased and the maximum possible supersaturation, the upper boundary of the metastable region, is attained. When this boundary is exceeded, the rate of nucleation rapidly increases and the crystallization process becomes essentially uncontrolled and large numbers of small crystals are formed. Thus, in order to have a controlled crystallization process, it is necessary to maintain the solution within the metastable region—bounded on one side by the saturation concentration and on the other by the upper boundary of the metastable region.

The kinetics of crystallization can be usefully divided into two stages: formation of new crystal nuclei or nucleation and crystal growth proper. Both stages occur simultaneously, but they can to a greater or lesser extent be separated and considered independently in discussions of the crystallization process.

7.4.2. Nucleation Processes

Depending on the crystallization process being employed, it is usual to divide nucleation into two types: primary nucleation, in which crystals begin to form in the absence of solid particles of the crystallized substance; and secondary nucleation, which requires the presence of "seed" crystals of the substance of interest. Primary nucleation can be further subdivided into homogeneous and heterogeneous nucleation. In the former, spontaneous nucleation occurs without the intervention of any solid phase, whereas for the latter, the presence of a foreign surface such as dust, colloids, or vessel walls acts as a catalyst to initiate crystal formation. Secondary nucleation can also be subdivided into true, apparent, and contact nucleation. Those topics will be left for the interested reader to pursue.

While there are a number of theories that attempt to predict crystal growth patterns or habits and growth rates based on thermodynamic and kinetic principles, only one will be mentioned here. The theory of limiting faces is of particular interest because it is relatively simple, it is directly related to the question of the surface energy of a surface, and it provides a way of estimating the energy of a given crystal face based on the shape of the crystal.

The Theory of Limiting Faces. The theory of limiting faces is the oldest of the thermodynamic approaches to explaining crystal growth and habit. It is based on the derivation of the relationship between the crystal habit and the surface energy of the various crystal faces that go to produce a particular shape. If one assumes that a crystal is in equilibrium with its mother liquor (or vapor phase), then one can argue that the Gibbs condition for the most favorable crystal shape has been achieved; that is, the overall surface energy is a minimum for a given temperature and volume. If the surface energy of the ith face is defined as the product of the specific surface energy, σ_i, and the area of that face, A_i, then the Gibbs equilibrium condition can be expressed as

$$\Delta\sum_i^n \sigma_i A_i = 0 \qquad (V, T = \text{constant}) \tag{7.2}$$

where the summation is carried out over all n crystal planes. Equation (7.2) leads directly to the Wulff theorem, which is given as

$$\frac{\sigma_1}{h_1} = \frac{\sigma_2}{h_2} = \ldots = \frac{\sigma_i}{h_i} = \ldots = \frac{\sigma_n}{h_n} \tag{7.3}$$

where h_i is the distance of the ith plane from the center of the crystal, the Wulff point (Fig. 7.6). What the theorem says, in brief, is that a crystal in equilibrium with its mother liquor contains a point whose distance from the various crystal faces (not in contact with a foreign surface) is proportional to the specific surface free energy of that face.

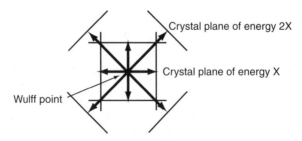

FIGURE 7.6. A Wulff construction for a hypothetical, two-dimensional crystal with surface energies X and $2X$, from which the "ideal" geometric shape of the crystal can be predicted. The arrows emanating from the common point are proportional to the surface free energy of the intersecting crystal faces.

If equilibrium is disturbed, the system will attempt to reestablish its condition of minimum energy by precipitating (or dissolving) more substance on (from) those crystal faces from which the greatest amount of energy will be released. Those faces will then correspond to areas of greater linear growth rate (or shrinkage). Thus faces with smaller linear growth rates will become larger and those with greater growth rates will become smaller until they eventually disappear. The relationships among the growth rates of the individual faces can be modified (changes in temperature, pressure, additives, etc.) to achieve, in principle at least, various desired results.

Crystal Growth Modification. As a practical example of crystal habit modification, one can consider the growth of ice crystals in ice cream. If the crystals grow too large or attain certain shapes, the organoleptic or perceived quality of the product will be reduced significantly—the ice cream becomes "sandy." In practice, the crystallization phenomenon is controlled by the addition of various natural gums (e.g., locust bean gum) that presumably adsorb on specific crystal faces and retard or prevent further deposition of water molecules.

Another less tasty but potentially more important example is the control of the growth of ice crystals in biological systems. In cold polar seas, for example, where water temperatures may be well below 0°C, fish swim through water thick with ice. The fish themselves, however, are protected from freezing by a natural antifreeze: compounds of protein and sugar that keep the liquids in a fish's body from freezing.

Just how the "antifreeze" works is still something of a mystery, although it is assumed that the protein compounds act as a vigilant defense force, homing in on ice crystals as they begin to form and quickly adsorbing onto specific crystal faces. They apparently do not actually lower the freezing point of the fluids significantly, but inhibit the growth of crystals by adsorption onto the preferred crystal face, thereby slowing growth at temperatures where it would normally be very rapid.

Medical researchers are interested in retarding ice crystal growth in organs destined for transplantation. Temperatures that halt or retard deterioration in the organs may also cause intercellular fluids to freeze, a process that tears the tissues and ruptures cell walls. If the "fish antifreeze" or related substances could be modified to keep human organs free from ice crystals at, say, $-10°C$, the viability of potential transplant organs might be extended.

7.4.3. Amorphous Solid Surfaces

From a practical standpoint, the "plastic" nature of our modern existence carries with it important questions concerning the surface characteristics and interactions of primarily amorphous (i.e., noncrystalline) polymeric surfaces. Because of the molecular size, polydispersity, and generally random nature of polymeric solids (and their surfaces), many of the principles applied to studying and modeling ordered crystalline surfaces are no longer valid. Like crystals, polymer surface energies can be history-dependent; however, since polymers generally have relatively low surface energies, complications due to rapid adsorption and contamination are somewhat reduced.

Dynamics in Polymer Surfaces. The concepts of solid surfaces discussed above assumed that the surfaces in question were effectively rigid and immobile. Such assumptions allow one to develop certain models and mathematical relationships useful for estimating and understanding surface energies, surface stresses, and specific interactions, such as adsorption, wetting, and contact angles. It is assumed that the surfaces themselves do not change or respond in any specific way to the presence of a contacting liquid phase, thereby altering their specific surface energy. Although such assumptions are (or may be) valid for truly rigid crystalline or amorphous solids, they more often than not do not apply strictly to polymeric surfaces. Glass, for that matter, has been shown to undergo interactions with liquids such as water that lead to specific alterations in its surface properties. And glass is popularly considered to be an inert, rigid solid.

The structure and characteristics of polymeric surfaces, like other solids, is generally time- and environment-dependent. The reason, of course, is that polymers in general are composed of mixtures of long, very anisotropic molecules of a variety of molecular weights (polydispersity) that will seldom have the opportunity to achieve their equilibrium condition. Solid polymer surfaces, therefore, are inherently nonequilibrium structures and exhibit a variety of time- and condition-dependent properties that may change dramatically with those variables. Although often recognized in bulk polymer problems, such character changes are often ignored in the context of surface properties, sometimes much to the dismay and detriment of researchers, manufacturers, and users.

Because of their large molecular size, complex bonding patterns, the presence of side chains, and other characteristics, polymers exhibit a number of phenomena in the solid state that are much less common in crystalline solids. In the study of bulk polymers, the time, temperature, and other variable-related characteristics have come to be classed as either relaxations or transitions. As a general definition, a relaxation can be considered a time-dependent motion in a polymer system in which the molecules return to an equilibrium from which they have been displaced by the action of some external force. For example, if a polymer sample is compressed under some external load that forces the molecules to rearrange to attain a new equilibrium state and the force is then removed, the material will, with time, relax or return to its original state (before compression).

A transition in a polymer system is considered to be a temperature-dependent process. In a crystalline material we are familiar with the transition from the solid to the liquid phase which will normally occur at some relatively sharp, well-defined temperature. Similar processes occur in polymers, but because of their nature, they are seldom sharply defined, but rather occur over a relatively broad temperature range. The melting process is ideally an equilibrium process and occurs essentially independently of time. In amorphous polymers and glasses, there exists a temperature range over which the system undergoes a dramatic change in its physical properties. It changes from a rigid, possibly brittle system to a viscous liquid. Its tensile strength, elasticity, and other properties change dramatically. It undergoes a transition to an essentially distinct class of material.

Polymer molecules in or near a surface are also found to undergo relaxations and transitions similar to those found in the bulk. However, those motions in a surface are somewhat different because of the different environment encountered there. They no longer interact only with other polymer molecules, but also with the surrounding phase. Like liquids and solids, surface polymer molecules will, given sufficient time, orient themselves at the surface so as to attain the configuration of minimum surface energy. The reorientations will, of course, be time/temperature dependent and correspond to related bulk phenomena. At low temperatures, the transitions may require long equilibration times to become evident, while at higher temperatures the effects may become apparent in short order.

In contact with condensed phases, especially liquids, surface relaxations and transitions can become quite important. Even basically hydrophobic, rigid polymers such as poly(methylmethacrylate) which contain somewhat hydrophilic ester side chains will, in contact with water, undergo surface molecular reorientation, due to the interaction of water with the ester groups. The interfacial region may become plasticized (roughly put, softened) because the water–ester interaction liberates to some extent the side chains (or lubricates the interchain interaction region) and increases their mobility; see Fig. 7.7). The important point is that these surface interactions can dramatically

FIGURE 7.7. Rigid, hydrophobic polymers such as poly(methylmethacrylate) that contain somewhat hydrophilic ester side chains may, in prolonged contact with water, undergo surface molecular reorientation due to the interaction of water with the ester groups. The interfacial region may become plasticized or softened as a result of the water–ester interaction, liberating to some extent the side chains (or "lubricating" the interchain interaction region) and increasing their mobility.

change the interfacial characteristics of a polymer with possibly important consequences in a particular application. And since the processes are time-dependent, the changes may not be evident over the short span of a normal experiment. For critical applications in which a polymer surface will be in contact with a liquid phase, it is not only important to know the surface characteristics (e.g., coefficient of friction, adhesion, adsorption) under normal experimental conditions but also to determine the effects of prolonged (equilibrium) exposure to the liquid medium of interest.

As a practical example, take the use of a polymer in some biomedical applications such as an implant device, in which the polymer surface will continually contact blood or other body fluids. Classic surface studies using contact angle measurements, wetting phenomena, X-ray photoelectron spectroscopy, or other analytical techniques may indicate that the material should be biocompatible and not cause problems such as blood platelet deposition and clot formation and immune responses. Typical surface analyses, however, are not or cannot normally be carried out under conditions of use. Under such conditions, surface transitions and relaxations may occur with time that will transform the polymer surface into one that is no longer biocompatible from the standpoint of blood or other body fluid interactions. The result could be catastrophic for the recipient of the transplant or implant made of such material.

It is therefore important for biomedical as well as many other applications that the surface characteristics of a material of interest be determined under conditions that mimic as closely as possible the conditions of use and over extended periods of exposure to those conditions, in addition to the usual characterizations.

Much more can be said about solid surfaces, such as their formation, energetics, theoretical models, adsorption characteristics, and reactivities. Later chapters will address specific subjects related to interfacial phenomena.

PROBLEMS

7.1. If a solid has a large surface energy, would one expect that to correspond to an increase or decrease in solubility with decreasing particle size? Why?

7.2. How many nearest neighbors are there for a sphere in the surface of a hexagonal close-packed array when the sphere is (a) part of a terrace, (b) part of a monotonic step, (c) is adjacent to a kink in a step, and (d) is isolated atop a terrace.

7.3. What must be the particle size of a colloidal silica if approximately 25% of the silicon atoms are to be on the surface? What is the approximate surface area per gram? Assume a density of 2.3 g cm^{-3}.

7.4. A spray-cooled, organic powder (m.p. 80–82°C) composed of monodisperse spherical particles is prepared and stored in a cold storage room at −10°C for 3 months. At the end of that time the powder remains free-flowing with no signs of clumping. The powder is then shipped across the southern United States in a nonrefrigerated truck in August, the trip taking 8 days. When the powder arrives and is opened, it is found to have formed a solid, nonflowing crust that must be mechanically broken up. Give a qualitative explanation for the observed result of the trip.

7.5. While it is almost universally found that the surface tension of a liquid decreases with an increase in temperature, it has been observed that some lead glasses actually show an increase in σ as the temperature is increased. Propose an explanation for such a phenomenon.

7.6. A fresh mica surface is prepared under three sets of conditions—in air, under argon, and in a vacuum—and the surface energy determined. Will the surface energies determined be equal? If not, rank them in order of increasing value and give your reason for the order chosen.

7.7. In an industrial process, product crystallization is initiated by the careful addition of a nonsolvent to a solution. On a particular day, the nonsolvent pump broke down after approximately one half of the specified amount of material was added. Although crystallization occurred, the product was found to be outside specifications and had to be reprocessed. Were the rejected crystals too small or too large? Explain.

7.8. A large, perfect crystal prepared aboard the space shuttle Discovery was found to have three distinct faces. The ratio of distances of each face from the center of the crystal was determined to be $1:2.5:4$. Given that the specific surface energy of the shorter face is 45 mJ m^{-2}, what are the energies of the other two faces?

7.9. If crystal growth modifiers can be used to inhibit the production of large ice crystals (in stored ice cream, for example), could a similar approach be used to bring about the production of large crystals? Explain.

7.10. A particular polymeric material is being evaluated for use in prosthetic devices. Initial in vitro tests showed the material to cause no apparent problems of blood compatibility. Long-term animal tests, however, resulted in the formation of dangerous blood clots in the region of the implant. Suggest an explanation for the observed results.

8 Liquid–Fluid Interfaces

Liquids have several distinct characteristics that differentiate them from solid and gas phases. One of the more important ones (from the point of view of surface chemistry, at least) is that, unlike a gas, liquids have a relatively high density and fixed volume, while they possess a mobility at the molecular level that is many orders of magnitude greater than that in solids. As a result of that mobility, interfaces involving liquids and another fluid generally (but not always) behave as though homogeneous and therefore lack many of the complications encountered when considering solid surfaces. The not always qualification applies primarily in viscous liquids and those multicomponent systems containing surface-active materials that must diffuse to the surface. In such cases, rapid measurements of newly formed surface may produce surprising results, as described later.

8.1. THE NATURE OF A LIQUID SURFACE: SURFACE TENSION

As pointed out in Chapter 2, it is common practice to describe a liquid surface as having an elastic "skin" that causes the liquid to assume a shape of minimum surface area, its final shape being determined by the "strength" of that skin relative to other external factors such as gravity. In the absence of gravity, or when suspended in another immiscible liquid of equal density, a liquid spontaneously assumes the shape of a sphere. In order to distort the sphere, work must be done on the liquid surface, increasing the total surface area and therefore the free energy of the system. When the external force is removed, the contractile skin then forces the drop to return to its equilibrium shape.

While the picture of a skin like a balloon on the surface of a liquid is easy to visualize and serves a useful educational purpose, it can be quite misleading, since there is no skin or tangential force as such at the surface of a pure liquid. It is actually an imbalance of forces on surface molecules pulling into the bulk liquid and out into the adjoining vapor phase that produces the apparent contractile skin effect. The forces involved are, of course, the same van der Waals interactions that account for the liquid state in general and for most physical interactions between atoms and molecules. Because the liquid state is of higher density than the vapor, surface molecules are pulled away from the surface and into the bulk liquid, causing the surface to contract spontane-

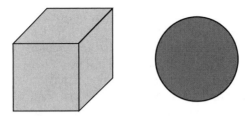

FIGURE 8.1. The unbalanced, inward pull of bulk liquid molecules on those at the surface results in the phenomenon observed as surface tension. The drive to reduce the surface area to a minimum produces the observed tendency of liquids to form spherical drops (in the absence of gravity)—the geometry of minimum surface area for a given volume of material.

ously (Fig. 8.1). For that reason, it is more accurate to think of surface tension (or surface energy) in terms of the amount of work required to increase the surface area of the liquid isothermally and reversibly by a unit amount, rather than in terms of some tangential contractile force.

As will be seen later in the chapter, the same basic ideas that are used to describe the liquid–vapor interface apply to the liquid–liquid interface. However, since a second liquid phase is much more dense than a vapor phase, the various attractive interactions among units of the two phases across the interface, which depend on the number density of interacting units (see Chapter 4), are significantly greater. For a given increase in liquid–liquid interfacial area, the excess surface energy of each unit (and therefore the total energy) will be lower. In other words, the net work required to increase the interfacial area, the interfacial tension, will be reduced relative to the liquid-vapor case. Table 8.1 lists the surface tensions of several typical liquids and their corresponding interfacial tensions against water and mercury.

TABLE 8.1. Typical Liquid Surface and Interfacial Tensions at 20°C (mN m-1)

Liquid	Surface Tension	Interfacial Tension versus Water
Water	72.8	—
Ethanol	22.3	—
n-Octanol	27.5	8.5
Acetic acid	27.6	—
Oleic acid	32.5	7.0
Acetone	23.7	—
Carbon tetrachloride	26.8	45.1
Benzene	28.9	35.0 (357 vs. mercury)
n-Hexane	18.4	51.1 (378 vs. mercury)
n-Octane	21.8	50.8
Mercury	485	375

There are two quick observations that one may note from the data in Table 8.1: (1) the interfacial tension between a given liquid and water is always less than the surface tension of water; and (2) for an homologous series of materials such as the normal alkanes, the interfacial tension between the members of the series and water (or any other immiscible liquid) will change only slightly as a function of the molecular weight of the material. Those characteristics are a direct consequence of the nature of the interactions at the interface. Where the two liquids are highly immiscible, the interfacial tension will lie between the two surface tensions (e.g., water–alkane); if significant miscibility exists, the interfacial tension will be lower than the lower of the two surface tensions (e.g., water–octanol). The difference stems from the surface activity of the molecules of the miscible liquid (in water), a topic introduced in Chapter 3 that will be addressed again in Chapter 15.

Most commonly encountered room-temperature liquids have surface tensions against air or their vapors that lie in the range of 10–80 mN m^{-1}. Liquid metals and other inorganic materials in the molten state exhibit significantly higher values as a result of the much greater and more diverse interactions occurring in such systems. Water, the most important liquid commonly encountered in both laboratory and practical situations, lies at the upper scale of what are considered normal surface tensions, with a value in the range of 72–73 mN m^{-1} at room temperature, while hydrocarbons reside at the lower end, falling in the lower to middle 20s. Materials such as fluorocarbons and silicones may go even lower.

8.1.1. Surface Mobility

The common concept of interfacial tensions is simplistic in the sense that it implies that the surface or interface is a static entity. There is, in reality, a constant and for liquids and gases, rapid interchange of molecules between the bulk and interfacial region, and between the liquid and vapor phases. If it is assumed that molecules leave the interfacial region at the same rate that they arrive, it is possible to estimate the exchange rate, β, of an individual molecule from the relationship

$$\beta = \alpha(2\pi mkT)^{1/2} p_0 \tag{8.1}$$

where α is a so-called sticking coefficient (i.e., the fraction of molecules striking the surface that actually becomes part of it), p_0 is the equilibrium vapor pressure of the liquid, m the mass of the molecule, and k Boltzmanns constant. Assuming α to lie in the range of 0.03–1.0, a water molecule at 25°C will have an average residence time of ≤ 3 ms at the air–water interface. The corresponding residence time for a mercury atom would be roughly 5 ms, while that for a tungsten atom (obviously not in the liquid state) would be 10^{37} s at room temperature.

With such molecular mobility, it is clear that the surface of a pure liquid offers little resistance to forces that may act to change its shape. That is, there will be very little viscous or elastic resistance to the deformation of the surface. An important consequence of that fact is that a pure liquid does not support a foam for more than a small fraction of a second (see Chapter 12). A similar situation exists at the liquid–liquid interface. As we shall see in later chapters, the highly mobile nature of liquid interfaces has significant implications for many technological applications such as emulsions and foams, and forms the basis for many of the most important applications of surface-active materials or surfactants.

8.1.2. Temperature Effects on Surface Tension

Because of the mobility of molecules at fluid interfaces, it is not surprising to find that temperature can have a large effect on the surface tension of a liquid (or the interfacial tension between two liquids). An increase in surface mobility due to an increase in temperature will clearly increase the total entropy of the surface and thereby reduce its free energy, ΔG. Since the surface tension has been thermodynamically defined as

$$\sigma = \frac{\Delta G}{\Delta A} \tag{8.2}$$

one would expect to encounter a negative temperature coefficient for s. While that is the case for most normal liquids, including most molten metals and their oxides, positive coefficients have been encountered. The reason for that phenomenon is not entirely clear, although it probably results from some change in the actual atomic composition of the surface as the temperature is increased.

At temperatures near the critical temperature of a liquid, the cohesive forces acting between molecules in the liquid become very small and the surface tension approaches zero. That is, since the vapor cannot be condensed at the critical temperature, there will be no surface tension. A number of empirical equations that attempt to predict the temperature coefficient of surface tension have been proposed; one of the most useful is that of Ramsey and Shields:

$$\sigma\left(\frac{Mx}{r}\right)^{2/3} = k_s(T_c - T - 6) \tag{8.3}$$

where M is the molar mass of the liquid, ρ its density, x the degree of association, T_c the critical temperature, and k_s a constant.

8.1.3. The Effect of Surface Curvature

Because many practical situations involve surfaces and interfaces with high degrees of curvature, it is important to understand the effect of curvature on interfacial properties. As pointed out in Chapters 2 and 6, there will develop a pressure differential across any curved surface, with the pressure being greater on the concave side of the interface. In other words, the pressure inside a bubble will always be greater than that in the continuous phase. The Young–Laplace equation

$$\Delta p = \sigma\left(\frac{1}{r_1} + \frac{1}{r_2}\right) \tag{8.4}$$

relates the quantities of interest in this situation, in which Δp is the drop in pressure across a curved interface, r_1 and r_2 are the principal radii of curvature, and s is the surface (or interfacial) tension. For a spherical surface where $r_1 = r_2$, the equation reduces to

$$\Delta p = \frac{2\sigma}{r} \tag{8.5}$$

For a very small drop of liquid in which there is a large surface : volume ratio, the vapor pressure is higher than that over a flat surface of equal area. The movement of liquid from a flat interface into a volume with a curved interface requires the input of energy into the system since the surface free energy of the curved volume increases. If the radius of a drop is increased by dr, the surface area increases from $4\pi r^2$ to $4\pi(r + dr)^2$, or by a factor of $8\pi r\, dr$. The free energy increase is $8\pi\sigma r\, dr$. If during the process δn moles of liquid are transferred from the flat phase with a vapor pressure of p_0 to the drop with vapor pressure p_r, the free energy increase also is given by

$$\Delta G = \sigma n RT \ln\left(\frac{p_r}{p_0}\right) \tag{8.6}$$

Equating the two relationships leads to the expression

$$RT \ln\left(\frac{p_r}{p_0}\right) = \frac{2\ \sigma M}{r\rho} = \frac{2\sigma V_m}{r} \tag{8.7}$$

known as the Kelvin equation. In Equation (8.7), ρ is the density, M the molar mass, and V_m the molar volume of the liquid. It can be shown that extremely small radii of curvature can lead to the development of significant pressure differences in drops. For a drop of water with a radius of 1 nm, the partial pressure ratio from (8.7) is about 3. Obviously, the condensation of liquid molecules to form very small drops will be retarded by a relatively high energy barrier due to curvature. Understanding the consequences of the Kelvin equation helps in explaining the ability of many liquid–vapor systems to become supersaturated, when logic dictates that condensation should readily occur. It is the input of energy by scratching, agitation, or other perturbations, or the provision of a heterogeneous nucleation site, thereby reducing the effective radius of curvature of the incipient drops, that brings about the rapid condensation or crystallization of a supersaturated system.

8.1.4. Dynamic Surface Tension

For a pure liquid in equilibrium with its vapor, the number density and orientation of molecules at the surface will be different from that of bulk molecules (Fig. 8.2). When new surface is created, it is reasonable to assume that a finite amount of time will be required for new molecules to diffuse to the surface and to return the system to equilibrium. In that interim, as short as it may be, the measured surface tension of the system will be different from that of the system in equilibrium. The surface tension of such "new surface" is referred to as the dynamic surface tension.

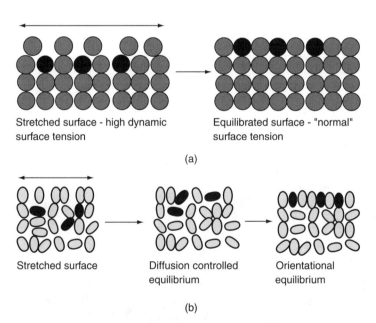

FIGURE 8.2. Dynamic surface tension in pure liquids: (*a*) for a liquid of isotropic molecular shape, dynamic surface tension effects are controlled by the rate of diffusion of molecules from the bulk to the new surface; (*b*) in polar or anisotropic liquids, the situation may be further complicated by the question of molecular orientation at the surface.

Qualitatively, it is assumed that the time required to attain equilibrium after formation of new surface is related to the time for diffusion of liquid molecules to the surface—that is, to the self-diffusion constant. Diffusion times are usually on the order of 10^{-6} cm^2 s^{-1}, which translates to times of milliseconds for the attainment of equilibrium. The accurate measurement of surface tensions over such short time frames is difficult at best, so that a great deal is still in question concerning the thermodynamics and kinetics of such fresh surfaces.

If a surface-active solute is present, it becomes rather easy to demonstrate the dynamic surface tension effect experimentally, although the problem is complicated by the questions of solute concentration and orientation (among others). While a number of theories exist concerning dynamic processes of adsorption at freshly formed surfaces, the uncertainties involved make them somewhat problematic for most practical purposes. As a general concept, however, it is usually assumed that the initial rate of adsorption at the new surface approximately equals the adsorption rate for molecules of the pure liquid. However, as surface adsorption occurs, the solution region just below the surface becomes depleted of solute, and diffusion is slowed until more solute diffuses into the region from the bulk (Fig. 8.3). Obviously, the rate of such movement is related to the solute diffusion constant; the smaller the constant, the longer the time required to attain equilibrium. For relatively large solute molecules, reliable data have been obtained that indicate that, in agreement with intuition, the attainment of equilibrium surface tension values takes longer for larger solute molecules and for lower bulk concentrations.

An additional complication in evaluating dynamic surface tensions may arise in terms of molecular orientation at the surface. For a symmetric molecule, orientation will not be a problem; however, for many systems, especially

 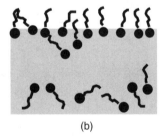

(a) (b)

FIGURE 8.3. Dynamic surface tension effects can be particularly important in certain applications of surfactant solutions. (*a*) In a quiescent solution, the surface concentration and orientation of surfactant molecules produces the equilibrium surface tension. (*b*) When new surface is formed rapidly, as in certain coating operations, diffusion of surfactant molecules to the new surface from the bulk will require a certain time, reflected as a dynamic surface tension, greater than the equilibrium value. If significant new surface is formed, the concentration of surfactant near the surface may be depleted, producing a greater dynamic effect.

FIGURE 8.4. Polar molecules in solution will, when possible, orient themselves at surfaces and interfaces in order to minimize the overall interfacial energy of the system. In aqueous solutions containing organic solutes that usually results in the nonpolar portion of the molecule "pointing" away from the aqueous phase, or at least lying along the surface.

those involving asymmetric (in both shape and chemical nature) molecules such as alcohols and other surface-active organic materials, the surface tension will be a function of the orientation of the molecule at the surface. For an aqueous solution of a long-chain alcohol, the equilibrium surface tension results when the adsorbed molecules are oriented with the alkyl chain pointing out toward the vapor phase and the hydroxyl group "buried" in the water (Fig. 8.4). Some finite time is required for such orientation to occur, so that two materials with essentially identical bulk diffusion coefficients may exhibit distinct dynamic surface tension characteristics due to differences in orientation rates. That effect is especially apparent in systems in which the solute is a macromolecule. It has been found that polymer solutions may take minutes, hours, or even days to attain their equilibrium surface tension, due primarily to the long times required for the chains to orient and accommodate themselves at the surface.

   A great deal of practical work has been done on the effects of dynamic surface tension in processes such as high-speed coating operations. Unfortunately, much of it has been in terms of specific industrial systems and the results remain buried in the never-never land of "proprietary information."

8.2. SURFACE TENSIONS OF SOLUTIONS

Because of differences between the shape, size, and/or chemical nature of a solute relative to a given solvent, the presence of the solute often results in the alteration of the surface tension of the solution relative to that of the pure liquid. Most commonly, the effect is to lower the surface tension, although the opposite effect is also found.

   Intuitively, the surface tension of a solution of two miscible liquids may be expected to be some mathematical average of that of the two pure components. The simplest such combination for a binary mixture would be an additive combination related to the quantity of each component in the mixture, such as mole fraction. Such a relationship may be written

$$\sigma_{mix} = \sigma_1 X + \sigma_2(1 - X) \tag{8.8}$$

where σ_{mix} is the surface tension of the solution, σ_1 and σ_2 are the surface tensions of the respective components, and X is the mole fraction of component 1 in the mixture. In ideal systems where the vapor pressure of the solution is a linear function of the composition, such relationships are found. Normally, however, there will be some positive or negative deviation from linearity; the latter are most commonly encountered. Some examples of the variation of the surface tension of mixtures with composition are shown in Figure 8.5.

In aqueous solution, when the second component is an inorganic electrolyte that requires significant solvation, the relationship between surface tension and composition may be quite varied, depending on the exact nature of the interaction. It is generally found, for example, that the addition of inorganic electrolyte to water results in an increase in the surface tension of the solution, although the effect is not dramatic and requires rather high salt concentrations to become significant (Fig. 8.6). The relative effectiveness of ions at increasing the surface tension of water generally follows the Hofmeister series: $Li^+ > Na^+ > K^+$, and $F^- > Cl^- > Br^- > I^-$, indicating that the effect results from a structuring of the water molecules at the surface due to solvation phenomena.

Unlike the inorganic electrolytes, the presence of an organic material in aqueous solution will result in a decrease in the surface tension of the system. The extent of such lowering depends upon a number of factors, including the relative miscibility of the system (or the solubility of the organic solute) and the tendency of the organic material to adsorb preferentially at the water-air interface. Liquids such as ethanol or acetic acid produce gradual decreases in

FIGURE 8.5. For an ideal mixture, the surface tension will be a linear function of the composition of the mixture and the surface tension of the pure components (light lines). In practice, most mixtures show significant deviation from ideality (bold lines).

FIGURE 8.6. Unlike organic solutes, many common inorganic electrolytes can produce increases in the surface tension in aqueous solution, although relatively high concentrations are usually required to produce significant increases. Typical values include those for LiCl (curve *a*), NaCl (curve *b*), and NaBr (curve *c*).

the surface tension of their aqueous solutions (as in Fig. 8.5, curve *a*), while longer-chain organic compounds such as butanol can produce more dramatic effects (Fig. 8.7). When the organic solute has a limited solubility in water, the effect on surface tension becomes characteristic of solutions of surface-active materials. For solutions of such materials, one encounters a steady decrease in surface tension with increased solute concentration. At some point, a minimum value of σ will be obtained as the solute concentration increases before surface saturation or some form of solute behavior change (precipitation, micelle formation, etc.) prevents further change in the surface tension.

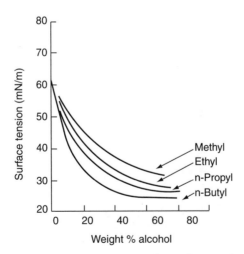

FIGURE 8.7. The effect of carbon number on the surface tension of water increases rapidly with the addition of each carbon unit, as illustrated for the series of short-chain alcohols.

A typical surface tension-concentration curve is shown in Figure 8.8. Surface-active agents or surfactants form a large class of materials whose importance to our existence (both literally and figuratively) cannot be overstated. Such materials, including classic soaps, detergents, and hundreds of other types of materials, and their production is part of a multibillion dollar industry world-wide. But that is only part of the story. Surfactants, because of their properties for lowering surface tension and altering related phenomena, are practically indispensable in an even wider range of processes and products, so that their actual economic value is many times that of their direct production value.

Of greater importance from an existential point of view, however, is the vital importance of surface-active molecules in our most important biological functions. The walls of the cell which make up our bodies are assemblies of surface-active molecules, plus other components. Vital functions such as respiration, blood transport phenomena, nutrition, and the functioning of our antibody systems all involve surface activity and surface interactions. In a nutshell, we are creatures of surface activity.

8.2.1. Surfactants and the Reduction of Surface Tension

The fundamental principle underlying our present understanding of surface activity is the Gibbs adsorption equation discussed in more detail in Section 9.1. Since the surface tension of a liquid is determined by the excess energy of the molecules in the interfacial region, the displacement of surface liquid molecules by adsorbed solute directly affects the measured value of σ, assuming a significant difference in the energies of adsorbed solute and solvent molecules. It is the relationship between the chemical structure of an adsorbing molecule and the rate and extent of adsorption under given circumstances that differentiates the various surfactant types and determines their utility in applications where surface tension lowering is of importance.

FIGURE 8.8. In a surfactant solution, the surface tension will typically change rapidly over a relatively small concentration range and then level off at some characteristic minimum value. The exact shape of the curve will depend on the molecular characteristics of the surfactant and its purity.

In aqueous solutions, the interface between the liquid and vapor phases involves interactions between relatively densely packed, highly polar water molecules, and relatively sparse, nonpolar gases. The result is a large imbalance of forces acting on the surface molecules and the observed high surface tension of water (72.8 mN m^{-1}). If the surface solvent molecules are replaced by adsorbed surfactant molecules with lower specific excess surface energy, the surface tension of the solution is decreased accordingly, with the amount of reduction being related to the surface excess concentration of solute and the nature of the adsorbed molecule.

A large body of literature has grown up around the question of how one can relate the chemical structure of a surfactant to its surface activity in various situations. At the present time, most such structure–property relationships are semiquantitative at best, but they can serve one well as a guide to the choice of the best surfactant for a given situation. Several works reviewing the literature in that area are cited in the Bibliography, and some specific examples will be noted in the appropriate contexts of later chapters.

8.2.2. Effects of Phase Densities

If the vapor phase is replaced by a condensed phase that has a higher molecular density and more opportunity for attractive interaction between molecules in the interfacial region, the interfacial tension will be reduced significantly. In the case of water, the presence of a liquid such as octane, which interacts only by relatively weak dispersion forces, lowers the interfacial free energy to 52 mN m^{-1}. If the extent of molecular interaction between phases can be increased by the introduction of polar groups that interact more specifically with the water, as, for instance, in octanol, the interfacial energy reduction will be even greater (to 8.5 mN m^{-1}). Clearly, any alteration in the nature of the molecules composing the surface would be expected to result in a lowering of the surface energy of the system. And therein lies the basic explanation for the action of surfactants in lowering the surface and interfacial tension of aqueous solutions.

The same qualitative reasoning also explains why most surfactants do not affect the surface tension of organic liquids—the molecular nature of the liquid and the surfactant are not sufficiently different to make adsorption a particularly favorable process. Moreover, if adsorption occurs, the energy gain is not sufficient to produce a significant change in the surface energy to be useful. The actions of fluorocarbon and siloxane surfactants are exceptions since the specific surface free energy of such materials may be significantly lower than that of most hydrocarbons. They will therefore be positively adsorbed at hydrocarbon surfaces and lower the surface tension of their solutions.

8.3. SURFACTANT ADSORPTION AND GIBBS MONOLAYERS

The basic concepts behind the factors governing the adsorption of surface-active molecules at interfaces has already been mentioned several times in terms of the Gibbs adsorption isotherm, which relates the surfaces excess concentration of the adsorbed species to the surface or interfacial tension of the system. Because of the broad range of areas in which such adsorption occurs and produces significant alterations in the surface chemistry of the systems involved, specific details of the phenomenon and resultant effects will be covered in chapters related to specific topics, such as emulsions, foams, solid–liquid interfaces, and wetting phenomena.

For now, the discussion will be limited to some general concepts related to adsorption at liquid–fluid interfaces, such as some general relationships between surfactant structure and the rate and effect of adsorption.

8.3.1. Efficiency, Effectiveness, and Surfactant Structure

When discussing the performance of a surfactant in lowering the surface tension of a solution it is necessary to consider two aspects of the process: (1) the concentration of surfactant in the bulk phase required to produce a given surface tension reduction and (2) the maximum reduction in surface tension that can be obtained, regardless of the concentration of surfactant present. The two effects may be somewhat arbitrarily defined as follows: the surfactant efficiency is the bulk phase concentration necessary to reduce the surface tension by a predetermined amount, for example, 20 mN m^{-1}. Its effectiveness is the maximum reduction in σ that can be obtained by the addition of any quantity of surfactant. The typical shape of the surface tension–concentration curve for aqueous surfactants is shown in Figure 8.8.

Because the extent of reduction of the surface tension of a solution depends on the substitution of surfactant for solvent molecules at the interface, the relative concentration of surfactant in the bulk and interfacial phases should serve as an indicator of the adsorption efficiency of a given surfactant and, therefore, as a quantitative measure of the activity of the material at the solution–vapor interface. For a given homologous series of straight-chain surfactants in water, $CH_3(CH_2)_n-S$, where S is the hydrophilic head group and n is the number of methylene units in the chain, an analysis based on the thermodynamics of transfer of a surfactant molecule from the bulk phase to the interface leads to the conclusion that the above-defined efficiency of adsorption is directly related to the length of the hydrophobic chain. The efficiency can be defined mathematically by the expression

$$-\log (C)_{20} = pC_{20} = n\,\frac{-A}{2.3RT} + \frac{-B}{2.3RT} + K \tag{8.9}$$

where A, B, and K are terms for the free energies of transfer of methylene, terminal methyl, and head groups, respectively, from the bulk solution to the interface; and C is the concentration of surfactant required to lower σ by 20 mN m^{-1}. For a given head group at constant temperature, solvent composition, and so on, the equation reduces to a direct dependence of efficiency on the length of the hydrocarbon chain *n,* as expected.

Since the surfactant efficiency is directly related to the thermodynamics of chain transfer from bulk to interface, it is reasonable to expect chain modifications that alter that characteristic, that is, changes in the hydrophobic character of the surfactant, to produce parallel changes in adsorption efficiency. For example, branching in the hydrophobic group results in a reduction in the hydrophobicity of a surfactant chain relative to that of a related straight-chain material with the same total carbon content. Carbon atoms located on branch sites contribute approximately two-thirds as much to the character of a surfactant molecule as one located in the main chain. Similar results are observed for surfactants with two or more short-chain hydrophobes of equal total carbon content (e.g., internal substitution of the hydrophilic group) and for the presence of unsaturation in the chain. A benzene ring usually contributes an effect equivalent to approximately 3.5 methylene groups.

If a surfactant possesses two polar groups, the methylene groups lying between the two polar groups contribute an effect equivalent to approximately one-half that found for such groups located in the main body of the hydrophobe.

In cationic surfactants, the presence of short-chain alkyl groups (fewer than four carbon atoms) attached to the nitrogen seem to have little effect on the efficiency of adsorption of the molecule. The dominant factor will always be the length of the primary hydrophobic chain. That effect is true regardless of whether the alkyl groups are attached to a quaternary ammonium group, an amine oxide, or a heterocyclic nucleus such as pyridine.

Within limits, the nature of the charge on an ionic surfactant has little effect on the efficiency of surfactant adsorption. Again, it is the nature of the hydrophobic group that predominates. Some increase in efficiency, however, is seen if the counterion is one that is highly ion paired, that is, one that is not highly solvated in the system and therefore produces a lower net electrical charge as the molecules are adsorbed at the interface, facilitating the movement of molecules into the interface. The addition of neutral electrolyte to an ionic surfactant solution produces a similar result in increasing the efficiency of adsorption by compression of the electrical double layer associated with the ionic head group.

Polyoxyethylene (POE) nonionic surfactants with the same hydrophobic group and an average of 7–30 OE units, exhibit adsorption efficiencies that follow an approximately linear relationship of the form

$$pC_{20} = A_{tr} + mB_{tr} \tag{8.10}$$

where A_{tr} and B_{tr} are constants related to the free energy of transfer of $-CH_2-$ and OE groups, respectively, from the bulk phase to the interface and m is the number of OE units in the POE chain. As is usually the case for POE nonionic materials, most data reported have been obtained using nonhomogeneous POE chains. The available data indicate that the efficiency of adsorption decreases slightly as the number of OE units on the surfactant increases.

To this point, we have seen that the efficiency of surfactant adsorption at the solution–vapor interface is dominated by the nature of the hydrophobic group and is relatively little affected by the hydrophilic head group. It is often found that the second characteristic of the adsorption process, the so-called adsorption effectiveness, is much more sensitive to other factors and quite often does not parallel the trends found for adsorption efficiency.

8.3.2. Adsorption Effectiveness

The choice of 20 mN m^{-1} as a standard value of surface tension lowering for the definition of adsorption efficiency is convenient, but, as mentioned, somewhat arbitrary. When one discusses the effectiveness of adsorption, as defined as the maximum lowering of surface tension regardless of surfactant concentration, the value of σ_{min} is determined only by the system itself and represents a more firmly fixed point of reference. The value of σ_{min} for a given surfactant will be determined by one of two factors: (1) the solubility limit or Krafft temperature (T_k) of the compound, or (2) the critical micelle concentration (cmc). In either case, the maximum amount of surfactant adsorbed is reached, for all practical purposes, at the maximum bulk concentration of free surfactant.

Because the activity of surfactants used below T_k cannot reach their theoretical maximum as determined by the thermodynamics of surfactant aggregation (see also Chapter 15), they will also be unable to achieve their maximum degree of adsorption at the solution–vapor interface. It is therefore important to know the value of T_k for a given system before considering its application. Most surfactants, however, are employed well above their Krafft temperature, so that the controlling factor for the determination of their effectiveness will be the cmc.

When one examines the shape of the surface tension–ln C curve for a surfactant, it can be seen that the curve becomes approximately horizontal at some concentration below the cmc. It can be shown that the effectiveness of the adsorption of a surfactant, $\Delta\sigma_{cmc}$, can be quantitatively related to the concentration of surfactant at which the Gibbs equation becomes linear, C_1, the surface tension attained at C_1, σ_1, and the cmc. The relationship has the general form

$$- \Delta\sigma_{cmc} = (\sigma_0 - \sigma_1) + 2.3\Omega RT\,\Gamma_m \log \frac{C_{cmc}}{C_1} \qquad (8.11)$$

where σ_0 is the surface tension of the pure solvent and Γ_m is the maximum in surface excess of adsorbed surfactant at the interface. The factor Ω in this equation is related to the number of molecular or atomic units that become adsorbed at the interface with the adsorption of each surfactant molecule; for nonionic surfactants or ionic materials in the presence of a large excess of neutral electrolyte, $\Omega = 1$; for ionic surfactants $\Omega = 2$, since one counterion must be adsorbed for each surfactant molecule giving a total of two species.

The effectiveness of a surfactant can be conveniently quantified by using a value of C_1 at which the surface tension has been reduced by 20 mN m^{-1}, assuming $\Gamma_{20} \approx \Gamma_m$. Application of Equation (8.11) allows the calculation of a standard quantity, cmc/C_{20}, which serves as a useful measure for evaluating surfactant effectiveness. For several surfactants of comparable cmc, for example, a larger value for cmc/C_{20} indicates that at a given concentration, the available free surfactant is being used more effectively at the surface. Some representative values that illustrate the effects of well-controlled changes in surfactant structure are given in Table 8.2.

It is often found that the efficiency and effectiveness of surfactants do not run parallel; in fact, it is commonly observed that materials that produce significant lowering of the surface tension at low concentrations (i.e., are more efficient) are less effective (i.e., have a smaller Γ_m). This follows from the complex relationship between adsorption at the interface and micelle formation in the solution.

On a molecular basis, the conflicting factors can be seen conceptually as arising from the different roles of the molecular structure in the adsorption process. Surfactant efficiency is related to the extent of adsorption at the interface as a function of bulk surfactant concentration. At a concentration well below that at which micellization becomes a factor, efficiency can be structurally related to the hydrophobicity of the surfactant tail and the nature

TABLE 8.2. Experimental Values of cmc/C_{20}, G20 ($\times$ 10^{10} mol cm^{-2}), and σ_{min} (mN m^{-1}) for Some Typical Surfactants in Aqueous Solution

Surfactant	Temperature (°C)	cmc/C_{20}	Γ_{20}	σ_{min}
$C_{12}H_{25}SO_4^-Na^+$	25	2.0	3.3	40.3
$C_{12}H_{25}SO_3^-Na^+$	25	2.3	2.9	40.8
$C_{16}H_{33}SO_4^-Na^+$	60	2.5	3.3	37.8
$C_{12}H_{25}C_6H_4SO_3^-Na^+$	70	1.3	3.7	47.0
$C_{12}H_{25}C_5H_5N^+Br^-$	30	2.1	2.8	42.8
$C_{14}H_{29}C_5H_5N^+Br^-$	30	2.2	2.8	41.8
$C_{12}H_{25}N(CH_3)_3^+Br^-$	30	2.1	2.7	41.8
$C_{10}H_{21}(POE)_6OH$	25	17.0	3.0	30.8
$C_{12}H_{25}(POE)_6OH$	25	9.6	3.7	31.8
$C_{16}H_{33}(POE)_6OH$	25	6.3	4.4	32.8
$C_{12}H_{25}(POE)_9OH$	23	17.0	2.3	36.8
$C_{16}H_{33}(POE)_9OH$	25	7.8	3.1	36.8
$C_{12}H_{25}(POE)_{12}OH$	23	11.8	1.9	40.8
$C_{16}H_{33}(POE)_{12}OH$	25	8.5	2.3	39.8
$C_{16}H_{33}(POE)_{15}OH$	25	8.9	2.1	40.8

of the head group. For a given homologous series of surfactants, it will be a function of the thermodynamics of transfer of the hydrophobic tail from the bulk to the surface phase. A plot of σ versus ln C for such a series will exhibit a relatively regular shift in the linear portion of the curve to lower concentrations (for a given σ) as methylene groups are added to the chain.

While the role of molecular structure in determining surfactant efficiency is primarily thermodynamic, its role in effectiveness is more directly related to the size of the hydrophobic and hydrophilic portions of the adsorbing molecules. When one considers the adsorption of molecules at an interface, it can be seen that the maximum number of molecules that can be fitted into a given area depends on the area occupied by each molecule. That area will, to a good approximation, be determined by either the cross-sectional area of the hydrophobic chain or the area required for the arrangement for closest packing of the head groups (Fig. 8.9), whichever is greater. For straight-chain 1:1 ionic surfactants, it is usually found that the head group requirement predominates, so that for a given homologous series, the surface tension minimum obtained varies only slightly with the length of the hydrocarbon chain.

Since the decrease in surface tension obtained is directly related to the surface excess adsorption of the surfactant by the Gibbs equation, a reduction in the amount of material that can be adsorbed in a given surface area reduces the ultimate surface tension lowering attained. The efficiency will, of course, change more or less regularly with the chain length. The sign of the charge on the ionic surfactant has only a minor effect on the ultimate surface tension attained, indicating that the geometric requirements (including electrostatic effects) are fairly constant from one head group to the next. In the presence of neutral electrolyte, of course, electrostatic repulsion between adjacent molecules are reduced, so that their effective areas are smaller. The net result is a slight increase in surfactant effectiveness.

While an increase in the hydrocarbon chain length in a series of normal alkyl surfactants between C_8 and C_{20} carbon chains will have a minor effect on the effectiveness of a surfactant, other structural changes can produce

(a)   (b)

FIGURE 8.9. The packing efficiency of a surfactant at the surface will be determined by the combined effects of the hydrophilic head group and the hydrophobic tail: (*a*) straight chains and large head groups (relative to the tail cross section) favor close, effective packing; (*b*) branched, bulky, or multiple hydrophobic chains hinder effective close packing at the interface.

much more dramatic effects. As will be seen in more detail in Chapter 15, structural features such as branching and multiple-chain hydrophobes will generally result in increases in the cmc of surfactants with the same total carbon content. Those changes seem to have a much smaller effect on the efficiency of the surfactant (C_{20}) than on its effectiveness.

The introduction of polar groups such as ethylenic unsaturation; ether, ester, or amide linkages; or hydroxyls located well away from the head group, usually results in a significant lowering of both the efficiency and effectiveness of the surfactant as compared to a similar material with no polar units. Such a result has generally been attributed to changes in orientation of the adsorbed molecule with respect to the surface due to interactions between the polar group and the water (Fig. 8.10). If the polar group is situated very near to the primary hydrophilic group, its orientational effect will be much less dramatic, although it may still have a significant effect on the cmc of the material.

Changes in the hydrophobic group in which fluorine atoms are substituted for hydrogen usually result in significant increases in the efficiency and effectiveness of the surfactant. The substitution of fluorine for hydrogen in a straight-chain surfactant results in a relatively small increase in chain cross-sectional area, as compared to a methyl branch, for example, so that the changes must be related to the chemical nature of the substitution. As has already been pointed out, fluorinated organic materials have a relatively low cohesive energy density and therefore little interaction with adjacent phases, or themselves, for that matter. They therefore have very favorable thermodynamic driving forces for adsorption (leading to high efficiency), as well as low surface energies. Their effectiveness is reflected in the very low surface tension values produced (as low as 20 mN m^{-1} in some instances).

There appear to be only relatively minor variations in effect from one head group to another in anionic surfactants. The difference in cross-sectional area between sulfate and sulfonate groups does not appear to influence greatly the activity of surfactants in lowering surface tensions, although some difference can be noted when differences in cmc are taken into consideration. The

(a)   (b)

FIGURE 8.10. The presence of a second polar group near the principle hydrophile, or further along the chain, will affect the surface area occupied by each surfactant molecule and, therefore, its surface characteristics.

role of the counterion can be important when changes result in significant alterations in the ion binding properties of the molecule. Tight ion binding, for instance, reduces the extent of electrostatic repulsion between adsorbed molecules, allowing for tighter packing of surfactant at the interface and, in general, increases in both the efficiency and the effectiveness of the surfactant. A similar result is obtained by the addition of neutral electrolyte.

While the head group may be of minor importance for hydrophiles closely related in size and charge character, alterations in those factors can produce significant changes in their activity at the solution–fluid interface. A class of surfactants well suited to the study of such effects is that of the quaternary ammonium salts in which three of the alkyl groups are short-chain units such as methyl, ethyl, and propyl. The substitution of the larger alkyl groups for methyl surfactants of the type $RN^+(CH_3)_3X^-$ results in a significant reduction in the efficiency of adsorption, while not affecting σ_{min} significantly. Presumably, the presence of bulkier alkyl groups on the head group greatly increases its area and therefore reduces its adsorption efficiency.

The practical effects of surface tension lowering have not been addressed here because they are generally more meaningful when presented in the context of related phenomena such as emulsification, foaming, wetting, and detergency. For further details on the subject of surface tension lowering and surfactant adsorption at fluid interfaces, the reader is referred to the works cited in the Bibliography.

8.4. INSOLUBLE MONOMOLECULAR FILMS

The preceding discussion of liquid surface tensions and the effects of adsorbed molecules on them was concerned primarily with systems in which the adsorbing species has a sufficiently large solubility in the solvent that, when the surface is saturated, there remains in solution a reservoir of dissolved surface-active molecules to "fill in the blanks" if more surface is created, or to participate in other phenomena such as micelle formation. Surface films of materials that do not have sufficient solubility to fit this model, but that also exhibit interesting and useful properties have also been recognized for centuries and have added much to our understanding of adsorption and surface phenomena in general. Such materials form the so-called insoluble monolayers—that is, monomolecular layers or films of adsorbed molecules that have very low solubility in the supporting liquid phase, so that they are essentially isolated on the surface.

A schematic comparison of the two situations is shown in Figure 8.11. On the left, a reservoir that has a movable barrier (B) is filled with pure water. The surface tension of the water on each side of the barrier will be the same (σ_0). If a quantity of surfactant is added to one side of the reservoir, time is allowed for the system to reach equilibrium, and the surface tension of each side is measured (σ_L and σ_R), one finds that $\sigma_L = \sigma_R$. Surfactant molecules

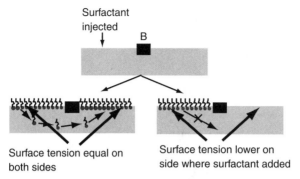

FIGURE 8.11. If a surface-active material is added to the surface of water in a container where the surface is divided by a barrier, two types of adsorption may occur: (*a*) if the material has significant solubility in water, diffusion through the water will occur and adsorption will result on both sides of the barrier to produce a Gibbs monolayer; (*b*) if the material is essentially insoluble in water, adsorption will occur on the side of the barrier to which it is applied, but not on the opposite side, leaving an insoluble monolayer.

placed on one side will dissolve in the supporting liquid and be adsorbed on the other side of the barrier. Addition can be continued until a saturated monolayer (a Gibbs monolayer) is produced; the important point is that the monolayer is formed on both sides of the barrier due to dissolution and diffusion processes.

Consider now the situation on the right in which an essentially insoluble, but still surface-active, material such as stearic acid is placed on one side of the barrier. Assuming that the barrier does not "leak," the system may be left for any practical period of time and when the surface tension of each side is measured, it will be found that σ for the side to which stearic acid has been added has been lowered, while the other side remains that of the pure water, σ_0. Stearic acid may be added until a monolayer is formed with no change in (for the "clean" side. The added stearic acid molecules, being insoluble in water, cannot be dissolved and transported through the water to be adsorbed on the other side of the barrier. The monolayer ultimately formed is an insoluble monolayer.

If the barrier B is made movable and connected to some indicating device (a torsion wire or some electronic transducer, for example) one will see that for the experiment on the left, at equilibrium, there is no change in the location of the barrier—the readout is zero. For that on the right, the barrier will be seen to move away from the side to which the stearic acid has been added, effectively increasing the area of surface containing stearic acid relative to clean surface (arrow). By analogy to a three-dimensional system, the system on the right behaves as if some pressure has been applied to the barrier, pushing it away from the added material.

8.4.1. Surface Pressure

The surface pressure of a monolayer film, π, is defined as the difference between the surface tension of the pure supporting liquid, σ_0, and that of the liquid with an adsorbed film, σ:

$$\pi = \sigma_0 - \sigma \tag{8.12}$$

The phenomenon of surface pressure has been studied since the late nineteenth century. Some consequences of insoluble monolayer formation were known (but not understood) as early as Biblical times and were of interest to the likes of Benjamin Franklin. For example, the pouring of oil on stormy waters was recognized as an effective measure to protect fragile ships in a storm.

The surface pressure, as defined by Equation (8.12) represents an expanding pressure exerted by the monolayer acting against the (contracting) surface tension of the pure liquid substrate (Fig. 8.12). Analogous to the pressure–volume $(P–V)$ curve of a three-dimensional bulk material, one can construct a pressure–area $(\pi–A)$ curve for a monolayer. Experimentally, either one can work with a fixed surface area and increase the pressure by incrementally adding more of the adsorbed material, or a known amount of material can be added to a surface and the pressure increased by slowly decreasing the available area. In practice, the latter approach, as represented by the classical Langmuir trough, is much preferred, since it is easier to measure area reproducibly than to measure the addition of small amounts of dilute solutions. (Typically, the adsorbed material is added as a dilute solution in a volatile solvent.)

With a known amount of material on the surface, the $\pi–A$ curve allows one to determine something about the physical nature of the film and some molecular characteristics of the adsorbed material.

FIGURE 8.12. The surface pressure [Eq. (8.12)] can be visualized as arising from the mutual "pushing" action of neighboring adsorbed molecules working against the "pull" of the surface tension of the liquid.

8.4.2. Surface Potential

It is a general fact that at an interface or phase boundary between two dissimilar materials, there exists a surface electrical potential that reflects differences in the electronic makeup of the two phases. Because almost all surface-active materials (for aqueous systems, at least) have a polar head group, when the molecules adsorb at the surface, the dipole moments of those groups become at least partially oriented with respect to the interface. As a result of the orientation of the dipoles (or charges), the potential difference across the interface will be altered. The surface film potential due to the monolayer, ΔV, is the change in the interfacial potential due to the presence of the monolayer.

If the monolayer is treated as a parallel plate condenser, the measured surface film potential can be used to deduce information about the orientation of the adsorbed molecules. While the calculation is only approximate, if n is the number of molecules in the adsorbed film (a known quantity), μ is the dipole moment of the head group (also known or accessible), and ε is the permittivity of the film (its dielectric constant $\times$ the permittivity of a vacuum), the approximate relationship

$$\Delta V = \frac{n\mu \cos \theta}{\varepsilon} \tag{8.13}$$

allows one to estimate the angle of inclination of the dipole to the normal to the surface, θ. Using known bond angles and distances, one can then deduce the orientation of the entire molecule with respect to the surface.

If a mixed monolayer is present (see below), the surface film potential can be used to estimate the homogeneity of the film or, for a homogeneous mixture, the film composition (assuming that the values for the pure films are known). The technique is particularly useful for studying the penetration of insoluble monolayer films by surface-active molecules injected just below the surface. For example, if a monolayer of a film analogous to a biological membrane is formed and another material of interest is injected below the film (a drug, for instance), surface potential measurements may indicate whether the injected material can penetrate the membrane, and if so, how fast. While the results of such experiments cannot be considered to be absolute in the sense of mimicking an in vivo system, they may be very useful as an aid in interpreting other data.

8.4.3. Surface Rheology

Because of the mobility of molecules in the surface of a pure liquid, such surfaces have very little elasticity. For that reason, pure liquids cannot support a foam. In the presence of an adsorbed monolayer film, however, the rheological properties of the surface can change dramatically. By analogy with bulk phases, the physical state of a surface film can be distinguished by its viscosity.

If talc powder is gently placed on a liquid surface, gently blowing on the surface causes the particles to move relatively freely. If the surface is covered with a low density monolayer film movement becomes more restricted, but relatively free movement is still evident. If the film pressure is increased (i.e., more molecules per unit area of surface), at some point the particles become fixed in place—the surface viscosity has increased substantially and the film behaves as if it is in a condensed state (see text below).

It is the increase in surface viscosity produced by adsorbed films (insoluble and Gibbs monolayers, adsorbed polymers, etc.) that leads to the production of persistent foams, helps stabilize emulsions, and explains the role of spread monolayers in dampening surface waves, among other important interfacial phenomena.

8.5. THE PHYSICAL STATES OF MONOLAYER FILMS

Like bulk materials, monolayer films exhibit characteristics that can (sometimes with a bit of imagination) be equated to the solid, liquid, and gaseous states of matter. For films, the equivalent states are roughly defined as

1. Condensed (solid) films, which are coherent, rigid (essentially incompressible), and densely packed, with high surface viscosity. The molecules have little mobility and are oriented perpendicular (or almost so) to the surface (Fig. 8.13a).
2. Expanded films, roughly equivalent to the liquid state, in which the monolayer is still coherent and relatively densely packed but is much more compressible than condensed films. Molecular orientation is still approximately perpendicular to the surface, but the tails are less rigidly packed (Fig. 8.13b).
3. Gaseous films, in which the molecules are relatively far apart and have significant surface mobility. The molecules act essentially independently,

(a)   (b)   (c)

FIGURE 8.13. Molecules in a monomolecular film are usually considered to exist in one of three principle "states." (a) The gaseous state is that in which the molecules are relatively far apart and have little mutual interaction; the film is compressible. (b) The liquid expanded state is that in which the head groups are relatively closely packed, but there is significant degree of tail mobility; the film is compressible to a limited extent. (c) The condensed state in which the molecules are closely packed and have very limited mobility; the film is essentially incompressible.

much as a bulk phase gas and molecular orientation will be random (Fig. 8.13c).

Each type of monolayer film exhibits its own special characteristics analogous to the corresponding bulk phase, as well as distinct "phase transitions," which are useful in characterizing the nature of the film in terms of its equation of state, molecular orientation, interfacial interactions, and other parameters.

8.5.1. Gaseous Films

An ideal (bulk) gas will have an equation of state given by the ideal gas law

$$PV = nRT \tag{8.14}$$

Similarly, an "ideal" gaseous monolayer should follow the corresponding law

$$\pi A = kT \tag{8.15}$$

which means that the π–A curve should be a rectangular parabola such as that in Figure 8.14. In fact, such ideal behavior is rare due to the finite size of the adsorbed molecules.

FIGURE 8.14. Each "state" of monomolecular films has a characteristic pressure–area (π–A) curve that can be used as a diagnostic tool for identifying the state of a given film: (a) the condensed film, being incompressible, has a sharp transition as the film area is decreased; (b) the liquid expanded film shows a more gradual transition as it is compressed, but the degree of compression is limited; (c) the gaseous film is relatively easy to compress and shows no sharp transition.

If molecular interactions are taken into consideration, Equation (8.15) can be modified to

$$\pi A = x k T \tag{8.16}$$

where x is a constant (usually between 1 and 2) that adjusts the system to allow for lateral interaction between molecules. Because the value of x must be determined independently for each material, Equation (8.16) is sometimes inconvenient to use. A more generally useful modification is to employ a constant A_0, which is essentially the area excluded (or occupied) by the adsorbed molecule:

$$\pi(A - A_0) = kT \tag{8.17}$$

It is tempting to equate A_0 directly with the cross-sectional area (in the plane of the surface) of the vertically adsorbed molecule. However, lateral interactions tend to reduce (or in cases of repulsion increase) its value, so that for many films, A_0 may be significantly less than or greater than the expected cross-section of the adsorbate molecule.

Of course, even more complicated equations of state such as a combination of Equations (8.16) and (8.17) can be devised, as, for example

$$\pi(A - A_0) = xkT \tag{8.18}$$

However, since fitting experimental data (which may likely contain some uncertainty) to isotherms and equations of state is sometimes quite subjective, the use of overly complicated relationships involving fitted parameters such as A_0 and x are seldom truly justified.

Gaseous films are common for soluble surfactants solutions (Gibbs monolayers) since solvent-adsorbed solute interactions tend to keep the adsorbed molecules "independent" of neighboring molecules. While they are also encountered in insoluble monolayers, many materials of interest are not so "well behaved" in that they do not exhibit the parabolic π–A curve of Figure 8.14.

8.5.2. Liquid Films

Liquid films are coherent in that they appear to involve some degree of cooperative interaction between portions of the adsorbed molecules, either head groups or tails. They exhibit characteristics of a fluid in that they appear to have no yield point, yet their π–A curve extrapolates to zero at molecular areas significantly larger than than that corresponding to the "theoretical" cross-sectional area (Fig. 8.14b). This indicates the presence of molecular interactions at relatively long distances—a coherent structure, albeit loose or disorganized.

It is sometimes found convenient to designate two subclasses of liquid films—liquid-expanded (L_1) and liquid condensed (L_2)—based on subtle differences in respective π–A curves. The L_1 curve is one which typically extrapolates to a limiting value of π (sometimes zero) at a molecular area of about 0.5 nm^2. In contrast to the bulk liquid analog, such films exhibit a significant degree of compressibility but show no signs of "island" or hemimicelle formation; that is, it appears to maintain the characteristics of a uniform phase. In many cases, L_1 films show a transition to a gaseous film at low pressures and perhaps to an L_2 film as the available area per molecule is decreased.

The L_2 films are characterized by the fact that they have considerably lower compressibility than L_1 films and that their π–A curve undergoes a gradual transition to linearity, reminiscent of the solid films. Such films are commonly viewed as having head groups that are close-packed, but that can, under pressure, be rearranged somewhat to give a still tighter packing arrangement. For example, in Figure 8.15, the head groups are shown schematically in a square (or cubic, in three dimensions) lattice at π_1. If π is increased ($\pi_2 > \pi_1$), the arrangement may change to a hexagonal structure, with tighter packing, at which point further significant changes in "crystal structure" are precluded and the π–A curve becomes linear and steep.

While L_2 films ultimately extrapolate to some limiting area at high pressures, that area is usually found to be some 20% larger than the cross-sectional area of a hydrocarbon chain taken from X-ray data, or 10% greater than the respective condensed film (0.22 versus 0.205 nm^2).

8.5.3. Condensed Films

Condensed films are composed of densely packed, highly oriented molecules with little mobility and low compressibility. Unlike the gaseous films, the π–A

FIGURE 8.15. As pressure is applied to a condensed film, the adsorbed molecules can rearrange to a small extent by a change in packing structure (e.g., cubic to hexagonal). Beyond that point added pressure will result in film "buckling."

curve for such materials as the saturated, straight-chain carboxylic acids (e.g., stearic acid) exhibit minimal change is pressure as the surface area is decreased until a critical area is reached, at which point the pressure will increase rapidly (Fig. 8.16). The interpretation of the form of the π–A curve is that in such cases, the cohesive interaction between adsorbed molecules is sufficient to cause the formation of clusters of molecules or hemimicelles on the surface. Because of the strong cohesion, as the available area is decreased, the clusters grow in size and/or number, while inter-cluster interaction remains small, and it is the interaction between floating clusters that is measured as changes in π. When the area is reduced to the point that the clusters are forced to interact (by physical contact), the pressure increases rapidly.

A typical π–A curve for stearic acid on water at 20°C, plotted as area per molecule, is shown in Figure 8.16. The curve is found to become very steep at an area per molecule of 0.205 nm^2. If the area is decreased further, the pressure suddenly falls, indicating a buckling or collapse of the film. That point is commonly referred to as the yield point. The critical molecular area of 0.205 nm^2 found for stearic acid is the same as that for palmitic, myristic, and the other members of the series with more than 12 carbons in the chain. X-ray diffraction data indicate that the cross-sectional area of the stearic acid molecule is 0.185 nm^2, suggesting that the critical limiting area represents the point at which the molecules become more efficiently packed, approximating the packing in the solid crystal.

With very careful experimental work, it is sometimes possible to identify various intermediate phase transitions occurring before the formation of the condensed film. For example, myristic acid spread on 0.1 N HCl at 14°C can (with extreme care) produce a curve similar to that in Figure 8.16, in which

FIGURE 8.16. A monomolecular film of a straight chain carboxylic acid such as myristic acid on distilled water will show a sharp transition in the π–A curve as the head groups become more closely packed. That kind of transition may be viewed as something like a reversed sublimation in which the film passes from gaseous to solid condensed without passing through the liquid expanded state. With very careful experimental work, it is sometimes possible to identify an intermediate liquid expanded phase as illustrated for myristic acid on 0.1 N HCl.

an intermediate liquid-expanded phase is found between the gaseous and condensed phases.

8.5.4. Some Factors Affecting the Type of Film Formed

The type of monolayer film formed by a given material will depend on a number of factors, both intrinsic and external, including the natures of the tails and head groups, the degree and nature of solvation of the head group, the nature of the substrate or supporting liquid phase, and temperature.

The Nature of the Tail. The first factor to consider is the nature of the hydrocarbon tail. For simple, straight-chain materials such as the saturated carboxylic acids, $CH_3(CH_2)_nCOOH$, solid or L_2 films will be favored. At a given temperature, long chain lengths (e.g., $n \geq 14$ carbons at 20°C) tend to favor solid films, while shorter chains ($10 < n < 14$) tend to produce L_2. For $n < 8$, the acids begin to have significant water solubility, so that gaseous films may result. Similar trends will be found for other classes of n-alkyl materials (alcohols, amines, etc.), although such factors as solvation of the head group may become important. Octadecanol (cetyl alcohol, $C_{18}OH$), for example, can be compressed to an L_2 film at 20°, but not to a condensed film, presumably due to the solvation of the $-OH$ group by hydrogen-bonded water. The carboxyl group, on the other hand, is much more strongly associated with "its own kind," and produces a condensed film.

If the tail is branched, the larger cross-sectional area of the molecule precludes the close packing and lateral cohesion required for the formation of condensed films and expanded films result. Similarly, molecules having two (or more) hydrocarbon groups such as esters of polyhydric alcohols, are limited in their lateral interaction and generally produce expanded films, although this depends somewhat on the length of the chains and the temperature.

Molecules containing two hydrophilic groups will exhibit characteristics curves reflecting the interaction of the second (and usually weaker) hydrophile with the substrate. Unsaturated and hydroxy carboxylic acids, some esters and amides, are similar, tend to lie more or less flat on the water surface at low pressures as a result of interactions between the water and the second hydrophilic group (Fig. 8.17). As the pressure is increased, energy is required to force those groups away from the surface so that for a given area, π will be greater than that for a normal chain material such as stearic acid. The process of standing up the tails in such materials is gradual, so that a curve characteristic of an expanded film will result.

For unsaturated carboxylic acids, the nature of the monolayer film will depend of the configuration of the double bond. For a *trans* double bond, the hydrocarbon chain will be more or less straight, so that lateral interactions and good packing efficiency may lead to the formation of a solid or an L_2 film. The corresponding cis isomer has a forced bent structure, reducing its

FIGURE 8.17. Surfactant molecules having two or more hydrophilic head groups will tend to lie more or less parallel to the liquid surface (*a*). As pressure is applied to the film, the molecules may be forced to "stand up" and assume a more vertical orientation (*b*). The energy requirements for such a process produces a π–A curve characteristic of a liquid-expanded film.

ability to pack tightly and leading to expanded (probably L_1) film formation (Fig. 8.18).

If hydrogen atoms on, for example, a long, straight-chain acid are substituted by fluorine or other halides, the film type gradually changes from solid to L_2, presumably due to packing difficulties imparted by the bulkier halogen atoms as well as the weaker cohesive interactions present in such materials. Irregular and complex molecules such as steroids, dyes, and polymers usually exhibit complex phase behavior in monolayer films and often defy clear classification. In such cases, one may say that a monolayer is "essentially" solid, fluid, etc., with the understanding that some allowance is being made for the nature of the beast.

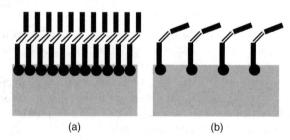

FIGURE 8.18. Unsaturation in the hydrophobic chain can have a significant effect on packing at the interface: (*a*) a *trans* double bond allows for a reasonably uniform close packed arrangement at the interface and a more condensed state; (*b*) a *cis* isomer, due to the inherent curved structure of the molecule, prevents close packing and produces a much more expanded film.

Effect of the Head Group. A second important factor in determining the type of film formed is the nature of the head group. A bulky head group that requires more surface area to accommodate it tends to keep the tails farther apart so that lateral cohesive interactions, and therefore efficient packing, are prevented. Such systems tend to form expanded rather than solid films. Charged head groups act similarly relative to uncharged species, in that the electrostatic repulsion between adjacent molecules will force them apart and reduce the interactions necessary for the formation of a tightly packed solid film.

Finally, the degree of solvation of the head group will affect its effective size; groups that are in fact relatively small, such as the hydroxyl group, but which require a significant amount of solvation, form more expanded films than a similar molecule with a larger, but less solvated group (e.g., $-COOH$).

Effect of Temperature. The discussion to this point has been couched more or less in terms of a fixed temperature (e.g., 20°C). Just like three-dimensional phases, however, monolayer phases and phase changes are sensitive to temperature. In general, as the temperature is lowered, the behavior of a given film goes from expanded to condensed or solid. The temperature at which the transition occurs depends on the specific molecule; however, for a homologous series of materials, it is usually found that the addition of one $-CH_2-$ group to the chain corresponds to an increase of 5°C in the temperature at which the transition from condensed to expanded film occurs.

For ionized or bulky head groups, the temperature at which the expanded to condensed film transition occurs will be lower than that for the corresponding unionized material. Slightly ionized salts of polyvalent cations, on the other hand, will have higher transition temperatures.

Effects of Changes in the Nature of the Substrate. It is often found that the nature of the film formed by a given type of molecule will depend greatly on the pH and other characteristics of the aqueous substrate (i.e., concentration and valence of solute ions). This is especially true for ionizable materials such as carboxylic acids and amines. In the first case, at low pH, the unionized acid will tend to form solid or condensed films, depending on other factors mentioned above. As the pH is raised, however, the degree of ionization of the head group will increase, leading to the expansion of the film for reasons already mentioned. For amines, the effect will be the opposite—lower pH leading to greater ionization and film expansion.

In the presence of polyvalent ions such as Ca^{2+} carboxylic acids tend to form metal soaps, which have significantly lower solubility in water than the corresponding acid or alkali salt. Because in such cases each cation becomes associated with two molecules of the adsorbate, the result is a tighter packing of the molecules and a transition to a solid or condensed film at higher temperatures. For ionized species, a similar result may be obtained with a substrate containing a relatively high concentration of neutral electrolyte (e.g.,

NaCl), which can reduce the electrostatic repulsion between head groups and thereby enhance packing efficiency. The result is the formation of a condensed or solid film at temperatures that produced expanded films on water alone.

8.5.5. Mixed-Film Formation

Brief reference has already been made to the formation of mixed monolayer films and some of their probable characteristics. For completeness, the basic ideas will be summarized again. In studies of mixed films of materials that form monolayers alone, the type of film formed may vary from an "ideal solution" film, through films involving specific compound formation, to essentially "immiscible" systems, all depending on the specifics of molecular interactions between the components. The simplest possibilities are illustrated schematically in Figure 8.19.

If two film components are structurally similar (e.g., two normal-chain carboxylic acids) the characteristics of the film produced by the mixture will lie between that formed by each separately (Fig. 8.19*a*). For example, if the two each form expanded films alone, the mixed film will also be of the expanded type. If, on the other hand, one is a condensed film and the other expanded, the mixture will be more condensed than the expanded film or more expanded than the condensed film.

If dissimilar materials are mixed that can undergo specific interactions (e.g., alcohols with carboxylic acids), interesting effects can be observed (Fig. 8.19*b*). For example, if an alcohol is added to an acid layer of the same chain length,

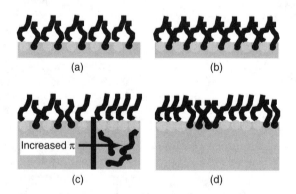

FIGURE 8.19. Mixed molecular films can have several possible structures. (*a*) An "idel" mixed film is one that has a homogeneous distribution of components throughout the film, but with no "special" interactions between components. (*b*) A synergistic mixed film or complex involves specific interactions between component molecules that produce characteristics different from those expected for an ideal film. (*c*) Immiscible components may produce the expulsion of the more soluble component of the mixture at high surface pressure. (*d*) Heterogeneous mixed films may form "islands" (two-dimensional micelles?) of the components.

the layer becomes more condensed than if the same amount of the acid is added to the film. The usual explanation is that the alcohol can undergo a strong specific interaction (hydrogen bonding) with the acid, reducing the effective molecular area of each molecule and essentially shrinking the area of the film (or reducing the film pressure at constant area). If the two materials are sufficiently different and no specific interactions occur (Fig. 8.19c), increasing the surface pressure may cause the complete expulsion of one component from the surface. In the absence of the effects mentioned above, two materials may form a heterogeneous film, with islands of one film "floating" in a sea of the other. One might think of it as a two-dimensional emulsion or dispersion (Fig. 8.19d).

A mixed-film phenomenon of particular interest in the biological and medical areas is that referred to as film penetration, in which a soluble surface-active material in the substrate enters into the surface film in sufficient quantity to alter its nature significantly, or to undergo some alternative physical or chemical process related to the surface (Fig. 8.20). Such penetration studies using films of biological materials have been used to mimic phenomena in biological systems (cell walls and membranes, for example) that cannot readily be studied directly. Of particular interest are such topics as cell surface reactions, catalysis, and transport across membranes.

A typical penetration experiment might involve the formation of an insoluble monolayer at a surface pressure π, after which a soluble surface-active material is injected below the monolayer and changes in surface pressure (at constant area) due to penetration or inclusion of the new material in the monolayer are monitored. Alternatively, one can study changes in surface area at constant π, changes in surface potential, or a combination of any or all.

8.5.6. Surface Films of Polymers and Proteins

High molecular weight polymers, including proteins, also form surface monolayer films. However, because of the length of the polymer molecules and the complex interactions involved in intra- and interchain interactions, the properties of such films are less distinct and more difficult to determine with

(a)   (b)

FIGURE 8.20. Film penetration is a potentially important process in which a monomolecular film (a) is modified by the insertion of molecules of a second component from the supporting solution (b). In the process, the second component penetrates the original monolayer film from solution to produce a mixed film with new properties.

any degree of confidence (quantitatively, at least). In order for a macromolecular film to reach its "true" equilibrium pressure, every unit of every chain must orient itself to provide the optimum thermodynamic situation. Such a process may involve hours (or more likely, days), which means that the problems associated with maintaining a stable, clean system are magnified enormously.

When carried out carefully, however, a great deal of interesting information about the macromolecule can be gained. In the case of proteins, for example, it has been shown that the secondary and tertiary structures of the spread protein are significantly different from those of the same material in its "native" solution configuration—it is denatured. In addition, at low pressures, it is generally found that the amino acid side chains are lying flat on the surface of the water, while higher pressures will cause them to stand up and point out into the air. Given time, the protein monolayer may even form a substantially rigid and strong gel or "skin" which can be physically removed from the surface as a unit (or at least in large units).

Nonprotein polymer films generally behave similarly to the protein films in terms of their $\pi–A$ curves, orientation, compressibility, and so on, depending on the nature of the side-chains and the possibilities of nonspecific and specific interactions between neighboring units.

8.5.7. Monolayer Films at Liquid–Liquid Interfaces and on Nonaqueous Liquids

With improvements in apparatus for doing monolayer film studies has come more interest in studying the nature of monolayer films at liquid–liquid interfaces, with one liquid usually being water or an aqueous solution. For monolayer films deposited at a water–nonpolar liquid interface, it is normally found that the area per molecule for alcohols and carboxylic acids is larger than the same material spread at the water–air interface. The accepted explanation is that the presence of the nonpolar liquid reduces the lateral cohesive interactions between adjacent tails, causing what may be termed a swelling of the monolayer film. A similar "swelling" effect can be noted for protein and other polymer films at the water–liquid interface. Since biologically important monolayer film models generally involve aqueous–oil type systems, the behavior of monolayer films at such interfaces can be of particular interest.

Studies of monolayer films on liquids other than water have been somewhat limited for various reasons, including experimental difficulties and possibly lack of obvious practical relevance. However, some work has been done using mercury, long-chain hydrocarbons, mineral oil, etc. Mercury, due to its high surface tension, adsorbs almost anything. That broadens the choice of monolayer material one might use and facilitates the formation of the film. However, it also creates the significant problem of assuring the presence of a clean surface before deposition of the film of interest. As a result, little quantitatively reliable information is available.

Silicone polymers and fully fluorinated surface-active materials have been found to be the best candidates for spread monolayer film studies on nonpolar liquids. Because nonpolar liquids are more difficult to manipulate in terms of their solvent properties (e.g., by changing pH, electrolyte content) it is often necessary to talk in terms of adsorbed Gibbs monolayers, rather than true insoluble monolayers. However, sometimes we must take what we can get from nature and make the most of it.

8.5.8. Deposited Monolayer and Multilayer Films

It was discovered early in the studies of insoluble monolayer films that the adsorbed monolayer could, with careful attention to detail, be transferred from the liquid surface to a solid substrate which was passed through the surface. The technique, commonly referred to as the "Langmuir–Blodgett technique," is illustrated schematically in Figure 8.21. The solid surface on which the layer is deposited is usually glass or metal, although any material that has a relatively strong affinity for one part of the monolayer material will serve.

If an insoluble polyvalent salt of a carboxylic acid is deposited on a glass slide (which would normally be completely wetted by water), the resulting surface exhibits a water contact angle equal to or greater than that of water on pure paraffin. The explanation is that the monolayer is transferred intact to the solid surface, producing a densely packed layer with the hydrocarbon tails oriented out toward the air (Fig. 8.21a). If the deposited monolayer is passed back down through a new monolayer, adsorption will occur with the reverse orientation ("back-to-back" or Y films; Fig. 8.21b), producing a bilayer film but this time one that is completely wetted again. The process can be repeated many times, always producing films of alternate wetting and nonwetting character.

With modifications in the dipping technique, it is possible to produce multilayer films with each layer having the same orientation (head-to-tail or X films; Fig. 8.21c). It is usually found that the Y films are more stable than

(a)   (b)   (c)

FIGURE 8.21. Insoluble monolayer films can, with careful work, be deposited onto solid substrates from the liquid surface. The so-called Langmuir–Blodgett films will usually have alternating molecular orientations: (a) first-layer deposition with the head (or tail) toward the solid substrate and (b) the second-layer head-to-head or tail-to-tail in a "Y" film orientation. Under some conditions it is possible to produce an "X" film with "head-to-tail" orientation (c).

their X film counterparts, a result that is intuitively satisfying considering the interfilm interactions involved.

The adsorption of deposited monolayer films can be very tenacious, especially when the solid substrate can react with the deposited material. Carboxylic acid monolayers deposited on metal or metal oxide surfaces, for example, almost certainly form metal soaps that are extremely difficult to remove by any means other than direct chemical etching. For example, such strongly adsorbed films are very useful in friction and wear studies of boundary lubrication (see Chapter 18).

Monolayer films on solid substrates need not necessarily be deposited from a monolayer system. They can also be deposited directly from solution or a melt of the material to be deposited, or by a vapor deposition process. Such films are generally referred to as self-assembled monolayers. There are several possible advantages to such monolayers, mostly associated with simplification of the process (no need to be concerned with depositing the monolayer on a liquid surface, maintaining a constant surface pressure, problems of surface contamination, etc.)

Deposited monolayer films have received a great deal of attention in recent years because of the possibility of using surface-active materials that can be reacted after deposition to produce very thin, strong films. If a multilayer (e.g., 50 molecules thick) of photosensitive material is deposited on the surface of a semiconductor, the resulting "microphotoresist" coating can be used to produce a circuit with a definition and resolution several orders of magnitude better than that possible using a normal photoresist coating. The result is the possibility of significantly higher packing of circuit information in a given area of semiconductor: greater miniaturization.

8.6. A FINAL COMMENT

The preceding discussion of liquid–fluid interfaces, adsorption, monolayer films, and so forth, was extremely limited and a great deal of interesting and useful information was excluded. Although the theories and techniques discussed may be considered to be old in comparison to much of modern science, one finds repeatedly that a great many modern technological problems can be understood and solved on the basis of a little understanding of the classical ideas of such interfaces.

PROBLEMS

8.1. Using logic based on molecular considerations and the Gibbs adsorption equation, explain why the surface tension of an aqueous surfactant will usually remain constant after the concentration of surfactant passes a certain level.

8.2. A cloud chamber operates on the principle that a supersaturated vapor phase is caused to condense into visible droplets by the passage of high energy particles. In a hypothetical experiment, an unknown particle induces the formation of 200 water droplets 2000 nm in diameter in a cloud chamber at 1 atm pressure and 25°C. The density and surface tension of water under the conditions of the experiment are 0.9971 g cm^{-3} and 72.49 mN m^{-1}, respectively. Calculate the free energy increase caused by the passage of the particle.

8.3. The surface tensions of a series of solutions of an unknown surfactant where found to be the following:

Concentration (mM)	σ (mN m^{-1})	Concentration (mM)	σ (mN m^{-1})
0.01	72.5	1.2	36
0.05	71.0	1.3	34
0.08	69.5	1.5	32
0.10	68.0	1.7	33
0.20	64	2.0	35
0.30	57	2.5	36
0.50	52	2.7	36
1.0	44	3.0	36
1.1	40	3.2	36

After repeated recrystallization of the unknown, the following surface tension results were obtained:

Concentration (mM)	σ (mN m^{-1})	Concentration (mM)	σ (mN m^{-1})
0.01	72.5	1.2	36
0.05	71.0	1.3	34
0.08	69.5	1.5	32
0.10	68.0	1.7	33
0.20	64	2.0	35
0.30	57	2.5	36
0.50	52	2.7	36
1.0	44	3.0	36
1.1	40	3.2	36

Plot the two sets of results (σ vs. ln C) and explain qualitatively using molecular considerations and the Gibbs equation the observed differences in the two sets of data.

8.4. Using the data for the purified material in Problem 8.3, calculate the surface excess and area per molecule for the material at surface saturation.

8.5. The material from Problem 8.3 was dissolved in a solution of 0.2 M sodium chloride and the surface tension measurements repeated. The results are given in the following table:

Concentration (mM)	σ (mN m^{-1})	Concentration (mM)	σ (mN m^{-1})
0.01	72	1.0	46
0.02	70	2.4	34
0.04	64	3.5	33
0.09	60	3.9	33
0.15	57	4.5	33
0.30	51	8.0	33

Calculate the surface excess and area per molecule for the material in 0.2 M salt solution. Compare the results with those from Problem 8.3 and explain any differences.

8.6. A 5.2×10^{-5}-g sample of hexadecanoic acid is prepared as a dilute solution in toluene and applied to the surface of a 0.2 M sodium chloride solution in a Langmuir trough. As pressure is applied a condensed film is formed at a area of 250 cm^2. Calculate the area per molecule occupied by the acid in a close packed monolayer.

8.7. The following data were obtained for the butanol–water system at 25°C:

[Butanol] (mol kg^{-1} water)	Activity (mol kg^{-1} water)	σ (mN m^{-1})
0.0033	0.0033	72.80
0.0066	0.0065	72.26
0.0132	0.0130	70.82
0.0264	0.0258	68.00
0.0532	0.0518	63.14
0.1050	0.0989	56.31
0.2110	0.1928	48.08
0.4330	0.3796	38.87
0.8540	0.7119	29.27

Prepare a plot of butanol adsorption versus activity. What is the area per butanol molecule adsorbed at surface saturation?

8.8. Write the Gibbs adsorption equation for a three-component system in which species (1) is the solvent, species (2) is the primary surface-active solute, and species (3) is an impurity in (2) present at a constant percent. It is found that in the dilute solution region a plot of surface tension versus the concentration of (2) goes through a minimum. Provide an explanation of what may be causing the minimum and what is the behavior of each term of the Gibbs equation in that region.

8.9. Explain the logic behind the statement that dissolution of one liquid in another might be considered as a situation in which the interfacial tension between the two is negative.

8.10. Knowing that ethyl alcohol is surface-active and that it evaporates more rapidly than water, explain why, in a glass of brandy, droplets or tears form on the glass above the liquid surface and drain back into the liquid.

8.11. Given the following surface and interfacial tensions at 20°C:

Interface	σ (mN m^{-1})	Interface	σ (mN m^{-1})
Air–water	72	Mercury–water	415
Air–valeric acid	25	Mercury–valeric acid	329
Air–hexane	18	Mercury–hexane	378
Air–mercury	485	Water–hexane	50

(a) Will valeric acid spread at the mercury–water interface? (b) If it spreads, will the polar group on the acid be oriented toward the water or the mercury? Why?

8.12. A monolayer of protein spreads at the air–water interface. The amount of protein spread is equivalent to 0.80 mg m^{-2} and gives a surface tension lowering of 0.035 mN m^{-1}. What is the molecular weight of the protein?

8.13. n-Octane spreads to form a duplex film on water provided that some carboxylic acid (e.g., myristic acid) is present in the oil. A duplex film containing 0.01 M acid has a surface pressure of 14 mN m^{-1}. How many milligrams of acid are present per square centimeter, assuming that the acid is present as a gaseous film at the oil–water interface? At 25°C the surface tension of n-octane is 25 mN m^{-1} and its interfacial tension with water 53 mN m^{-1}.

8.14. Data for the surface pressure, π, as a function of area per molecule of $C_{15}COOH$, a, are

π ($\times$10-3 mN m^{-1})	a (nm^2)	π ($\times$10-3 mN m^{-1})	a (nm^2)
53	0.75	86	0.22
60	0.50	128	0.21
85	0.30	312	0.20
86	0.25		

The density of $C_{15}COOH$ is 0.85 g cm^{-3} and its molecular weight = 256 g mol^{-1}. Calculate the cross-sectional area of the acid molecule at the interface.

8.15. Can the interfacial tension of two liquids ever be larger that the sum of their surface tensions? Larger than the largest of the two? Larger than the smallest?

8.16. How much work is done in distorting a liquid sphere 2000 nm in diameter into a prolate ellipsoid with an axial ratio of 5. The liquid surface tension is 65 mN m^{-1}.

8.17. What will be the ratio of the Laplace pressures of the two configurations in Problem 8.16?

9 Adsorption

9.1. INTRODUCTION

A topic of particular importance in surface and colloid science is that of the adsorption of atoms and molecules at interfaces. The process of adsorption is one of the principal ways in which high-energy interfaces can be altered to lower the overall energy of a system. While the phenomenon can be quite complex from the point of view of molecular theory, the classical approaches to understanding the various processes have been founded mostly on empirical observations and conceptual insight rather than fundamental first principles. Modern computers allow one to attack the problems more rigorously through complex mathematical modeling, although the "practical" gains from such modeling remain limited. In this and other areas covered in this text, reference is made to "practical" versus "model" results. The distinction is made because, in almost all cases, computer models are based on ideal conditions and consider the interactions of a few or a few hundred "units." Such models cannot take into consideration the sometimes small, but often very significant, deviations from ideally encountered in real systems, even under the most rigorously controlled laboratory conditions. As one might expect, correlations between model predictions and experimental results are most often qualitative or semi-quantitative, except in the simplest systems. For that reason the "user" of theoretical models and concepts should always be aware of possible differences and not over interpret experimental results in terms of preconceived models.

Adsorption can be most simply defined as the preferential concentration (i.e., location) of one component of a system at an interface, where the local (i.e., interfacial) concentration of one or more components of one or both phases is different from those in the bulk phases. "Adsorption" should be clearly differentiated from "absorption," in which physical penetration of one phase into another is involved, although the two may operate concurrently. Adsorption can occur at any type of interface, although the distinct characteristics of solid versus liquid interfaces make the analysis of each case somewhat different. For that reason, the discussion of each situation is best presented in the context of specific interfaces. In many practical systems, all four of the principle interfaces may be present, leading to complex situations that make complete analysis very difficult or impossible.

Where the interfacial concentration of the adsorbed species is greater than that in the bulk phase(s), one can refer to "positive" adsorption, although the positive aspect is usually assumed. It is possible, however, for negative adsorption to occur. In such a situation, the concentration of a system component in the region of the interface will be less than that in one or both bulk phases. The result of negative adsorption can be to increase the interfacial energy of a system relative to a defined standard state. While such situations are less frequently encountered, the possibility should not be discarded where experimental evidence cannot be explained by other means.

Interfaces containing only liquids and vapors generally exhibit simpler adsorption characteristics (in principle, at least) than those containing solid surfaces, because in liquid surfaces, the complications arising due to specific structures and surface heterogeneities can be ignored. In addition, multilayer adsorption (see below) can usually be ruled out. The adsorption characteristics of solid surfaces, on the other hand, can be very much history dependent and the possibility of such "historical" differences should always be kept in mind.

Under the general topic of adsorption phenomena, there are a few terms that must be kept clearly in mind to avoid confusion, particularly when solid surfaces are involved. When adsorption occurs on a solid, the solid is referred to as the adsorbent and the adsorbed material the adsorbate. In some cases absorption may also occur. It can be difficult to separate the effects of the two phenomena if they occur together. In fact, it is likely that one will affect the other and produce a situation even more difficult to analyze than that produced by one alone.

In the consideration of adsorption processes, there are two aspects that must be addressed: (1) thermodynamics—the effect of the adsorption process on the final equilibrium interfacial energy of the system, and (2) kinetics—the rate at which the adsorption process occurs. For the most part, the discussions to follow will be concerned only with equilibrium conditions, and dynamic processes will not be addressed. For many applications, such a restriction will not result in significant limitations to the validity of the concepts involved. For others, however, the kinetics of adsorption can play a very important role. Some of those situations will be addressed in later chapters.

9.1.1. The Gibbs Surface Excess

Before beginning a discussion of specific adsorption phenomena, it will be useful to introduce one of the concepts fundamental to "classical" adsorption theory—the Gibbs dividing surface and the Gibbs adsorption isotherm. While the following introduction is very superficial (excuse the pun), the concepts involved, once clearly understood, provide a good basis for understanding more complex models and approaches.

A consequence of the nature of the interfacial region is that the total concentration of a given component in a system of fixed volume and interfacial area will be determined by the shape of the concentration profile at the interface (Fig. 9.1). For example, if a component in a fluid–fluid system is positively adsorbed at the interface, its concentration profile may resemble

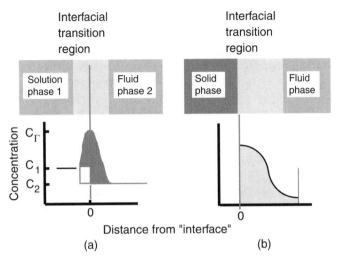

FIGURE 9.1. When a material is positively adsorbed at an interface, its adsorption profile will resemble (*a*) for a liquid–fluid interface or (*b*) for a solid–fluid interface.

that depicted in Figure 9.1*a*. In a solid–liquid system a situation similar to Figure 9.1*b* would be expected, assuming no dissolution of the solid in the liquid and no absorption of liquid. The concept of the concentration profiles illustrated have been around in theory for many years. Unfortunately, procedures for their direct quantitative verification eluded, until recently, even the best experimentalists. A number of indirect methods for obtaining the interfacial concentration profiles have been suggested, but the one most often used in that of Gibbs. Although much criticized, the approach of Gibbs has so far stood the test of time based on its ability to fit experimental data, its generality, and the relative ease of obtaining useful results.

The Gibbs approach to determining (or at least estimating) the concentration of components in the interfacial region is quite simple and straightforward. Consider a system containing a substance i in one or both of two phases α and β. If the unit concentration of i in phase α (C_i^α) is uniform throughout and that in β (C_i^β) is likewise uniform, for given volumes of a (V_α) and β (V_β), the total amount of i, n_i is given by

$$n_i = (C_i^\alpha V_\alpha + C_i^\beta V_\beta) \tag{9.1}$$

However, since the local value of C_i varies going through the interface (except in very unusual circumstances), there will generally be a different concentration of i present in the interfacial region than that indicated by Equation (9.1). That difference, defined as the surface excess amount of i (n_i^σ), is given by

$$n_i^\sigma = n_i - (C_i^\alpha V_\alpha + C_i^\beta V_\beta) \tag{9.2}$$

The quantity n_i^σ results from the presence of the interface and is dependent on the shape of the concentration profile of i in the transition region between α and β. From a practical standpoint, the surface excess can also be considered to be the amount of i adsorbed at the interface. The major theoretical problem with this approach is the question of exactly how to define or locate the interface between α and β. Gibbs' approach was to define the interface (or the "Gibbs dividing surface") as the region (really a two-dimensional plane) in which the concentration of one phase component, for example substance α, becomes zero. In a solution, it is standard practice to define α as the solvent. The concentration profiles shown in Figure 9.2 illustrate graphically the situation under consideration. Figure 9.2a gives the concentration profile for component α, the solvent. The shaded area represents the excess concentration of a in the region. The profile for i is given in Figure 9.2b and shows how, at $C_\alpha^\sigma = 0$, there is a relative excess amount of i with respect to α. When the system is limited to an interface of area A^σ, one obtains a concentration termed the surface excess concentration of i with respect to α, $\Gamma_i^{(\alpha)}$, where

$$\Gamma_i^{(\alpha)} = \frac{n_i^\sigma}{A^\sigma} \tag{9.3}$$

It should be remembered that the Gibbs approach is a model that facilitates the handling of data mathematically, and does not imply that the surface excess of i is actually physically located in the region of the Gibbs dividing surface. The dividing surface is a mathematical plane with zero dimension in the third direction (into phases α and β), while the units of i (i.e., atoms or molecules) are three-dimensional and cannot occupy such a mathematical plane.

As mentioned, it is extremely difficult to measure a surface excess quantity directly. Such measurements have been made at solid–liquid interfaces, but

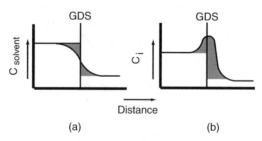

FIGURE 9.2. In the Gibbs approach to defining the surface excess concentration, the Gibbs dividing surface (GDS) is defined as the plane in which the solvent excess concentration becomes zero (the shaded area is equal on each side of the plane) as in (a). The surface excess of component i will then be the difference in the concentrations of that component on either side of that plane (the shaded area) (b).

liquid–liquid and liquid–vapor interfaces still pose extreme experimental difficulties. That is not to say, however, that the Gibbs approach is not applicable in such systems. On the contrary, fundamental thermodynamics allows one to use the Gibbs approach to determine surface excess concentrations in fluid systems from relatively simple and straightforward determinations of interfacial tensions via the Gibbs adsorption equation.

9.1.2. The Gibbs Adsorption Equation

In order to define unambiguously the state of a bulk phase, α, it is necessary to specify the values of several variables associated with that state. Those variables include the temperature, T^α, its volume, V^α, and its composition, n_i^α. Alternatively, the system pressure, P^α, may be fixed. In terms of the Helmholtz free energy F, the system can be specified by

$$F^\alpha = -S^\alpha T^\alpha - P^\alpha V^\alpha + \Sigma \mu_i^\alpha n_i^\alpha \qquad (9.4)$$

or in differential terms (with constant P)

$$dF^\alpha = -S^\alpha \, dT^\alpha - P^\alpha \, dV^\alpha + \Sigma \mu_i^\alpha \, dn_i^\alpha \qquad (9.5)$$

For a two-phase system, a similar equation can be written for the second or β phase. At equilibrium, the two phases will have the same temperature, T; the same pressure, P; and the same chemical potential, μ, for all components. The complicating factor in a system of two phases in contact is that the presence of the α–β interface that may be considered to be a third "phase," σ, and which will make a separate contribution to the overall energy of the system. The total energy, then, will be given by

$$F^T = F^\alpha + F^\beta + F^\sigma \qquad (9.6)$$

where F^σ is the interfacial excess free energy. For systems in which the interfacial area is small relative to the bulk volumes, the contribution of F^σ is usually ignored. However, in many contexts, such as adsorption, catalysis, membrane activity, and especially colloidal systems, the free energy of the interface may be the primary factor determining the overall molecular, microscopic, and macroscopic properties observed. Analogous to Equation (9.5), the derivative of the surface free energy may be given by

$$dF^\sigma = -S^\sigma \, dT + \sigma \, dA^\sigma + \Sigma \mu_i \, dn_i^\sigma \qquad (9.7)$$

where σ is the surface or interfacial tension between α and β, and μ_i has the same value as that in the bulk phases. The term $\sigma \, dA^\sigma$ replaces the $P \, dV$ term from eq. 9.5 because the interface (as the Gibbs dividing surface) is a mathematical plane having only two dimensions and is therefore an area rather than a volume term. Similarly, the sign of the $\sigma \, dA^\sigma$ term is positive rather than negative because σ is conceived of as a tension (pulling) rather than a pressure (pushing).

In bulk thermodynamics, one can derive the Gibbs–Duhem equation by integration of Equation (9.5) while holding the intensive properties T, P, and μ_i constant to give

$$F^\alpha = PV^\alpha + \Sigma \mu_i^\alpha n_i^\alpha \tag{9.8}$$

Differentiation then yields, at equilibrium

$$dF^\alpha = V^\alpha dP - S^\alpha dT + \Sigma n_i^\alpha \, d\mu_i^\alpha = 0 \tag{9.9}$$

Similarly, the interfacial "phase" yields the Gibbs adsorption equation

$$- A^\sigma d\sigma - S^\sigma dT + \Sigma n_i^\sigma \, d\mu_i = 0 \tag{9.10}$$

At constant temperature, Equation (9.10) reduces to

$$- d\sigma = \frac{\Sigma n_i^\sigma d\mu_i}{A^\sigma} \tag{9.11}$$

or, where $\Gamma_i = n_i^\sigma / A^\sigma$,

$$- d\sigma = \Sigma \Gamma_i \, d\mu_i \tag{9.12}$$

For a two-component liquid–vapor system where the Gibbs dividing surface is defined so that the surface excess concentration of the solvent is zero ($\Gamma^\alpha = 0$), the summation in Equation (9.12) is no longer necessary and a simple relationship between the surface tension of the liquid phase, σ, and the surface excess concentration of solute i, Γ_i, is obtained. It is therefore possible to employ experimentally accessible quantities such as surface tension and chemical potential to calculate the surface excess concentration of a solute species and to use that information to make other indirect observations about the system and its components.

A more general form of the Gibbs adsorption equation is

$$d\sigma = - \Gamma_2^{(1)} \, d\mu_2 \tag{9.13}$$

in which 2 designates a solute dissolved in bulk phase 1. At equilibrium, the chemical potential of each component is equal in all phases, so that μ_i at the interface can be taken as that value in either of the adjacent bulk phases. The chemical potential of 2, then, can be related to its concentration in either of the bulk phases by

$$d\mu_2 = RT \, d \ln a_2^{(1)} = RT \, d \ln x_2\gamma_2 \qquad (9.14)$$

where $a_2^{(1)}$ is the activity of 2 in (1), the bulk phase, x_2 is its mole fraction, and γ_2 its activity coefficient. These manipulations lead to the relationship

$$d\sigma = - RT \, \Gamma_2^{(1)} \, d \ln x_2\gamma_2 \qquad (9.15)$$

At equilibrium, the chemical potential of a species will increase with its concentration, although not necessarily in a linear fashion. As a result, an increase in Γ_2 with μ_2 (i.e., positive adsorption) will produce a decrease in the interfacial tension, σ.

A material that is strongly adsorbed at an interface may be termed a surface active material (a "surfactant"), and will normally produce a dramatic reduction in interfacial tension with small changes in bulk phase concentration. In dilute solution, it is assumed that the activity coefficient of a material, γ_2, can be approximated as unity so that the last term in Equation (9.15) can be substituted for by the molar concentration, c_2. The practical applicability of this relationship is that the relative adsorption of a material at an interface, its surface activity, can be determined from measurement of the interfacial tension as a function of solute concentration:

$$\Gamma_2^{(1)} = - \frac{1}{RT} \left[\frac{d\sigma}{d \ln c_2} \right] \qquad (9.16)$$

The preceding discussion of the Gibbs adsorption equation was referenced to a fluid–fluid interface in which the surface excess, Γ, is calculated based on a measured quantity, σ, the interfacial tension. For a solid–fluid interface, the interfacial tension cannot be measured directly, but the surface excess concentration of the adsorbed species can be, so that the equation is equally useful. In the latter case, Equation (9.16) provides a method for determining the surface tension of the interface based on experimentally accessible data.

The principles given above allow one to derive an expression relating theoretical concepts of surface excess concentration and adsorption to experimentally obtainable quantities. But what is the practical importance of those ideas? In fact, the phenomenon of adsorption at interfaces, tied to the resultant effects of such adsorption, carries with it a multitude of important consequences (some good and some bad) for many technological and biological processes.

Some of those effects will be discussed in detail in following chapters dealing with specific interfacial situations and interactions. For now, suffice it to say that the ability to understand and control interfacial tensions (or energies) through adsorption processes, and the secondary electrical and thermodynamic effects of the adsorbed species on a system constitute the bedrock of modern technology related to colloidal stability (and instability), wetting phenomena, emulsification and demulsification, foam formation and destruction, adhesion, lubrication, fluid displacement in capillary systems, and many more areas. Specifically, the areas of pharmaceuticals, cosmetics, food preparation, inks, paints, adhesives, lubricants, crude-oil recovery techniques, mineral ore separations, wastewater treatment, heterogeneous catalysis, lithographic and xerographic printing techniques, microelectronics fabrication, and photographic and magnetic recording media, live and die by the effective control of interfacial interactions through the preferential adsorption of surfactants, polymers, and colloidal solids. On a more personal level, the functioning of our most important life processes—cell structure, respiration, blood flow, muscle function, and many disease processes are based on interfacial phenomena. Small alterations in the functioning of those phenomena can literally be the difference between life and death. With so much riding on our ability to understand and control interfaces, it is important that the people involved in industrial and medical research and technology development have at least a basic working knowledge of the phenomena involved.

9.2. ADSORPTION AT THE SOLID–VAPOR INTERFACE

As already discussed in Chapter 7, the surface characteristics of a solid may be very history-dependent, its energy and adsorption characteristics depending on, among other things, the conditions of its formation and storage, the possibility of seemingly minor chemical reactions (e.g., surface oxide formation), the presence of seemingly insignificant (in both quantity and quality) potential contaminants, its handling during sample preparation, and other factors. For that reason, any detailed discussion of adsorption onto solid surfaces, and any effects attributed to such adsorption, must include knowledge of those "historical" elements along with the actual experimental results. As in many areas of interfacial and colloidal phenomena, a literal minefield of complicating factors awaits the unsuspecting adventurer.

A freshly formed, clean solid surface will often be of quite high surface energy (except for most polymer surfaces) so that there will exist a strong driving force for the reduction of the excess surface energy by whatever process may be available. In a liquid, some of that excess energy can be dissipated by spontaneously reducing the total interfacial area—the liquid forms a spherical drop (or as close to it as gravity and physical restraints allow). A solid does not have that option so that solid surfaces tend to adsorb materials that will not adsorb appreciably at liquid interfaces, namely gases such as nitrogen, oxygen, or carbon monoxide. Adsorption occurs because it reduces the imbalance of forces acting on the surface molecules of the solid or liquid. The energetics of adsorption will be essentially the same for any type of interface. However, a solid surface will almost certainly be heterogeneous in terms of the distribution of its excess surface energy, meaning that adsorption will not be a uniform process, while for liquids it is assumed to be so.

9.2.1. Energetic Considerations: Physical Adsorption versus Chemisorption

The forces involved in adsorption processes are the same as those encountered in every other interfacial process, and for that matter every chemical or physical phenomenon above the atomic level. They include nonspecific van der Waals forces, ionic or electrostatic forces, and specific forces involved in the formation of chemical bonds. Because the nonspecific interactions are orders of magnitude smaller than the specific forces, adsorption processes that involve only nonspecific interactions are generally referred to as "physical adsorption" while those in which stronger interactions occur are termed "chemisorption."

Since adsorption onto a solid surface by a vapor is a spontaneous process, the overall free energy change for the process must be negative. However, in the process, the adsorbing molecules lose a degree of freedom; that is, they become restricted to two instead of three degrees of freedom and their entropy decreases. From the thermodynamic relationship

$$\Delta G = \Delta H - T \Delta S \tag{9.17}$$

it is clear that for ΔG to be negative, ΔH must be negative; that is, adsorption must be an exothermic process. The situation may be different in the case of adsorption from solution due to the effects of changes in solvation and other factors.

For solid–vapor adsorption, because of its exothermic nature, the amount of gas adsorbed onto a solid will decrease as the temperature of the system is increased. That is why clean solid surfaces are more easily prepared by heating to high temperatures. In addition, high vacuum reduces the vapor pressure of the adsorbed gas and reduces its tendency to adsorb.

The heats of adsorption, ΔH, for gases onto a given solid can, in principle, be measured in a variety of ways and will, in reversible systems, adhere to the Clausius–Clapeyron equation

$$\left(\frac{\delta \ln p}{\delta T} \right)_v = -\frac{\Delta H_{ads}}{RT^2} \tag{9.18}$$

In systems where adsorption is strictly of the physical type, it is found that ΔH_{ads} is generally of the same magnitude as the heat of condensation for the vapor. In the case of nitrogen, for example, the heat of condensation is approximately -6 kJ mol^{-1}. The heat of physical adsorption of nitrogen is found to be in the range of -10 kJ mol^{-1} on iron, -12 kJ mol^{-1} on graphite, and -14 kJ mol^{-1} on TiO$_2$. In the case of chemisorption of nitrogen on iron the heat of adsorption rises to about $-150\,kJ\,mol^{-1}$, comparable in magnitude to the strength of chemical bonds. Obviously, chemisorption involves much stronger interactions.

In the absence of complicating factors such as capillary condensation and competitive adsorption, the process of physical adsorption has no activation energy; that is, it is diffusion-controlled and occurs essentially as rapidly as vapor molecules can arrive at the surface. The process will be reversible and equilibrium will be attained rapidly. Because the forces involved are the same as those involved in condensation, physical adsorption will generally be a multilayer process—that is, the amount of vapor that can be adsorbed onto a surface will not be limited simply by the available solid surface area, but molecules can "stack up" to a thickness of several molecules in a pseudoliquid assembly (Fig. 9.3). If the vapor pressure of the gas reaches saturation level, in fact, the condensation and adsorption processes overlap and become indistinguishable. The fact that physical adsorption can be a multilayer process is very important to the mathematical modeling and analysis of the process, as will be seen below.

Unlike physical adsorption, chemisorption involves very specific interactions between the solid surface and the adsorbing molecules, as illustrated by the much higher heats of adsorption. Another important result of that specificity is that chemisorption is by nature limited to the formation of a monomolecular adsorbed layer. Chemisorption processes will generally have some activation energy and may therefore be much slower than physical adsorption. They may also exhibit hysteresis; that is, they may not be readily reversible.

The energetic relationship between the two processes can be illustrated by analysis of the schematic energy diagrams shown in Figure 9.4. Curve 1 in the figure represents the energy diagram for physical adsorption. At large distances, there is essentially no attraction between the surface and the vapor molecule. As the vapor approaches the surface there develops an attraction due to van der Waals interactions leading to an energy minimum representing

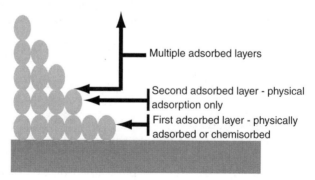

FIGURE 9.3. In multilayer adsorption on a solid surface the first adsorbed layer may be physically adsorbed or chemisorbed. Subsequent layers will be physically adsorbed.

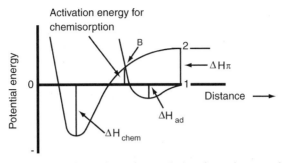

FIGURE 9.4. Although physical adsorption and chemisorption may be energetically quite different, the stronger chemisorption phenomenon must be preceeded by a physical adsorption process.

the heat of adsorption, ΔH_{ads}. At some distance, in this case the molecular radius of the vapor, there begins to develop some overlap between electron clouds, leading to the development of a repulsive interaction (Born repulsion).

Curve 2 represents the process of chemisorption. Because chemisorption involves specific interactions between the adsorbent and the adsorbate, the process must also involve some specific change in the molecular structure of the adsorbate molecule. For example, for a diatomic molecule A_2 that involves π bonding, chemisorption may involve the rupture of a π bond between the two atoms. The starting point for the isolated process (at large distances) must be the energy of the bond being broken, ΔH_π, which would represent the activation energy of the process. (One may think in terms of an activated molecule, A_2^*.) Because of the magnitude and nature of the specific interactions involved, the energy minimum for chemisorption (ΔH_{ch}) will be much deeper than ΔH_{ads} and will occur at shorter distances.

If the two adsorption processes were mutually exclusive, the activation energy of chemisorption would be so high that it would occur only under rather vigorous conditions. However, if the two occur in a cooperative way, it becomes clear that physical adsorption is an important component of the overall chemisorption process. Referring again to Figure 9.4, one can see that if physical adsorption of the gas molecule occurs first, it can approach the solid surface along a pathway of much lower energy than that of chemisorption alone. At the point where the two curves intersect (B), the energy difference between the physically adsorbed molecule and the activated molecule is greatly reduced. That difference is the activation energy for chemisorption. The activation energy clearly depends on the shapes of the two energy curves and will vary greatly from one system to another.

Both physical adsorption and chemisorption have very important practical applications and implications. Because physical adsorption is a rapid and reversible process, it can be considered to occur to some extent on almost all solid surfaces (with the possible exception of very low energy surfaces such as perfluorinated materials) except under extreme conditions of high temperature and high vacuum. If the physically adsorbed molecules are such things as oxygen or nitrogen from the air, their presence on the surface may cause no practical problems. However, if the atmosphere contains materials such as oil from fans and motors, which can be just as easily or more easily adsorbed, the nature of the solid surface may be altered to the extent that it no longer functions in an expected or desired way.

For example, a freshly cleaned silicon surface intended for use in the manufacture of microelectronic devices must be coated with a so-called micro-resist, a photosensitive polymer that allows the manufacturer to transfer the microcircuit design from a large mask or negative to the surface of the wafer. Since the size of the circuit components may be in the micron range, the resist coating must be free of defects. For a clean surface (or at least one with only atmospheric gases adsorbed), the microresist solution will uniformly wet and spread to give the quality of surface necessary for the production of usable microcircuits (Fig. 9.5). If the surface has organic materials adsorbed (e.g., "spots" of oily material) the coating will not spread uniformly, and an irregular, useless layer will result. More details on the effects of adsorbed materials on wetting and spreading are given in Chapter 17. It may be said, then, that random physical adsorption (which is practically unavoidable in any case except for an almost perfect vacuum) may be innocuous or detrimental in practical situations, depending on the nature of the adsorbed species and the sensitivity of the process involved.

9.2.2. Chemisorption and Heterogeneous Catalysis

Chemisorption, while generally slower, less ubiquitous, and more easily avoided than physical adsorption in most cases, has some very important practical aspects. Perhaps the most important of those is its role in heterogeneous catalytic processes. As pointed out above, chemisorption involves such strong interactions between adsorbent and adsorbate that one may assume in

(a)   (b)

FIGURE 9.5. Even a monolayer of physically adsorbed material can have a disastrous effect on some technological processes, especially critical coating processes in which absolutely smooth, uniform coating is required such as the microelectronics industry. (*a*) An uncontaminated surface should produce a smooth, uniform coated layer. (*b*) Surface contamination by a lower-energy material (such as oil molecules) may result in the formation of "repellency" spots or other defects, producing less than optimum or even useless finished products.

some cases the formation of chemical bonds between the two. Even if that is not the case, it has been shown repeatedly that just the "forceful presence" of the solid surface can alter the electronic structure of an adsorbed molecule sufficiently to alter its electronic and vibrational spectra, and therefore its chemical reactivity. In addition, if two species are chemisorbed on a surface, their close proximity in the "activated" chemisorbed state may lead to chemical reaction between the two.

The number of important chemical processes that involve the chemisorption of gases at solid surfaces is quite large and cannot be even partially covered here. However, a few general examples of reaction types include

1. Combination (hydrogen + alkene → alkane)
2. Decomposition (ethanol → ethylene + water)
3. Polymerization (ethylene → polyethylene)
4. Depolymerization or cracking (higher alkanes → lower-molecular-weight materials)
5. Isomerization (n-alkane → branched alkane)
6. Various other mixed reactions, including photolytic and photosynthetic processes

The complications of heterogeneous catalytic reactions are such that each specific case must be considered individually, and few if any are really fully understood. There are, however, a few fundamental aspects that can be considered in order to get an overall picture of what individual molecular processes may be involved (Fig. 9.6). Those fundamentals include (1) the initial physical adsorption process, (2) possible surface diffusion of the adsorbed species, (3) chemisorption processes (e.g., bond breaking, if it is involved), (4) chemical reaction between adsorbed species; and (5) desorption of the product. All or only some of those steps (or perhaps others not mentioned) may be involved in a given catalytic process. Any one of them may be the rate-determining step.

To illustrate the processes involved, consider a hypothetical heterogeneous oxidation of compound A with molecular oxygen

$$O_2 + 2\,A + catalyst \rightarrow 2\,AO + catalyst$$

for which it is known that the actual oxidizing species is atomic oxygen. The first step in the process must be the physical adsorption of the reactants onto the catalyst surface, followed, perhaps, by chemisorption of O_2 and dissociation to form two oxygen atoms.

A + catalyst → A (adsorbed)	(rate constant = k_1)
O_2 + catalyst → O_2 (adsorbed)	(rate constant = k_2)
O_2 (adsorbed) → O_2 (chemisorbed) → 2O·	(rate constant = k_3)

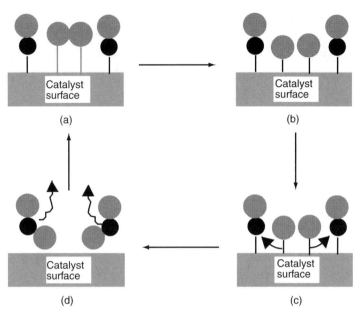

FIGURE 9.6. The heterogeneous catalysis process involves several adsorption and desorption steps before and after chemical reaction. The process for the catalytic conversion of carbon monoxide to carbon dioxide is the following (approximately): (*a*) carbon monoxide is adsorbed strongly through the carbon atom—molecular oxygen is adsorbed more weakly at both atoms; (*b*) the oxygen molecule is activated by splitting into two oxygen atoms; (*c*) the oxygen atoms react with neighboring adsorbed CO molecules to produce CO_2, which is only weakly adsorbed; (*d*) the CO_2 desorbs, diffuses away, and liberates the catalyst site for further reaction.

Once atomic oxygen has been formed, it must encounter an adsorbed molecule of A in order for reaction to occur. That process may require surface diffusion with an accompanying rate (i.e., diffusion) constant k_4. The reaction between O and A will obviously have its own rate constant k_5 to be thrown into the soup. Finally, the overall rate of the process will depend on the availability of surface sites for adsorption, so that as product AO is formed, it must be desorbed to free up space for further desired reaction (k_6). Clearly, in order to understand such a catalytic process, one must understand a variety of independent but interrelated processes. It is easy to see, therefore, why the subject of heterogeneous catalysis is so complex and in many cases poorly understood.

Chemisorption and heterogeneous catalysis also occur at the solid–liquid interface. The basic concepts mentioned above remain valid in such systems, although the situation will naturally be complicated by such factors at solvation.

9.2.3. Catalytic Promoters and Poisons

It is well known in the science and art of heterogeneous catalysis that the presence of small amounts of certain materials can greatly improve or disastrously ruin a catalytic reaction. Where improvement is found the additive is referred to as a "promoter" (a catalyst "catalyst"). When disaster results, the additive is termed a "poison."

The exact role of promoters is not very well understood in many cases, but it is now generally accepted that it is related to the formation of specific electronic surface states necessary for the given catalytic reaction. It apparently does not matter how that electronic state is produced; that is, whether it is formed in the preparation of the "native" catalyst surface or by the presence of some other component which "induces" the necessary state. As an example, the presence of small amounts of aluminum and potassium oxides on iron–iron oxide catalyst in the Haber ammonia synthesis greatly improves its activity. Either promoter alone has no significant effect on the process. Why? Such questions remain as fodder for further industrial or graduate research.

Catalyst "poisons" are materials that significantly alter, reduce, or completely destroy the activity of a given catalyst. Such materials generally function by binding strongly and (effectively) irreversibly to the specific surface sites necessary for the functioning of the desired process. Particularly troublesome materials in that sense are sulfur-containing compounds, especially thiols and thioethers. For example, the catalytic converters used to oxidize hydrocarbon residues in automobile exhausts will rapidly lose their effectiveness if exposed to such materials.

On the other hand, "poisoned" catalysts can have their uses. In some hydrogenations of organic molecules, for example, it may be desirable to produce a reaction between hydrogen and one functional group in the molecule, while leaving untouched another functionality that would normally react as well. By selectively poisoning the catalyst, surface states necessary for the desired reaction may be left untouched while those for the unwanted reaction are blocked.

Clearly, chemisorption and related catalytic processes are quite complex and remain a relatively poorly understood area of surface science. Modern surface analytical techniques have added much to our understanding of the molecular processes involved, but much remains in the realm of art (or perhaps black magic).

9.3. SOLID–VAPOR ADSORPTION ISOTHERMS

Study of the adsorption of gases onto solid surfaces has a long and illustrious history, with some of the fundamental aspects of physical adsorption being recognized early in the nineteenth century. It was known as early as 1814, for example, that the amount of a gas adsorbed by a given amount of a particular solid under "standard" conditions (room temperature and atmospheric pressure) was directly related to its ability to be condensed under those conditions (e.g., $NH_3 > H_2S > CO_2 > N_2 > H_2$). It was also recognized that heat was evolved when gas was adsorbed onto a solid. With the more detailed and quantitative analysis of adsorption process in modern times, it became obvious that adsorption processes were far from simple; that they were, in fact, quite varied and sometimes distinct, depending on the specific system under consideration.

Some systems, for example, were found to be completely reversible while others exhibited hysteresis, depending on from which side of the equilibrium an approach was made. The effects of temperature and gas pressure on the amount of gas adsorbed was noted. And in some cases it was noted that the gases given off as vacuum was applied were chemically different from those that were originally adsorbed. It was a confusing situation that required some innovative conceptual and mathematical modeling in order to satisfy the need to have a clear theoretical picture of the process.

The general relationship between the amount of gas (volume, V) adsorbed by a solid at a constant temperature (T) and as a function of the gas pressure (P) is defined as its adsorption isotherm. It is also possible to study adsorption in terms of V and T at constant pressure, termed isobars, and in terms of T and P at constant volume, termed isosteres. The experimentally most accessible quantity is the isotherm, although the isosteres are sometimes used to determine heats of adsorption using the Clausius–Clapeyron equation. In addition to the observations on adsorption phenomena noted above, it was also noted that the shape of the adsorption isotherm changed with temperature. The problem for the physical chemist early in the twentieth century was to correlate experimental facts with molecular models for the processes involved and relate them all mathematically.

9.3.1. Classification of Adsorption Isotherms

A useful model for a simple adsorption isotherm should take into consideration all phenomena involved in the process, including the initial monomolecular adsorption process at both low and high coverage, multilayer adsorption, if present, and accompanying (and complicating) phenomena such as chemisorption and capillary condensation. In a given situation, some or all of those factors may be important. As a result, there exists a wide variety of isotherm types, five of which are generally considered to be important in solid–vapor adsorption processes. The theoretical details behind each of the five major isotherm types will not be given here; however, each will be described qualitatively in terms of the relationship between shape and the (presumed) molecular processes involved. Because of their practical experimental utility, the mathematical foundations of three classical adsorption isotherms—the Langmuir, Freundlich, and Brunauer–Emmett–Teller (BET) isotherms—will be given in the following sections.

The five isotherms to be considered are shown schematically in Figure 9.7. The type I isotherm, usually termed the Langmuir type, is characterized by a fairly rapid initial rise in the amount of gas adsorbed with increasing gas pressure until some limiting value is reached. That limiting value is usually identified with the attainment of a complete monolayer coverage. Such an isotherm would be expected, for example, in chemisorption, where the system is limited to a monolayer. It may also be found for systems in which there is a strong nonspecific attractive interaction between the adsorbate and adsorbent, but weak attraction between the absorbate molecules themselves. Such isotherms may also be encountered in systems in which the solid has a very fine microporous structure.

Type II isotherms are typical of physical adsorption on nonporous solids. In contrast to type I, the adsorbate molecules in these cases also have relatively strong mutual interactions, which leads to the tendency for multilayer formation. The initial rapidly rising part of the isotherm corresponds to the equivalent type I adsorption. Point B on the curve is identified with complete monolayer coverage. Multilayer formation then begins which may lead to surface condensation. Type II isotherms are sometimes encountered for microporous solids, in which case point B would correspond to both completion of monolayer coverage and filling of the micropores by capillary condensation. The rest of the curve would then correspond to normal multilayer formation.

Types III and V isotherms are relatively rare and correspond to systems in which the interaction between adsorbate molecules is stronger than that between adsorbate and adsorbent. In these cases, the uptake of gas molecules is initially slow until surface coverage is sufficient so that interactions between adsorbed and free molecules begins to dominate the process. One might say that the processes are autocatalytic in terms of the adsorption process.

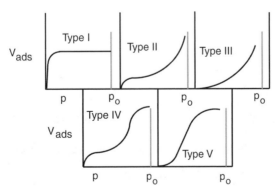

FIGURE 9.7. Adsorption isotherms are generally divided into five main types depending on the degree of adsorption (monolayer or multilayer), the mechanism of adsorption (physical or chemisorption), the nature of the adsorbent surface (porous or nonporous), and the relative strengths of adsorbate–adsorbent interactions.

Type IV isotherms are obviously similar to type II and usually correspond to systems involving capillary condensation in porous solids. In this case, however, once the pores have become filled, further adsorption to form multilayers does not occur and the terminating plateau region results. This would indicate a relatively weak interaction between the adsorbate molecules. While more complex isotherm classifications are available, they generally represent combinations and extensions of the five basic types described above.

9.3.2. The Langmuir Isotherm

Because of its simplicity and wide utility, the Langmuir isotherm has found wide applicability in a number of useful situations. Like many such "classic" approaches, it has its fundamental weaknesses, but its utility generally outweighs its shortcomings. The Langmuir isotherm model is based on the assumptions that adsorption is restricted to monolayer coverage, that adsorption is localized (i.e., that specific adsorption sites exist and interactions are between the site and a specific molecule), and that the heat of adsorption is independent of the amount of material adsorbed. The Langmuir approach is based on a molecular kinetic model of the adsorption–desorption process in which the rate of adsorption (rate constant k_A) is assumed to be proportional to the partial pressure of the adsorbate (p) and the number of unoccupied adsorption sites ($N - n$), where N is the total number of adsorption sites on the surface and n is the number of occupied sites, and the rate of desorption (rate constant k_D) is proportional to n.

At equilibrium, the rates of adsorption and desorption will be equal so that

$$k_A p(N - n) = k_D n \qquad (9.19)$$

Ignoring entropy effects, the equilibrium constant for the process will be $K_{eq} = k_A/k_D$, so that

$$K_{eq} = \frac{n}{p(N - n)} = \exp \frac{-\Delta H°}{RT} \qquad (9.20)$$

where $\Delta H°$ is the heat of adsorption per mole at temperature T and standard pressure P_{st}.

The fraction of the adsorption sites occupied at a given time, q, is given by

$$q = \frac{n}{N} \qquad (9.21)$$

so that Equation (9.20) can be written

$$q = \frac{K_{eq}p}{1 + K_{eq}p} \tag{9.22}$$

or

$$q = \frac{p}{p + K_{eq}^{-1}} = \frac{p}{p + P_{st} \exp(\Delta H^\circ / R)} \tag{9.23}$$

A useful characteristic of the Langmuir isotherm is that it can be rearranged to the linearized form

$$n^{-1} = N^{-1} + (K_{eq}Np)^{-1} \tag{9.24}$$

so that a plot of n^{-1} versus p^{-1} should be linear and yield values of K_{eq} and N from the slope and intercept. If the plot is not linear, then the Langmuir model does not fit the adsorption process in question.

The assumptions accompanying the derivation of the Langmuir isotherm mentioned above are rather rigorous. Although those conditions are almost never met in practice, the ability of the model to accurately express a significant amount of adsorption data in a mathematically simple and accessible way makes it invaluable as a basis for adsorption studies.

9.3.3. The Freundlich Adsorption Isotherm

It is often found that the simple Langmuir isotherm does not adequately describe adsorption systems of theoretical and practical interest. Another classic isotherm that has found application in describing adsorption, especially at moderate pressures, is the so-called Freundlich adsorption isotherm

$$V = kp^{1/a} \tag{9.25}$$

where V is the volume of adsorbed gas and k and a are constants, a usually being greater than 1. Equation (9.25) can be linearized by taking the logarithm of each side to give

$$\ln V = \ln k + \frac{1}{a \ln p} \tag{9.26}$$

Obviously, a plot of $\ln V$ versus $\ln p$ should give a straight line. Although originally derived empirically, the Freundlich equation can also be derived theoretically using a model in which it is assumed that the heat of adsorption is not constant but varies exponentially with the extent of surface coverage, a condition that is probably closer to reality than the Langmuir assumption in most cases.

9.3.4. The Brunauer–Emmett–Teller (BET) Isotherm

A major assumption of the Langmuir isotherm model is that adsorption stops at monolayer coverage. However, since the van der Waals forces leading to physical adsorption are the same as those involved in the formation of the liquid state, it should not be surprising to find that the limitation of adsorption to a single monolayer is unrealistic in many cases. As mentioned previously, such a case is expected only when the interactions among adsorbate molecules are very much weaker than those between adsorbate and adsorbent.

Assuming the formation of a multilayer of adsorbed molecules, Brunauer, Emmett, and Teller (BET) modified the Langmuir approach of balancing the rates of adsorption and desorption for the various molecular layers. The BET model assumes that the adsorption of the first monolayer has a characteristic heat of adsorption ΔH_A, but that subsequent layers are controlled by the heat of condensation of the vapor in question, ΔH_L.

The BET equation will not be derived here, but the most common linearized form of the final equation is

$$\frac{p}{V(p_0 - p)} = \frac{1}{V_m c} + \frac{(c - 1)p}{V_m c p_0} \tag{9.27}$$

where V is the volume of adsorbed vapor at STP, V_m is the monolayer capacity at STP, p is the partial pressure of the adsorbate, p_0 is the saturation vapor pressure of the adsorbate, and

$$c \approx \exp \frac{(\Delta H_A - \Delta^H_L)}{RT} \tag{9.28}$$

The BET isotherm was developed primarily to describe the commonly encountered type II isotherm shown in Figure 9.7, such as is found for the adsorption of relatively inert gases (N_2, Ar, He, etc.) on polar surfaces ($c \approx 100$). However, it reduces to the Langmuir isotherm (type I) when restricted to monolayer coverage ($\Delta H_A \gg \Delta H_L$), and describes type III isotherms in the unusual situation where the adsorption of the first monolayer is less exothermic that that of the subsequent layers (e.g., $c < 1$) resulting in low adsorption at low values of p/p_0.

9.3.5. Surface Areas from the BET Isotherm

One of the most common uses of the BET isotherm is for determining the surface area of finely divided solids by physical adsorption. Such information can be of great importance in a number of areas including heterogeneous catalysis and various sorption applications. While the BET model for multilayer adsorption contains several potential sources of error due to the assumptions of the absence of lateral interactions between adsorbed molecules, the constancy of the heat of adsorption (after the first monolayer), and solid surface homogeneity, it generally produces useful results at pressures, p, between 0.05 p_0 and 0.35 p_0. It must be used with caution, however, for porous solids that show adsorption hysteresis, or when point B on the isotherm corresponding to V_m cannot be accurately determined.

The monolayer capacity V_m is of primary interest because it allows for the calculation of the surface area based on the area occupied by each adsorbed gas molecule. According to Equation (9.27), a plot of $p/[V(p_0 - p)]$ versus p/p_0 should be linear over the pressure region noted above. From the slope of the line, $S = (c - 1)/V_m c$ and the intercept $I = 1/V_m c$ one can calculate the monolayer capacity V_m of the solid and thereby its specific surface area A_s:

$$V_m = \frac{1}{S + I} \qquad (9.29)$$

$$A^s = \frac{V_m k}{\text{sample weight}} \qquad (9.30)$$

$$k = \frac{N_a A}{M_v} \qquad (9.31)$$

where N_a is Avogadro's number, A is the area per molecule of the adsorbed gas, and M_v is the gram molecular volume of gas (22.400 L at STP).

The adsorbate most commonly employed for BET surface area determinations is nitrogen, which has an effective area per molecule, A, of 0.162 nm^2 at liquid nitrogen temperature (77 K). Nitrogen produces good results because it generally gives a well-defined value of B (i.e., $- \Delta H_A \gg - \Delta H_L$), while not having c so large that adsorption becomes localized. Other useful gases include argon ($A = 0.138$ nm^2) and krypton ($A = 0.195$ nm^2). Other vapors can be used so long as A is known. Effective molecular areas can in principle be determined by using a solid of known specific surface area, as determined by nitrogen adsorption, for example, for calibration. The potential problem with many other adsorbed gases is that, due to specific interactions with the surface sites, adsorption may become localized in the first monolayer. It is not uncommon, therefore, to find that A for a given molecule will vary significantly from one solid to another.

9.4. ADSORPTION AT SOLID–LIQUID INTERFACES

Interactions between solid surfaces and solutions are of fundamental importance in many biological systems (joint lubrication and movement, implant rejection, etc.), as well as in mechanics (lubrication and adhesion), in agriculture (soil wetting and conditioning and pesticide application), in communications (ink and pigment dispersions), in electronics (microcircuit fabrication), in energy production (secondary and tertiary oil recovery techniques), in foods (starch–water interactions in bakery dough), and in paint production and application (latex polymer and pigment dispersion stabilization).

With so many applications in theory and in practice, a good basic understanding of adsorption phenomena at solid–liquid interfaces can be of great importance to both the theoretician and the practitioner. As was the case for solid–vapor adsorption, the proper interpretation of experimental observations will almost always require a sound conceptual idea of the processes (possibly) involved, even though a specific model may not exactly describe the results.

9.5. THE ADSORPTION MODEL

Like all processes in this family, the adsorption of molecules at a solid–liquid interface creates a transition region on the order of molecular dimensions in which the composition of the system changes from that of the bulk solid to that of the bulk liquid. A "typical" system consisting of a solid surface and a liquid phase is illustrated in Figure 9.8. For a pure liquid, in the interfacial region it will be noted that the concentration of liquid molecules (the black circles in Fig. 9.8a) is apparently greater near the surface than in the bulk of the liquid phase indicating positive adsorption or solvation of the solid surface. If specific interactions occur between the liquid molecules and the solid, the adsorbed molecules may undergo a specific orientation, resulting in a change in the density, dielectric constant, or other physical (or even chemical) characteristics of the liquid near the interface. Except for the most delicate experimental work or in the context of catalytic processes, such effects are usually of little practical concern. It is the adsorption of a second component of the liquid phase (solute molecules) at the solid–solution interface that is generally of practical concern.

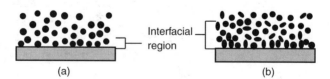

FIGURE 9.8. At a solid–liquid interface two adsorption "situations" may be encountered: (*a*) in the case of a pure liquid, the molecular distribution will be approximately uniform, although some molecules may be adsorbed to produce a small surface excess (black circles); (*b*) for a solution of surface-active solute (black ovals), extensive adsorption will occur, producing a significant interfacial region of excess solute concentration.

For a solution, a higher concentration of the solute near the interface is an indication of positive adsorption of the solute molecules. A typical concentration profile of the adsorbate is shown in Figure 9.8b. From both theoretical and practical standpoints, it is of interest to know the characteristics of such adsorption profiles for a given system in order to understand the mechanism of the adsorption process, as well as its consequences.

We have already seen (and will continue to see) the Gibbs adsorption equation which can be conveniently written as

$$\Gamma_{2(1)} = -\frac{1}{RT}\frac{d\sigma}{d\ln c} \tag{9.32}$$

As will be seen in later sections concerned with liquid–fluid systems, this equation normally employed to determine the amount of adsorbed material at an interface as a function of interfacial tension, σ, in the case of solid surfaces, it is difficult or impossible to determine σ directly. It is, however, relatively easy to determine the amount of adsorbed material directly and use that information to calculate a value of the interfacial tension. Such exercises are of great theoretical importance in understanding why and how molecules are adsorbed at an interface, and of even greater practical importance for understanding how such adsorption affects the characteristics of the interface and its interaction with its surroundings, especially in the context of colloidal stability and wetting phenomena.

Some degree of adsorption will occur at any solid–liquid interface, although it may be so small as to be effectively negligible. In fact, the adsorption may even be negative; that is, the concentration of the "adsorbed" component may be lower near the interface than in the bulk. Such situations, however, are rather rare. Important exceptions are in the context of electrical double-layer theory and some polymer solutions. Of more interest are systems in which one or more components of the liquid phase are strongly (and positively) adsorbed at the interface, bringing about a significant lowering of the interfacial tension and, in some cases, a significant change in the nature of the interface altogether. The effects of such strong adsorption are of great practical importance and allow us to manipulate solid–liquid interfaces to our own best advantage. A number of examples will be encountered in later chapters.

When the adsorption of a molecule from solution onto a solid surface is considered, there are several quantitative and qualitative points that are of interest, including (1) the amount of material adsorbed per unit mass or area of solid, (2) the solute concentration required to produce a given surface coverage or degree of adsorption, (3) the solute concentration at which surface saturation occurs, (4) the orientation of the adsorbed molecules relative to the surface and solution, and (5) the effect of adsorption on the properties of the solid relative to the rest of the system. In all the above, it is assumed that such factors as temperature and pressure are held constant.

9.6. QUANTIFICATION OF SOLUTE ADSORPTION

The conventional method for determining the above quantities in a given system is by way of the adsorption isotherm, using the basic concepts already introduced. The basic quantitative equation describing the adsorption of one component of a binary solution onto a solid substrate can be written as

$$\frac{n_0 \Delta x_1}{m} = n_1 \sigma x_2 - n_2 \sigma x_1 \tag{9.33}$$

where n_0 is the total number of moles of solution before adsorption, $x_{1,0}$ the mole fraction of the adsorbing component 2 in solution before adsorption, x_1 and x_2 the mole fractions of components 1 and 2 at adsorption equilibrium ($\Delta x_1 = x_{1,0} - x_1$), m the mass of solid adsorbent present, in grams, and n_1 and n_2 the number of moles of components 1 and 2 adsorbed per gram of solid at equilibrium.

In the case of a dilute solution where the solute (2) is much more strongly adsorbed than the solvent (1), the equation simplifies to

$$n_2 \sigma = \frac{\Delta n_2}{m} = \frac{\Delta C_2 V}{m} \tag{9.34}$$

where $C_{2,0}$ is the molar concentration of 2 before adsorption, C_2 the molar concentration of 2 at equilibrium, $\Delta C_2 = C_{2,0} - C_2$, and V the volume of the liquid phase in liters.

For relatively dilute systems containing surface-active materials (i.e., surfactants; see Chapter 3), the concentration of adsorbed material can be calculated from the known amount of material present before adsorption and that present in solution after adsorption equilibrium has been reached. A wide variety of analytical methods for determining the solution concentration are available, and almost all have been used at one time or other. In surfactant systems, the use of the Gibbs equation and measurements of σ are experimentally simple and straightforward (with proper precautions, of course). The utility of a specific method will depend ultimately on the exact nature of the system involved and the resources available to the investigator.

9.6.1. Adsorption Isotherms in Solid–Liquid Systems

The experimental evaluation of adsorption from solution at solid–liquid interfaces usually involves the measurement of changes in the concentration of the solute in the solution after adsorption has occurred. The usual method for evaluating the adsorption mechanism is through the adsorption isotherm. The important factors to be considered are (1) the nature of the interaction(s) between the adsorbate and the adsorbent, (2) the rate of adsorption, (3) the shape of the adsorption isotherm and the significance of plateaus, points of inflection, and so on, (4) the extent of adsorption (i.e., monolayer or multilayer formation), (5) the interaction of solvent with the solid surface (solvation effects), (6) the orientation of the adsorbed molecules at the interface, and (7) the effect of environmental factors such as temperature, solvent composition, and pH on these factors.

Just as in the case of solid–vapor adsorption, interactions between the adsorbent and adsorbate may fall into two categories: relatively weak and reversible physical adsorption, and stronger and sometimes irreversible specific adsorption or chemisorption. Because of the varied possibilities of adsorption mechanisms, a variety of isotherm shapes have been determined experimentally. Although most are found to fall into two main categories, a satisfying theory of adsorption must encompass the complete range of forms. A general classification of isotherms with various shapes that have been justified theoretically is shown in Figure 9.9. The classification system gives four fundamental isotherm shapes based on the form of the isotherm at low concentrations; the subgroups are then determined by their behavior at higher concentrations.

The L-class (Langmuir) isotherm is the most common and is identified by having its initial region (L1) concave to the concentration axis. As the concentration of adsorbate increases, the isotherm may reach a plateau (L2), followed by a section convex to the concentration axis (L3). If the L3 region attains a second plateau, the region is designated L4. This family of isotherms can be compared to types I, II, and IV for solid–vapor adsorption shown in Figure 9.7.

In the S class of isotherms, the initial slope is convex to the concentration axis (S1) and is often broken by a point of inflection leading to the characteristic S shape (S2). Further concentration increases may then parallel those of the

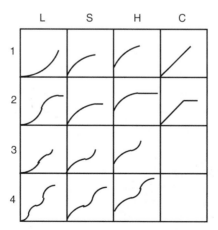

FIGURE 9.9. Various isotherm shapes are encountered in adsorption from solution. The more common shapes are illustrated here.

L class (see types III and V, Fig. 9.7). The H or high-affinity class of isotherm occurs as a result of very strong adsorption at low adsorbate concentrations. The result is that the isotherm appears to have a positive intercept on the ordinate. Higher concentrations lead to similar changes to those found in the L and S classes.

The final type of isotherm is the C class. Such systems exhibit an initial linear portion of the isotherm, indicating a constant partitioning of the adsorbate between the solution and the solid. Such isotherms are not found for homogeneous solid surfaces but occur in systems in which the solid is microporous. This classification system has proved very useful in providing information about the mechanism of adsorption.

9.6.2. Adsorption and Modification of the Solid–Liquid Interface

The adsorption of surface-active materials onto a solid surface from solution is an important process in many situations, including those in which we may want to remove unwanted materials from a system (detergency), change the wetting characteristics of a surface (coating and waterproofing), or stabilize a finely divided solid system in a liquid where stability may otherwise be absent (dispersion stabilization). In these and related applications, the ability of solute, usually a surfactant, to adsorb at the solid–liquid interface with a specific orientation and produce a desired effect is controlled by the chemical natures of the components of the system: the solid, the surfactant, and the solvent. The following discussions will summarize some of the factors related to chemical structures that significantly affect the mechanisms of surfactant adsorption and the orientation with which adsorption occurs.

The generally heterogeneous nature of solid surfaces (in both a physical and a chemical sense) has already been mentioned. As in the case of adsorption at solid–vapor interfaces, the exact nature of the adsorption process will depend to a great extent on the nature of the surface and its potential for interaction with the contacting solvent and dissolved species. Those interactions, as noted above, are normally studied and interpreted in terms of adsorption isotherms.

9.6.3. Adsorption and Nature of the Adsorbent Surface

The nature of the solid surface involved in the adsorption process is a major factor in determining the mode and extent of solute adsorption. When one considers the possible nature of an adsorbent surface, three principal groups readily come to mind: (1) surfaces that are essentially nonpolar and hydrophobic, such as polyethylene; (2) those that are polar but do not possess significant discrete surface charges, such as polyesters and natural fibers such as cotton; and (3) those that possess strongly charged surface sites. Each of these surface types will be discussed, beginning with what is probably the simplest, type 1.

Nonpolar, Hydrophobic Surfaces. Adsorption onto nonpolar surfaces occurs as a result of dispersion force interactions. From aqueous solution, it is obvious that the orientation of the adsorbed molecules will be such that the more hydrophobic portion of the adsorbing solute (the tail in case of surfactant) will be associated with the solid surface with the hydrophilic group directed toward the aqueous phase. In the early stages of adsorption of, for example, a surfactant with a linear hydrocarbon structure, it is likely that the hydrophobe, assuming a more or less linear hydrocarbon structure, will be lying on the surface much like trains or L's (Fig. 9.10*a,b*). As the degree of adsorption increases, however, the molecules will gradually become oriented more perpendicular until, at or near saturation, an approximately close-packed assembly will result (Fig. 9.10*c*). For surfactants, it is generally found that surface saturation is attained at or near the critical micelle concentration (cmc) for the surfactant under the specific conditions of the experiment (see Chapter 15). In many cases the isotherm is continuous, while in others an inflection point may be found. The existence of the inflection point is usually attributed to a relatively sudden change in molecular orientation—from train- or L-shaped to the more perpendicular arrangement. Because the orientation of the adsorbed molecules is with the hydrophilic group directed outward from the solid surface, there will normally be no inclination for the formation of a second adsorbed layer; that is, the process will usually be limited to monolayer formation.

The adsorption of surface-active agents onto nonpolar surfaces from non-aqueous solvents has been much less intensively studied than aqueous systems. Such studies have generally been limited to carbon black and crosslinked polymer dispersions in hydrocarbon solvents. The orientation of the adsorbed molecules in those cases appears to remain more or less parallel to the surface, although the exact details will depend on the history of the carbon surface (e.g., the presence of oxide layers, or charges) and the mode of preparation of the polymer dispersion (i.e., polar or ionic catalyst residues in the surface).

Polar, Uncharged Surfaces. Polar, uncharged surfaces include many of the synthetic polymeric materials such as polyesters, polyamides, and polyacrylates, as well as many natural materials such as cotton and silk. As a result of their surface makeup, the mechanism and extent of adsorption onto such materials is of great potential technological importance, particularly in terms of dyeing processes, waterproofing, and detergency. The mechanism of adsorption onto these surfaces can be much more complex than that of the nonpolar case discussed above, since such factors as orientation will be determined by a balance of several forces.

The potential forces operating at a polar surface include the ever-present dispersion forces, dipolar interactions, and hydrogen bonding and other acid–base interactions (see Chapter 4). The relative balance between the dispersion forces and the uniquely polar interactions is of supreme importance in determining the mode of adsorption. If dispersion forces predominate, adsorption

FIGURE 9.10. When a surfactant adsorbs on a solid surface, it may assume one (or more) of several orientations, depending on the natures of the surface and the molecule. For nonpolar surfaces the options will include (*a*) "trains" lying more or less flat on the surface; (*b*) "L"'s, in which significant portions of the molecule remain adsorbed parallel to the surface; and (*c*) vertical or perpendicular in which the major portion of the chain has no direct contact with the solid surface.

will occur in a manner essentially equivalent to that for the nonpolar surfaces (Fig. 9.10). If, on the other hand, polar interactions dominate, adsorption may occur in a reverse mode; that is, the surfactant molecules will orient in such a way that the hydrophilic head group will be at the solid surface and the hydrophobic group oriented toward the aqueous phase (Fig. 9.11). Obviously, the net result of the two adsorption modes will be drastically different. In aqueous systems, the final orientation will also be affected by the relative strength of solvent–adsorbent and solvent–adsorbate interactions. In marginal cases, the mode of adsorption may be reversed by small, subtle changes in the nature of the solvent (e.g., pH, electrolyte content, presence of a cosolvent.)

Surfaces Having Discrete Electrical Charges. The final class of adsorbent surface is the most complex of the three for several reasons. From the standpoint of the nature of the surface, these materials are capable of undergoing adsorption by all of the previously mentioned mechanisms. Possibly more important, however, is the fact that adsorption involving charge–charge interactions is significantly more sensitive to external conditions such as pH, neutral electrolyte, and the presence of nonsurface-active cosolutes than are the other mechanisms.

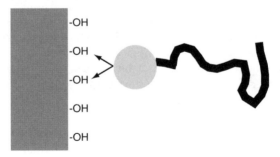

FIGURE 9.11. In adsorption in which the solid surface has a significantly polar character, the surfactant may orient itself with the polar head group toward the surface and the tail toward the solution.

Materials possessing charged surfaces include almost all of the inorganic oxides and salts of technological importance (silica, alumina, titania, etc.), the silver halides, latex polymers containing ionic comonomers, and many natural surfaces such as proteins and cellulosics. It is very important, therefore, to understand the interactions of such surfaces with surfactants or other adsorbates in order to optimize their effects in such applications as paint and pigment dispersions, papermaking, textiles, pharmaceuticals, and biomedical implants.

Because of the large number of possible interactions in systems containing charged surfaces and ionic surfactants or other solutes, it is very important to closely control all the variables in the system. As adsorption proceeds, the dominant mechanism may go from ion exchange through ion bonding to dispersion or hydrophobic interactions (Fig. 9.12). As a result, adsorption isotherms may be much more complex than those for the simpler systems.

Studies of the adsorption from aqueous solution of charged molecules onto surfaces of opposite charge have resulted in the identification of three principle regions of adsorption in which the rates vary due to changes in the mechanism of adsorption. It is generally assumed that such adsorption patterns involve three mechanisms as illustrated in Figure 9.12. In the early stages (region 1), adsorption occurs primarily as a result of ion exchange in which adsorbed

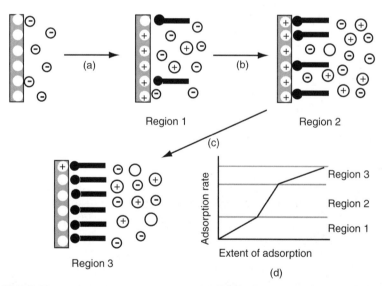

FIGURE 9.12. In the adsorption of an ionic surfactant onto a surface of opposite charge several stages will be encountered: (*a*) ion exchange, in which the native surface counterion will be exchanged 1:1 for surfactant molecules; (*b*) ion pairing, in which excess surfactant molecules are adsorbed (over the number of native ions bound to the surface); (*c*) charge neutralization, in which all accessible surface charges are neutralized by surfactant ions. Each stage of the process will have a characteristic adsorption profile as illustrated schematically (*d*).

counterions are displaced by surfactant molecules. During that stage the electrical characteristics (i.e., the surface charge or surface potential) of the surface may remain essentially unchanged. As adsorption continues, ion pairing may become important (region 2), resulting in a net decrease in surface charge. Such electrical properties as the surface and zeta potentials (see Chapter 5) will tend toward zero during this process. It is often found that in region 2 the rate of adsorption will increase significantly. The observed increase may be due to the cooperative effects of electrostatic attraction and lateral interaction between hydrophobic groups of adsorbed surfactants as packing density increases.

As the adsorption process approaches the level of complete neutralization of the native surface charge by adsorbed solute, the system will go through its zero point of charge (ZPC), where all the surface charges have been paired with adsorbed molecules (region 3). For systems containing surfactant, lateral interactions between adjacent groups may become significant, often leading to the formation of aggregate structures or hemimicelles. If the interaction between adjacent groups is weak (due to short or bulky structures) or if electrostatic repulsion between charged groups cannot be overcome (due to the presence of more than one charge of the same sign or low ionic strength), the enhanced adsorption rate of region 2 may not occur and hemimicelle formation may be absent. An additional result of the onset of dispersion force-dominated adsorption may be the occurrence of charge reversal as adsorption proceeds, which will be covered below.

9.6.4. Environmental Effects on Adsorption

Surfaces possessing charged groups in aqueous solvents are especially sensitive to environmental conditions such as electrolyte content and the pH of the aqueous phase. In the presence of high electrolyte concentrations, the surface of the solid may possess such a high degree of bound counterions that ion exchange is the only mechanism of adsorption available other than dispersion or hydrophobic interactions. Not only will the electrical double layer of the surface be collapsed to a few nanometers' thickness, but attraction between unlike charge groups on the surface and the surfactant, and repulsion between the like charges of the surfactant molecules, will be suppressed. The result is often an almost linear adsorption isotherm, lacking any of the mechanisms described above.

An increase in electrolyte content will generally cause a decrease in adsorption of surfactants onto surfaces of opposite charge and an increase in adsorption of like charged molecules (Fig. 9.13a). An important exception to that rule is the case of ionic surfactants in which the added cation has a specific interaction with the adsorbing solute (i.e., surfactant) that reduces its solubility in the solution leading to enhanced adsorption. A typical example is the addition of polyvalent cations to carboxylic acid soaps.

The presence in the solution of polyvalent cations such as Ca^{2+} or Al^{3+} generally increases the adsorption of anionic surfactants onto like-charged surfaces. Such ions are characteristically tightly bound to a negatively charged surface, effectively neutralizing charge repulsions. They also can serve as an efficient bridging ion by association with both the negative surface and the anionic surfactant head group (Fig. 9.13b).

Adsorption onto solid surfaces having weak acid or basic groups such as proteins, cellulosics, and many polyacrylates can be especially sensitive to variations in solution pH. As the pH of the aqueous phase is reduced, the net charge on the solid surface will tend to become more positive. That is not to say that actual positive charges will necessarily develop; instead, ionization of the weak acid groups (e.g., carboxylic acids) will be suppressed. The net result will be that the surface may become more favorable for the adsorption of surfactants of like charge (e.g., anionic surfactants onto carboxyl surfaces) and less favorable for adsorption of surfactants of opposite charge (Fig. 9.14). For surfaces containing weak basic groups such as amines, the opposite would be true. Thus, lowering the pH will lead to ionization of surface basic groups, increased adsorption of oppositely charged (negative) molecules, and decreased interaction with materials of the same charge.

An increase in temperature usually results in a decrease in the adsorption of ionic surfactants, although the change may be small when compared to those due to pH and electrolyte changes. Nonionic surfactants solubilized by hydrogen bonding, which usually have an inverse temperature–solubility relationship in aqueous solution, generally exhibit the opposite effect. In other words, adsorption will increase as the temperature increases, often having a maximum near the Krafft point of the particular surfactant.

(a)   (b)

FIGURE 9.13. In systems of negatively charged surfaces and anionic surfactants, the presence of polyvalent cations such as Ca^{+2} and Al^{+3}, can produce "bridging" phenomena that result in strong surface–surfactant interactions that would not occur in the presence of monovalent ions such as Na^{+} or K^{+}.

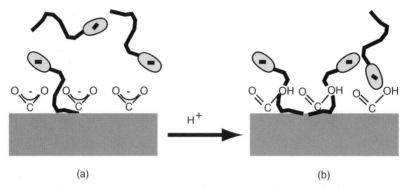

(a)   (b)

FIGURE 9.14. When the charge on a surface resulting from the ionization of a weak acid or base is neutralized, the propensity of a surfactant of like charge to adsorb on the surface is increased: (*a*) with charge, a surfactant of like charge can adsorb only by dispersion interaction, if at all; (*b*) with a neutralized charge, adsorption becomes more favorable, even to the extent of adsorbing through head group interactions.

9.6.5. Effects of Adsorption on the Nature of the Solid Surface

When a surfactant is adsorbed onto a solid surface, the resultant effect on the character of that surface will depend largely upon the dominant mechanism of adsorption. For a highly charged surface, if adsorption is a result of ion exchange, the electrical nature of the surface will not be altered significantly. If, on the other hand, ion pairing becomes important, the potential at the Stern layer will decrease until it is completely neutralized (see Fig. 9.5). In a dispersed system stabilized by electrostatic repulsion, such a reduction in surface potential will result in a loss of stability and eventual coagulation or flocculation of the particles (Chapter 10).

In addition to the electrostatic consequences of specific charge–charge interactions, surfactant adsorption by ion exchange or ion pairing results in the orientation of the molecules with their hydrophobic groups toward the aqueous phase; therefore, the surface becomes hydrophobic and less easily wetted by that phase. Once the solid surface has become hydrophobic, it is possible for adsorption to continue by dispersion force interactions (Fig. 9.15).

(a)   (b)

FIGURE 9.15. The interaction of an ionic surfactant with a surface of opposite charge will lead to charge neutralization (*a*) followed, in many cases, by charge reversal (*b*). Counterions are omitted for clarity.

When that occurs, the charge on the surface will be reversed, acquiring a charge opposite in sign to that of the original surface, because the hydrophilic group will now be oriented toward the aqueous phase. In a system normally wetted by water, the adsorption process reduces the wettability of the solid surface, making its interaction with other less polar phases (e.g., air) more favorable. Industrially, the production of a hydrophobic surface by the adsorption of surfactant lies at the heart of the froth flotation process for mineral ore separation. Because different minerals possess varied surface charge characteristics, leading to differences in adsorption effectiveness and efficiency, it becomes possible to obtain good separation by the proper choice of surfactant type and concentration.

Although surfactant adsorption and its effect on solid surface properties is often discussed in terms of colloidal systems, the same results can be of technological importance for macro surfaces, especially in the control of the wetting or nonwetting properties of materials (in waterproofing), detergency, lubrication (with cutting oils and other lubricants), the control of fluid flow through porous media (crude oil production), and corrosion control. Almost any process or product that involves the interaction of a solid and a liquid phase will be affected by the process of surfactant adsorption; thus the area represents a major segment of the technological application of surfactants. More detail on the effects of surfactant adsorption of wetting phenomena will be given in Chapter 17.

PROBLEMS

9.1. Assume that an aqueous solute adsorbs at the mercury–water interface according to the Langmuir equation:

$$\frac{x}{x_m} = \frac{bc}{1 + bc}$$

where x_m is the maximum possible amount adsorbed and $x/x_m = 0.5$ at $c = 0.2$ M. Neglecting activity coefficient effects, estimate the value of the mercury–solution interfacial tension when $c = 0.1$ M. The limiting molecular area of the solute is 0.2 nm^2. Assume that the temperature is 25°C.

9.2. The adsorption of a vapor on a solid is found to be rapid with a heat of adsorption (ΔH_{ad}) of approximately 50 kcal mol^{-1}. The vapor can be desorbed under vacuum at high temperatures. What kind of adsorption process is probably involved?

9.3. The amount of a vapor adsorbed on a surface is found to increase as the temperature is raised at constant concentration. Which of the following situations is suggested: (a) the process has a high activation energy;

(b) chemisorption is occurring; (c) there is competitive adsorption occurring; (d) all of the above; (e) none of the above.

9.4. In a situation of competitive adsorption of two adsorbates, A and B, from solution A will be more readily adsorbed than B if (a) the molecular weight of A is greater than that of B; (b) A is more soluble than B; (c) in the gas phase A is more readily adsorbed than B; (d) all of the above; (e) none of the above.

9.5. A spreading monolayer of camphor can be used to propel a toy boat through the water. The motion produced by the effect of camphor on the water surface tension at the rear of the boat is a result of (a) a permanent increase in σ; (b) a permanent decrease in σ; (c) a transient increase in σ; (d) a transient decrease in σ.

9.6. A compound is found to adsorb onto a glass surface in such a way that the resulting adsorbed layer may be either hydrophilic or hydrophobic, depending on the concentration of adsorbate, time of adsorption and temperature. In all probability, the process(es) involved is (are) (a) monolayer adsorption; (b) random multilayer adsorption; (c) oriented multilayer adsorption; (d) all of these; (e) none of these.

9.7. Data for an adsorption isotherm for N_2 (v_{ads} = cm^3 N_2 at 25°C and 1 atm) is as follows:

P/P_0	v_{ads}	P/P_0	v_{ads}
0.025	0.036	0.30	0.178
0.05	0.062	0.375	0.208
0.10	0.094	0.45	0.249
0.175	0.129	0.55	0.307
0.25	0.158	0.65	0.405

Plot the data as a linearized BET isotherm and determine the total surface area, assuming that the area occupied by an adsorbed nitrogen molecule is 0.162 nm^2.

9.8. A solution of 1 g of dye in one liter of water was treated with 14 g of activated carbon. The final concentration of the dye in solution after treatment was determined to be 0.3 g L^{-1}. How much activated carbon would be needed to reduce the dye concentration of two liters of 0.5 g L^{-1} of dye to 0.3 g L^{-1}?

9.9. Four (4) grams of activated carbon was added to 2 L of a 1.0 g L^{-1} aqueous dye solution causing the concentration of dissolved dye to drop to 0.60 g L^{-1}. Addition of a further 4 g of carbon dropped the dye concentration to 0.40 g L^{-1}. How much carbon must be added to bring the dye concentration down to 0.20 g L^{-1} if the adsorption follows the

Langmuir isotherm? How much if the Freundlich isotherm applies? Repeat the calculation for a final dye concentration of 0.04 g L^{-1}.

9.10. Explain qualitatively why a minimum in the entropy of adsorption may be expected at or near monolayer coverage.

9.11. Methylene blue dye has an adsorbed area per molecule of 1.75 nm cm^{-2}. A 100 ml aqueous solution of dye has an optical density (absorbance) of 0.65 at a certain wavelength. After the solution is equilibrated with 25 mg of an activated carbon, the supernatant solution is found to have an absorbance of 0.20. Estimate the specific surface area of the carbon.

9.12. When adsorption of solute from solution follows the Langmuir equation, what should be the form of the variation of surface film pressure with solute concentration?

9.13. Using the tabulated data for the adsorption of N_2 on TiO_2 at 75 K, use the BET equation to calculate v_m, C, and specific surface area for the adsorbent. For the system P_o = 572.48 torr, P is in torr, and v is in cm^3 STP per gram.

P	v	P	v	P	v
1.17	600.06	310.2	1547.37	471.2	2694.67
14.00	719.54	341.2	1654.15	477.1	2825.39
45.82	821.77	368.2	1766.89	482.6	2962.94
87.53	934.68	393.3	1890.11	487.3	3107.06
127.7	1045.75	413.0	2018.18	491.1	3241.28
164.4	1146.39	429.4	2144.98	495.0	3370.38
204.7	1254.14	443.5	2279.15	498.6	3499.13
239.0	1343.74	455.2	2418.34	501.8	3628.63
275.0	1441.15	464.0	2561.64		

10 Colloids and Colloidal Stability

In Chapter 1 the importance of the various classes of colloidal systems to modern science and technology was indicated in a general way. Because of the wide variety of colloidal systems one encounters, each having certain unique features that distinguish it from the others, it is convenient to discuss each major classification separately. For that reason, chapters have been devoted to specific systems such as solid dispersions, aerosols, emulsions, foams, lyophilic colloids (i.e., polymer solutions), and association colloids. There is a great deal of overlap in many aspects of the formation, stabilization, and destruction of those systems, and an effort will be made not to repeat more than is necessary. However, for purposes of clarity, some repetition is unavoidable.

The following discussion, while being general in nature, is intended to apply primarily to "classic" colloidal systems—dispersions of small solid particles in a liquid medium (also commonly referred to as "sols"). A minimum of information that gives a general conceptual understanding of the phenomena is presented below. It is intended to serve as a useful lead into more detailed information as needed.

10.1. THE IMPORTANCE OF COLLOIDS AND COLLOIDAL PHENOMENA

The importance of colloids and colloidal phenomena to our modern technological society cannot be overstated, even though it is quite commonly overlooked. An abbreviated list of some important products and processes involving colloids is given in Table 10.1. Considering the limited nature of that compilation, it should be obvious that the principles discussed here are basic to a large variety of practical areas. The listing also omits "natural" colloidal systems, which are of vital importance in biology, medicine, meteorology, and to some extent even cosmology. Unfortunately, experience indicates that few individuals working in those areas have more than a very cursory understanding of the subject and therefore may overlook important concepts related to their specific area of interest. While ignorance may be bliss in some situations, in this case, a little bit of knowledge properly applied may pay practical dividends many times the value of the effort expended to gain that knowledge. Previous chapters have covered the main concepts behind what is generally considered "surface chemistry." We now turn our attention to the basics concepts of

TABLE 10.1. Illustrative Examples of Practical Applications of Colloids and Colloidal Phenomena

Application	Principles Involved
Pharmaceuticals, cosmetics, inks, paints, foods, lubrication, food products, dyestuffs, foams, agriculture chemicals	Formation and stabilization of colloids for end-use products
Photographic products, ceramics, paper coatings, magnetic media, catalysts, chromatographic adsorbents, membrane and latex film, electrophotographic toners	Formation of colloids for use in subsequent manufacturing processes
Wettings of powders, enhanced petroleum recovery detergency, mineral ore flotation, purification by adsorption, electrolytic coatings, industrial crystallization, chemical waste control, electrophotography, lithography	Direct application of colloidal phenomena to processing
Pumping of slurries, coating technology, caking, powder flow, filtration	Handling properties of colloids, rheology, sintering
Water purification, sewage disposal, disperal of aerosols, pollution control, fining of wines and systems beers, radioactive waste disposal, breaking of unwanted emulsions and foams	Destruction of unwanted colloidal phenomena

"colloid chemistry"—what colloids are and how they are formed, stabilized, used, and destroyed.

10.2. COLLOIDS: A WORKING DEFINITION

Any attempt to define the term "colloid" rigorously will usually be found to be unsatisfactory in that it will restrict the range of systems excessively, especially in that systems which operate according to the "rules" of colloidal behavior may be excluded by some rather arbitrary factor, such as size. A working definition may reasonably be arrived at if one understands the "nature of the beast" in the context of the "normal" bulk states of matter: solids, liquids, and gases. We normally have few conceptual problems with respect to those three states since they possess certain characteristics that can readily be associated with each, including, for example, rigidity, fixed volumes under specified conditions, and characteristic phase transformations such as melting, boiling, and sublimation. We also understand a great deal about atoms, molecules, and solutions, and the forces controlling interactions at that level. Yet understanding of the "twilight zone" between bulk phases and the molecular level continues to be a mystery to most.

In homogeneous solutions, there exists a mixture of distinct species that are intermixed or dispersed as individual molecules (where the molecular size of the two materials are comparable in size). However, between pure bulk materials and molecularly dispersed solutions lies a wide variety of important systems in which one phase is dispersed in a second, but in units which are much larger than the molecular unit (e.g., a classical sol) or in which the molecular size of the dispersed material is significantly greater than that of the solvent or continuous phase (a macromolecular or polymer solution). Such systems are generally defined as colloids, although there may be accepted limitations on the unit size of the dispersed phase, beyond which other terminology may be used. The importance of many systems falling into the "colloidal" category has been recognized and recorded for thousands of years, although a reasonable understanding of the forces and concepts involved has lagged behind other areas such as chemistry, physics, biology, and medicine.

To define a colloid more or less adequately one must consider two aspects of the system: structure—how the components of the system are put together or mixed—and size—what the dimensions are of the dispersed units in the system.

10.2.1. Colloid Structure

In general, a colloid is a system consisting of one substance (the dispersed phase: a solid, liquid, or gas) finely divided and distributed evenly (relatively speaking) throughout a second substance (the dispersion medium or continuous phase: a solid, liquid, or gas). Commonly encountered examples of colloids are milk (liquid fat dispersed as fine drops in an aqueous phase), smoke (solid particles dispersed in air), fog (small liquid droplets dispersed in air), paints (small solid particles dispersed in liquid), gels (polymer molecules which, when dissolved in solvent, impart a semisolid structure to the solution), and bone (small particles of calcium phosphate dispersed in a solid matrix of collagen).

The different types of dispersed systems are classified depending on the nature of the dispersed phase and the continuous phase. A solid or liquid dispersed in a gas is termed an "aerosol"; smoke is a common example of the solid-in-air system. A liquid-in-air system is a mist or fog. Mayonnaise is an emulsion in which a liquid (vegetable oil) is dispersed in another liquid (water). Paints and inks are colloidal sols or dispersions and consist of solid particles dispersed in a liquid.

A second class of colloids of particular importance in the general context of surface chemistry, but less familiar to most people, is that of the so-called association colloids. Association colloids consist of aggregates or units of a number (sometimes hundreds or thousands) of molecules that associate in a dynamic and thermodynamically driven process leading to a system that may be simultaneously a molecular solution and a true colloidal system. As we shall see, the formation of association colloids involving specific substances will often depend on various factors such as concentration, temperature, solvent composition, and specific chemical structure. Many biological systems, including cell membrane formation, certain digestive processes, and blood transport phenomena, involve various forms of associated colloidal structures. This class of colloidal materials will be discussed in more detail in Chapter 15.

In addition to the colloids composed of insoluble or immiscible components, there are the lyophilic colloids which are in reality solutions, but in which the solute molecules (i.e., polymers) are much larger than those of the solvent. Lyophilic colloids are somewhat unique in that they have been able to "cross over" into another major area of science—polymer science—and thereby gain a great deal more general attention than more typical colloidal systems.

A fourth class of colloids often encountered is that of the network colloids. Such systems are difficult to define exactly because they consist of two interpenetrating networks, which make it hard to specify exactly which is the dispersed phase and which is the continuous phase. Classic examples of network colloids would be porous glass (air–glass), opal glass (solid–solid dispersion), and many gels. Practical examples of many of the colloids mentioned above are given in Table 10.2.

The examples of colloids listed in Table 10.2 may be considered "simple" colloids because they involve one fairly distinct type of dispersed and continuous phase. In practice, many colloidal systems are much more complex in that they contain a variety of colloidal types, such as a sol, an emulsion (or multiple emulsions), an association colloid, macromolecular species, plus the continuous phase. Such systems are often referred to as complex or multiple colloids. As we shall see, even the simplest colloids can be quite complex in their characteristics. It should be easy to understand, then, why the difficulty of understanding a multiple colloid increases dramatically with the number of components present.

TABLE 10.2. Examples of Commonly Encountered Colloidal Systems

System	Type	Dispersed Phase	Continuous Phase
Fog, mist	Liquid aerosol	Liquid	Gas
Smoke	Solid aerosol	Solid	Gas
Shave cream	Foam	Gas	Liquid
Styrofoam	Solid foam	Gas	Solid
Milk	Emulsion	Liquid (fat)	Liquid (water)
Butter	Emulsion	Liquid (water)	Solid (fat)
Paint	Dispersion	Solid	Liquid
Opal	Dispersion	Solid	Solid
Jello	Gel	Macromolecules	Liquid
Liquid soaps and detergents	Micellar solution	Micelles of detergent molecules	Liquid

10.2.2. Colloid Size

Working with the above general concept of what is meant by a "colloid" in structural terms, it is still necessary to specify what is meant in terms of the size of the dispersed units. The general definition refers to the dispersed phase as being "finely divided," but what exactly do we mean by that phrase? Experience over many years has shown that special "colloidal" properties are usually exhibited by systems in which the size of the dispersed phase falls in the range of 1–1000 nm, although those limits are far from rigid. It would be tempting, then, to limit the term to systems that fall into that size range. In practice, however, many systems with dimensions beyond that range, in particular most emulsions, paints, and aerosols, must also be included under the colloid umbrella, since their characteristics allow no other realistic option. Other colloidal systems, such as fibers, clays, and thin films, may "qualify" as colloids because one or two dimensions fall into the designated range, and their properties adhere to the "rules" of colloidal behavior. That concept is illustrated schematically in Figure 10.1. Ultimately, the most useful definition to use is that if it looks like a colloid and acts like a colloid, it is a colloid, regardless of other more restrictive limitations.

10.2.3. Some Points of Nomenclature

In addition to the general definitions given above, there are a number of additional terms related to colloids—their nature, characteristics, and stability—which should be clearly defined (where possible) before beginning the discussion of the topic. Some of the more important include:

> *Coagulum:* an aggregate of colloidal particles having a relatively tight, dense structure formed as a result of the inability of the colloidal system to

FIGURE 10.1. A colloid is basically defined by its dimensions. While one may set a size limit of, for example, 0.01 mm as an arbitrary upper limit to what may be called a colloid, the reality is that many systems with larger dimensions are considered "classic" colloids (e.g., clays) because at least one dimension falls into the limiting size range. To be considered a colloid, then, a system may have one dimension in the range as in a flat plate (*a*), two dimensions as in a cylinder (*b*), or three dimensions as in a drop or particle (*c*).

maintain its dispersed state. Such aggregates are normally formed irreversibly; that is, they cannot be returned to the colloidal state without significant input of work.

Coagulation: the process of forming coagulum.

Creaming: the separation of coagulum or flocs from the continuous phase, where the aggregate is less dense than that phase. (Slightly different definitions of creaming and sedimentation are presented in Chapter 11 for application in the context of emulsions.)

Floc: an aggregate of individual colloidal particles related to a coagulum but generally having a rather loose, open structure. Flocs may sometimes be formed reversibly and returned to the dispersed state with minimal energy input.

Flocculation: the process of forming flocs.

Monodisperse: having all particles in the colloidal system of approximately the same size (i.e., a narrow size distribution).

Polydisperse: having a rather broad range of particle sizes.

Sedimentation: as in creaming except that the aggregates are more dense than the liquid and settle to the bottom.

10.3. MECHANISMS OF COLLOID FORMATION

Having defined "colloids" (approximately, at least) another question of interest is how one may prepare systems of particles or drops of the proper size and stability. Since colloids represent a range of unit sizes intermediate between molecules and macroscopic bulk phases, it seems reasonable to expect that the problem can be attacked from one of two directions: by breaking down large pieces to the size required, known generally as comminution or by starting with a molecular dispersion and build up the size by aggregation, that is, by condensation. Special situations such as the formation of association colloids and lyophilic colloids will be addressed in other chapters.

10.3.1. Comminution or Dispersion Methods

It was shown in Chapter 2 that when a bulk material is divided into two pieces and separated to infinite distance there is a characteristic (theoretical) energy change per unit area that is termed the "specific surface free energy." As the subdivision of the sample is continued, the total free energy change will be the product of the surface energy (or surface tension) of the material, σ, and the total new surface area produced by the process. If carried to its ultimate, that is, reduction to the molecular level, the theoretical work required would be equal to the energy of evaporation, sublimation, or dissolution, depending on the situation. In practice, of course, a great deal more than the theoretical energy will be required due to the various irreversible, heat-generating processes that naturally accompany any grinding or pulverizing process.

It is reasonable to assume that the work required to reduce a given material to colloidal size varies directly with the surface energy of the material; higher-surface-energy materials require more work input. In addition, the natural tendency of subdivided particles is to reduce the total surface area by some aggregation process, especially if produced in a vacuum or inert atmosphere. Such processes of flocculation, coagulation, or sintering are very important and represent serious technological problems. As will be seen below, the attractive interaction between particles can be reduced by the introduction of an intervening medium, usually a liquid. The liquid medium will have two positive effects on the process: (1) it will reduce the surface energy of the system by adsorption on the new surface and (2) it will usually reduce the van der Waals attraction between the particles by "averaging" its Hamaker constant with that of the particulate material (see text below). For those reasons (plus a reduction in environmental problems such as dust formation), most industrial comminution processes are carried out in the presence of a liquid.

While the presence of the liquid normally facilitates the comminution process, the resulting dispersion still may not exhibit the stability necessary to make the process viable. That is, the dispersed particles may begin to flocculate or coagulate rapidly once the comminution process is halted (Fig. 10.2). The solution for the second problem is normally the addition of new components (surfactant, polymer, small particles, etc.) that adsorb at the solid–liquid interface and provide an electrostatic or steric barrier that retards or prevents "sticky" collisions between particles, thereby making the dispersion more stable. In the case of particulate colloids such additives are termed dispersing aids or agents. The mechanisms of action of such materials will be discussed in more detail in the context of specific colloidal phenomena.

The comminution of liquid phases is a special case of the above and is generally referred to as "emulsification." Because of the nature of liquid systems, emulsification has a number of additional variations not generally available for the formation of dispersions. These include spontaneous emulsification, electroemulsification, and spontaneous microemulsion formation.

Some important processes for the formation of sols involve first the formation of an emulsion or a liquid aerosol. In suspension or dispersion polymerization, a monomer or monomer mixture is emulsified to a drop size approximately the same as that of the final desired particle. Polymerization is then initiated using an initiator soluble in the monomer, so that chain growth occurs within each individual drop. The result (with luck) is a dispersion of polymer particles with the same average size as the original monomer emulsion. Normally, some type of stabilizer system is employed in the emulsification stage—a surfactant or very small particles of some material such as silica.

Another process involves the formation of a liquid aerosol or mist of the precursor to the solid. This is sprayed into a vapor atmosphere containing a

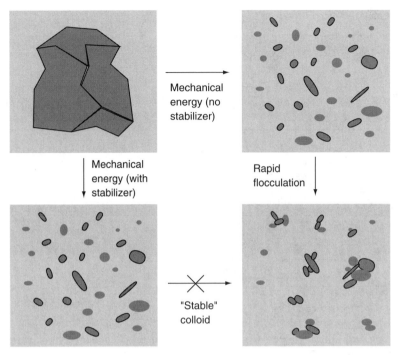

FIGURE 10.2. In the preparation of colloids by comminution, it is usually necessary to include a surfactant or polymeric stabilizer to inhibit the rapid flocculation of the newly formed particles.

reagent that induces the reaction leading to solid formation. In the case of polymer particles, for example, monomer drops are sprayed into a vapor containing an initiator producing spherical polymer particles. Particles of some inorganic materials can be prepared similarly by, for example, spraying a mist of a reactive precursor [e.g., titanium(IV) ethoxide] into water vapor to produce spherical particles of the corresponding oxide.

10.3.2. Condensation Methods

Approaching the formation of colloids from the other end of the size range involves one of several growth mechanisms. Such processes are commonly employed for the production of dispersions and aerosols, and less commonly in the production of emulsions. Typical examples of important condensation processes include fog formation (both water and chemical), silver halide "emulsions" (really dispersions) for use in photographic products, crystallization processes, colloidal silica, latex polymers, etc.

In emulsion polymerization, a monomer or mixture of monomers is emulsified in a liquid phase (usually water) in the presence (except under very special

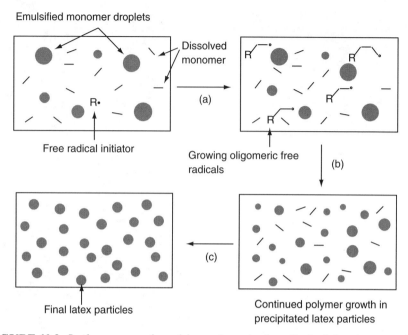

FIGURE 10.3. In the process of emulsion polymerization, the incipient latex particle begins as a free-radical-initiated dimer or oligomer in solution (*a*). As polymerization proceeds, the growing chain precipitates and continues to grow, fed by new monomer taken from the reservoir of emulsified material (*b*). Polymerization continues until all available monomer is consumed (*c*).

circumstances) of a surfactant, polymer or other stabilizer and a soluble free radical initiator. Particle formation and growth occur initially in the solution phase and then continues in precipitated particles (Fig. 10.3). Under proper conditions, particle size and dispersity can be closely controlled to meet specific needs. More details of some of those process are introduced in Chapters 11 and 13.

10.4. THE "ROOTS" OF COLLOIDAL BEHAVIOR

Earlier chapters have emphasized the fact that atoms or molecules at a surface have properties distinct from those located in a bulk phase or in solution. In normal bulk-phase chemistry and physics, the relative number of molecules located at a surface or interface is so small compared to that in the bulk that any surface effects will be easily overshadowed by bulk effects. However, as the bulk phase is subdivided into finer and finer particles, the relative ratio of surface to bulk molecules will increase until the effect of specific surface properties will begin to become significant, or even dominate the characteristics of the system.

For example, the amount of surface exposed by a given mass of material is inversely proportional to the linear dimension of the divided units. For unit mass of material, the "specific surface area" is given by $6/\rho d$, where ρ is the density and d particle dimension (diameter for a sphere or edge length of a cube). If the units (atoms or molecules) in the material have a linear dimension x and a molecular volume of x^3 ($\frac{4}{3}\pi x^3$ for a sphere), the fraction of molecules at the surface will be approximately equal to $6(x/d)$. Thus, for a material with $x = 0.4$ nm a cube with $d = 1$ cm will have roughly 0.000025% of its molecules at the surface, or about 1 in 5 million. If the cube is divided into small cubes with 1000-nm edges, the number of surface molecules increases to about one in 400 ($\approx 0.25\%$). If further divided to a dimension $d = 10$ nm, the system now contains approximately 25% of its units on the surface. Because of the higher energy, and therefore higher reactivity in many cases, of surface molecules, the observable effects of the surface begin to be significant as the colloidal size range is approached and usually becomes dominant very rapidly as the smaller end of the spectrum is approached.

Because the lower limit of the colloidal range is just larger than the size of some molecules and solvated species it is difficult to determine exactly where the distinction between "surface" and "bulk" ends and a molecularly dispersed system begins. For macromolecular systems, of course, the molecular size is such that even a "molecular dispersion" or solution easily falls into the size range of colloids. For that reason, primarily, such systems are referred to as "lyophilic colloids," even though the properties of such systems are governed for the most part by phenomena distinct from the "classic" surface interactions considered in lyophobic colloids. It is no trivial matter, therefore, to decide just where surface effects end and the characteristics of the individual free, solvated units begin.

10.5. GROUND RULES FOR COLLOIDAL STABILITY

Keeping in mind what we mean by "colloid," the next step is to define what is meant by a "stable" colloid, or "stability" in general. Basic thermodynamics tells us that any system, left to its own devices, will spontaneously tend to alter its condition (chemical and/or physical) in an effort to attain the condition of minimum total free energy. It says nothing about how fast such a transformation will occur—that is the province of kinetics—nor necessarily whether the system will "stop along the way" in the form of some metastable configuration. We only know that given a viable mechanism, it will occur.

The idea of such changes in energy can be illustrated by analogy to a game of golf (Figure 10.4). The free energy of a golf ball lying on a flat putting green (in this case its gravitational potential energy) relative to the bottom of the cup is given by

FIGURE 10.4. In colloids, as in golf, the system (the player) is seeking a minimum in energy (point C). While thermodynamics may be the ultimate controlling factor (the difference in height, h), inherently unstable systems (points A and B) may appear stable (to the golfers dismay) as a result of kinetic or other factors that prevent the system from readily reaching its final goal.

$$\Delta G_g = mg\ \Delta h \qquad (10.1)$$

where m is the mass of the ball, g is the gravitational constant, and Δh is the distance of the green surface above the bottom of the cup. Let us assume, in order to enhance the analogy, that the "lip" of the cup in the green is slightly higher than the level of the green. In position A the ball has a higher free energy than in position C; however, there is no available mechanism for it to reduce that energy spontaneously; it is metastable. When the player strikes the ball (assuming a certain degree of skill), it will roll toward the cup and up the lip. Going up the lip will increase the energy of the ball, but if it is struck with the proper force, the excess kinetic energy will allow it to go over the lip and fall into the state of lower energy (for the ball and the player); it will be stable relative to the cup. If not struck with sufficient force the ball will not be able to surmount the lip and it will roll back down to some energy level still higher than that in the cup but lower than that at the top of the lip. Under some circumstances, the ball may stop exactly at the top of the lip (position B), in which case it will have an even higher energy than before, as will the player, in all likelihood. In such a situation, which may be viewed as a reactive transition state, it will require only a very small force to cause the ball to roll in one direction back onto the green or the other into the cup.

The same general scheme is used to describe the situation in a chemical or physicochemical system in the form of a reaction coordinate. In that case, the height of the lip above the level of the green corresponds to the activation energy for the system which must be overcome in order for a given reaction or transformation to occur. If elements of a physicochemical system have sufficient energy (kinetic, electronic, vibrational, etc.) to overcome the barrier, reaction will occur. If not, the system will remain in the metastable state. In golf, of course, there is only one ball involved (we assume!) while chemical systems may involve on the order of 10^{13} (for typical "model" colloidal systems) or more. In such complicated systems, where there exists some statistical distribution of energies, it may be that a certain number of atoms or molecules may overcome the barrier, while the majority cannot. Likewise, some may have sufficient energy to pass in the reverse direction (like a golf ball that falls into the cup with such force that it bounces back out again). Such a distribution of energies leads to the establishment of an equilibrium in which individual units are constantly changing their energy "state," but in which the overall situation is constant. If the reverse reaction requires an energy that is essentially unavailable (i.e., the barrier is very high), the system may slowly "bleed" off until all of the units are in the stable state. Such a process, however, may require such a long time that it is imperceptible, or insignificant in the time frame of the observations.

In terms of colloids, we can say the following: interfacial energy considerations dictate that the "position" of lowest energy for a given system (assuming a positive interfacial free energy) will be that in which there is a minimum in the interfacial area of contact between phases. Put another way, in the absence of other factors, colloids should be unstable and rapidly revert to a state of complete phase separation. However, Nature has designed things in such a way that we (or she) can impose barriers of various types between metastable and stable states so that useful (and vital) colloidal systems can be made to exist and persist for enough time so that they can serve a useful function (like make up a significant part of our biological systems!)

10.5.1. A Problem of Semantics

At this point a problem of semantics begins to rear its ugly head, specifically in the operational use of the terms "stable," "unstable," and "metastable." The definitions given above refer to energetic states, without regard to time (i.e., kinetics). For practical purposes, however, it is often inconvenient and confusing to use the term "metastable," since its meaning can be somewhat ambiguous. As a general practice one commonly describes a colloid as "stable" if it remains in the energetically metastable state for some arbitrary or functionally determined length of time. It would be termed "unstable" if the system begins to lose its colloidal properties (e.g., its degree of dispersion or size) before that predetermined time has passed.

Such designations of stable and unstable colloids are very relative and must be made in the context of the application in question. It may be, for example, that a colloid that maintains its characteristics for two days would be considered stable in one application, while another would require that a minimum of two years pass without change. Obviously, then, one must be careful when discussing colloidal stability and instability. From this point on in the discussion, unless otherwise indicated, the kinetic (rather than energetic) "definition" of stability will be employed in its most general sense, it being assumed that all (or almost all) colloids are in reality metastable systems. Also, it must be kept in mind that stability in the present context is used in terms of lyophobic colloids, since lyophilic and association colloids will be inherently stable unless perturbed.

10.5.2. Mechanisms of Stabilization

Clearly, from what has been said so far, colloids must be considered to be metastable, in that surface forces "demand" that a state of fine dispersion represents a high-energy situation. The fact that colloids exist and that some samples have been kept for over 100 years tells us that energy barriers can be imposed which will make the transition from the metastable to the stable energy state difficult or impossible. On the other hand, if a colloid is metastable, and, for example, unwanted, changes in the system which lower existing energy barriers to a critical level will allow the system to spontaneously pass from metastable to stable.

In the analogy of the golf ball, the energy necessary for the transformation of the energy state of the ball was supplied by the player putting. In colloids, that energy is normally supplied by random collisions between the colloidal particles and molecules of the dispersion medium—that is, by Brownian motion. The average translational energy imparted to a colloidal particle due to such a mechanism is $\frac{3}{2}kT$ per particle, where the terms k and T have their usual significance. At room temperature (298 K), then, the energy of an average particle will be on the order of 5×10^{-21} J. For two colliding particles, the energy involved in the collision will be about 10^{-20} J. That will be for average particles; however, the actual energy of a given particle in a given collision may be much smaller or much larger, since the distribution of energies is in accordance with the Maxwell–Boltzman distribution.

If there exists an energy barrier between particles of some multiple, n, of kT, the probability of collisions with sufficient energy to overcome that barrier becomes smaller; the larger the value of n, the smaller is the probability of the two particles being able to make contact. Operationally, a value of $n = 10$ is usually cited as representing a condition in which a colloid can be considered stable. (That value did not, however, come down from the mountain cast in stone, so that for a specific situation, other conditions may prevail!) Conversely, in a stable (but perhaps unwanted) colloid, if the barrier can be reduced to some value of n closer to one, instability may be induced. Several mechanisms exist for lowering the barrier, including changes in temperature, which may or may not be significant depending on the nature of the barrier (see text below); changes in solvent properties of the continuous phase; pressure changes; or changes in electrolyte content.

10.5.3. A Review of Basic Intermolecular Forces

According to the theories introduced in Chapter 4, the force of attraction between two molecules as a function of the distance of separation is given by

$$F_{att} = -Ar^{-7} \tag{10.2}$$

where A is a constant defined by eq. 4.31. The work required to separate two molecules from a distance d to infinity will be

$$\Delta W = -\int_r^\infty F_{att} dr = A\int_r^\infty (r^{-7})dr = 6Ar^{-6} = A'd^{-6} \tag{10.3}$$

where $A = A'/6$.

Employing the convention that attractive energy is negative, at infinite separation the energy of the system will be zero, so that the free energy of attraction at distance r will be

$$\Delta G_{att} = -\Delta W = -A'r^{-6} \tag{10.4}$$

The value of A was previously defined by Equation (4.31) as

$$A = \tfrac{3}{4} h\nu\alpha^2 \tag{10.5}$$

For two dissimilar materials, 1 and 2, the value of A is given by

$$A_{12} = \tfrac{3}{4} h \left(\frac{\nu_1 \nu_2}{\nu_1 + \nu_2} \right) \alpha_1\alpha_2 \tag{10.6}$$

According to Equation (10.4), the attractive forces between two molecules increases (i.e., the energy becomes more negative) continuously as their distance of separation decreases. The relationship is shown graphically in Figure 10.5, curve 1. At some point in their approach, the electron clouds of the two

FIGURE 10.5. A "typical" interaction energy curve for two interacting units (curve 3) is the sum of the attractive potential (curve 1) and the repulsive contribution (curve 2).

molecules must begin to interact and ultimately overlap. If covalent bonding is not possible, that gives rise to a repulsion (the Born repulsion term) and consequent increase in free energy, which becomes effectively infinite when interpenetration occurs (Fig. 10.3, curve 2). As a good approximation, the Born repulsion term has the form

$$\Delta G_{rep} = B' \, d^{-12} \tag{10.7}$$

The total free energy will be the sum of the attractive and repulsive terms

$$\Delta G = \Delta G_{rep} + \Delta G_{att} = B' \, d^{-12} - A' d^{-6} \tag{10.8}$$

which is usually known as the Lennard–Jones 6–12 potential. The resulting energy relationship is shown as curve 3, Figure 10.5.

10.5.4. Fundamental Interparticle Forces

Given the relationships for interactions between molecules described above, how can we convert that information into interactions between large groups of molecules, specifically, colloidal particles? Mathematically, the simplest situation to analyze is that involving two hard, flat, nonpolar, effectively infinite surfaces separated by a distance H in a vacuum. Hamaker showed that the free energy of attraction per unit area in such a case is given by

$$\Delta G_{att} = \frac{-A_H}{12\pi H^2} \tag{10.9}$$

where A_H is the Hamaker constant. The value of A_H is related to A' by

$$A_H = \tfrac{3}{4} h\nu\alpha^2\pi^2 q^2 = A'\pi^2 q^2 \tag{10.10}$$

where q is the number of atoms or molecules in a unit volume of the phase. For two identical spheres of radius a, where $H/a \ll 1$, a similar type of approximate equation is

$$\Delta G_{att} = \left(\frac{-A_H a}{12H}\right)\left[1 + \frac{\tfrac{3}{4}H}{a} + \text{higher terms}\right] \tag{10.11}$$

In most practical instances, it is safe to neglect all of the higher terms.

A comparison of Equations (10.4) and (10.10) shows that the free energy of attraction between two surfaces falls off much more slowly than that between individual molecules. This extended range of bulk interactions plays an important role in determining the properties of systems involving surfaces and interfaces.

Due to the fact that interactions between surfaces fall off much more slowly with distance than those for individual atoms or molecules, a significant complicating factor enters into the quantum mechanical evaluation of the attractive forces. The quantum-mechanical effects leading to the London–van der Waals interactions occur close to the speed of light, yet even at the short distances involved in colloids, relativistic effects can be significant.

At close distances, the interactions resulting from fluctuating dipoles is effectively instantaneous. However, as the distance between interacting units increases, the time required for the electromagnetic "signal" of one unit to travel to its neighbor (at the speed of light), polarize the local electron cloud, and receive the return "signal" of said polarization, is long relative to the "lifetime" of the original dipole. The result is that over "long" distances, the net attractive interaction is reduced from that which would be expected otherwise (i.e., it is retarded). As mentioned in Chapter 4, this effect is termed the "retardation effect" and can be generally ignored for atoms and molecules. For larger units such as colloidal particles, however, which interact over greater distances, the retardation effect becomes significant and causes the attractive interactions to fall off faster than they would otherwise (H^{-3} vs. H^{-2}). While the retardation effect is important in quantitative theoretical discussions of surface interactions, from a practical standpoint, it is usually insignificant compared to other factors.

When both attractive and repulsive terms are taken into account, the interaction curve for particles resembles that shown in Figure 10.3, curve 3. In terms of colloidal stability, the key element in such a curve is the height of the so-called primary maximum indicated as ΔG_{max} on the curve. Later we will see from whence comes that energy maximum.

One must keep in mind that the preceding discussion was couched in terms of interactions in a vacuum or other inert environment, which is not a very practical situation for most applications. In order to understand "real" colloidal systems, one must take into consideration the effects of an intervening medium, the continuous phase, on the above interactions.

10.5.5. Attractive Interactions in Nonvacuum Media

The equations for particle interactions given above were derived for the situation in which the interacting units were separated by a vacuum. "Real life" dictates that in all except a few situations, interacting units are separated by some medium that itself contains atoms or molecules that will impose their own effects on the system as a whole. How will the relevant equations be modified by the presence of the intervening medium?

Surfaces interacting through an intervening fluid medium will experience a reduced mutual attraction due to the presence of the units of the third component. The calculation of interactions through a vacuum involve certain simplifying assumptions; therefore, it should not be surprising to find that models for three-component systems (where each particle is considered a

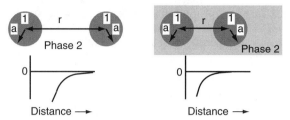

FIGURE 10.6. For two particles of phase 1 interacting in a medium of phase 2, as the nature of phase 2 more closely resembles that of phase 1, the attractive van der Waals interactions between the two particles is reduced, as is the distance over which they act. If the Hamaker constant of phase 2 becomes equal to that of phase 1, the system will be thermodynamically stable.

component) are even more theoretically complex. Although a number of elegant approaches to the theoretical problem have been developed over the years, for most purposes a simple approximation of a composite Hamaker constant is found to be sufficient. Consider, for example, two particles of material 1 dispersed in medium 2 (Fig. 10.6). The effective Hamaker constant (A_H^{eff}) for the two particles can be approximated by

$$A_H^{eff} = \left[A\,{}^{1/2}_{H(10)} - A\,{}^{1/2}_{H(20)} \right]^2 \tag{10.12}$$

where $A_{H(10)}$ is the Hamaker constant for component 1 in a vacuum, and the same for $A_{H(20)}$. An important result of the relationship in Equation (10.12) is that as the vacuum Hamaker constants for 1 and 2 become closer in value, A_H^{eff} ends toward zero, and the free energy of attraction between the two particles tends toward zero. Thus, the force of attraction is decreased and any repulsive term necessary to maintain an adequate value of ΔG_{max} is also reduced. As we will see, such a reduction in the attractive forces due to an intervening medium gives one a handle on ways to manipulate the stability of colloids. Since Equation (10.12) involves the square of the difference between the square roots of the Hamaker constants for components 1 and 2, the same effect applies for surfaces of component 2 separated by a continuous medium of 1. The form of the interaction curve for the above situation will be the same as that for the vacuum case, although the shape and values will differ because of the different value of the effective Hamaker constant.

10.6. SOURCES OF COLLOIDAL STABILITY

Knowing where the attractive interactions between colloidal particles stem from, we must now address the question of where the free energy maximum comes from, which gives a system what stability it may have. In general there are two practical mechanisms for stabilizing lyophobic colloids—electrostatic repulsion between electrical double layers and steric or entropic stabilization. The stabilizing effect of two phases having the same Hamaker constant already mentioned, while theoretically interesting, is not very practical. The earliest recognized, and in some ways the most easily understood, is that arising from electrostatic interactions, so we will begin the discussion in that area.

10.6.1. Charged Surfaces and the Electrical Double Layer

Chapter 5 presented a capsule summary of the primary sources of electrical charges at surfaces and an introduction to the energetics of such systems, including the concept of the electrical double layer (EDL). We now turn our attention to the question of how such interactions tend to provide stability to a colloidal system. If we first consider the situation of two isolated, similarly charged particles (or flat surfaces, for that matter) with their associated electrical double layers, it is relatively easy to understand the basic concepts involved in the electrostatic stabilization of colloids. In the context of (kinetic) stability, one can say that a system is stable so long as the individual particles maintain their identities—that is, as long as flocculation and/or coagulation do not occur. In order for aggregation to occur, of course, two particles must collide, and do so with sufficient force that the collision will be effective or "sticky." There are therefore two primary criteria which must be considered in the discussion of colloid stability: the number or frequency of particle collisions and the effectiveness of those collisions.

With reference to Figure 10.7, as the two charged particles approach, the two charge clouds or electrical double layers will begin to interact. Since the EDLs are of the same sign, their interaction will be repulsive, leading to an increase in the electrical potential between the particles. In simple terms, the EDL can be viewed as a spring located between the two particles. As they approach, the spring is compressed and develops a force pushing the particles apart. The magnitude of the repelling force will depend on the force constant of the spring (i.e., the magnitude of the EDL). The larger the constant, the greater the resistance to mutual approach and the more difficult a sticky collision leading to flocculation. The interaction of the respective electrical double layers, therefore, represents an energy barrier in the total interaction curve.

As the two particles approach, there will be two (at least) types of interaction: the repulsive interaction just described and the relentless van der Waals attractive interactions, which make most colloids inherently unstable. The total interaction energy for the system under consideration will be the sum of the two energies

$$\Delta G_{total} = -\Delta G_{att} + \Delta G_{rep} \tag{10.13}$$

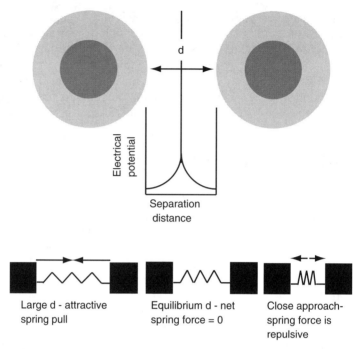

FIGURE 10.7. The interactions between two colloidal particles with electrical double layers can be visualized as two blocks connected by a spring. At large distances of separation the spring is stretched and applies a net force pulling the blocks together (attractive van der Waals interactions). At close approach, the spring is compressed producing a net repulsive force pushing the blocks apart (electrostatic repulsion). At some intermediate distance, the forces will be in equilibrium.

A general expression for the repulsive interaction between the electrical double layers around two spherical particles is quite complex and does not warrant discussion here. A simple and relatively good approximate equation derived by Reerink and Overbeek is

$$\Delta G_{rep} = \frac{Bek^2T^2a\gamma^2}{z^2} \exp{(-\kappa H)} \qquad (10.14)$$

where H is the distance between spheres of radius a, B is a constant equal to $3.93 \times 10^{39} A^{-2}\gamma^{-2}$, z is the charge on the counterion, e in the unit electrical charge, and

$$\gamma = \frac{\exp{(ze\psi_s/2kT)} - 1}{\exp{(ze\psi_s/2kT)} + 1} \qquad (10.15)$$

As pointed out in Chapter 5, the effective electrical potential of importance will be that at the Stern layer ψ_s rather than that actually at the surface, ψ_o.

The attractive interaction is given by Equation (10.11) so that the total interaction will be

$$\Delta G^{\text{total}} = \left[\left(\frac{Bek^2T^2a\gamma^2}{z^2} \right) \exp\,(-\kappa H) \right] - \left[\frac{A_H}{(12\pi H^2)} \right] \qquad (10.16)$$

Curves illustrating the individual and summed interactions for various electrolyte concentrations are shown in Figure 10.6. It should be fairly clear that the key element in determining the height of the energy barrier imposed by the electrical double layer is the concentration and valence of electrolyte in the system. An increase in the electrolyte concentration reduces the repulsive electrostatic interaction, reducing the energy barrier and facilitating effective particle collisions—the system is less stable. A good approximation to the point at which the system will begin to undergo rapid coagulation (indicating a loss of stability) is that at which $\Delta G_{\text{total}} = 0$ and $d\,\Delta G_{\text{total}}/dH = 0$.

10.6.2. Some Complicating Factors

In general, the analysis according to Equation (10.16) overestimates the attractive interactions due to van der Waals forces, because it does not take into consideration the retardation effect mentioned above. The total energy of interaction has some interesting characteristics that warrant pointing out. For example, the repulsive potential energy function is an exponential function of the distance of separation of the particles H, usually a distance of the same order of magnitude as the thickness of the electrical double layer, while the attractive potential decreases as the inverse square of H. As a result, the total interaction will be controlled primarily by the attractive potential at small and large distances, while the double-layer repulsion potential may dominate at intermediate distances, depending on the relative values of the two.

In summary, then, if the total potential-energy maximum is large compared to the average thermal energy of the particles, say 10 kT, then the system should be stable; if not, the system will flocculate or coagulate. The height of the barrier will be determined by the surface potential at the Stern layer ψ_s and the thickness of the double layer, $1/\kappa$.

10.7. STERIC OR ENTHALPIC STABILIZATION

The mechanism of stabilization discussed above refers only to those systems in which the dispersed particles carry a surface electrical charge, so that the interaction of their respective double layers provides the necessary energy barrier for kinetic stability. Another stabilizing mechanism has been known for centuries but has until recently been much less thoroughly studied and understood. That mechanism involves the presence of a lyophilic colloid, which adsorbs onto the particle surface and provides what is termed steric or entropic stabilization. (While there exists some controversy as to whether steric and entropic stabilization are the same phenomenon, the two terms will be used interchangeably in the following discussion.) India ink, carbon particles dispersed in an aqueous solution of a natural gum, is a good example of an old, well-known colloid stabilized (at least partially) in that way. Many other instances in which workable (i.e., stable) colloids can be prepared with the help of "protective action" provided by added lyophilic colloids (protective colloids) are known.

It has long been recognized that steric "protective agents" need not carry an electrical charge in order to be effective. However, they do have certain requirements, such as (generally) a dual chemical nature with respect to their solubility characteristics and a relatively high molecular weight. Some more details of the necessary characteristics are given below, but in general one can say that a steric stabilizer must have one portion of its molecule that exhibits relatively low solubility in the dispersion medium and/or a high tendency to adsorb onto the particle surface. The net result must be the formation of a relatively thick adsorbed layer that can impose a barrier to close particle approach, which will, of course, improve the stability of the colloid. Since most effective steric stabilizers are macromolecules, the following discussion tends to concentrate on the specific aspects related to their performance. More general information related to the adsorption of polymers at interfaces is provided in Chapter 14.

10.7.1. The Mechanism of Steric Stabilization

Referring to Figure 10.7, if two colloidal particles have an adsorbed layer of a lyophilic polymer, as they approach each other, those layers must begin to interpenetrate. Such interpenetration can have two effects: an osmotic effect due to an increase in the local concentration of the adsorbed species between the two particles, and an entropic or volume restriction effect because the interacting species begin to lose certain degrees of freedom due to crowding. In both cases, the local system will experience a decrease in entropy, which will, of course, be unfavorable, while the osmotic effect may be accompanied by an unfavorable enthalpic effect due to desolvation of the more closely packed units. In order to regain the lost entropy, the particles must move apart allowing them more freedom of movement, while solvent moves in to "resolvate" the units. The result is an energy barrier retarding the approach of particles and providing an effective mechanism for stabilization (another spring so to speak). The process is illustrated schematically in Figure 10.8. In a pure sterically stabilized system (i.e., no electrical charges involved) the net energy of interaction will be the sum of the attractive van der Waals forces and the repulsive steric interactions

$$\Delta G_{total} = \Delta G_{steric} - \Delta G_{att} \qquad (10.17)$$

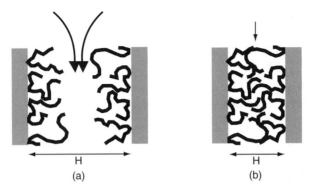

FIGURE 10.8. As two surfaces having adsorbed polymer approach, two phenomena occur that produce a net repulsive force between the surfaces. (*a*) At relatively close approach, but before actual interpenetration of the layers, the local concentration of polymer chains (between particles) increases above the "normal" equilibrium value giving rise to an osmotic pressure effect; solvent molecules move into the area between the surfaces and pushes them apart. (*b*) At distances where layer interpenetration occurs, the polymer chains begin to lose degrees of freedom (an entropy decrease) and thermodynamic factors introduce a second repulsive term.

As noted, polymeric protective agents or steric stabilizers must be such as to be strongly anchored to the particle surface at a minimum of one point, or even better, several points. If single-point attachment is involved, the result will be a system with a free-swinging "tail" projecting into the solution providing the protective action (Fig. 10.9*a*). If two or more points are involved the result will be the formation of various loops, and possibly some tails as well (Fig. 10.9*b*). For a given polymer chain length, one can see intuitively that in a system of "tails" the distance the protective layer extends into the solution will be greater than a comparable system of loops. From that one might assume that such a single-point attachment would provide better protection.

FIGURE 10.9. A polymer may be adsorbed at one end of the chain to produce "tails" (*a*) or in "loops" and "tails" as in (*b*). The situation in (*a*) has some advantage in terms of producing a thicker adsorbed layer for a given polymer chain length, but (*b*) will usually produce a stronger adsorption and a higher chain density.

On the other hand, for loops, once interpenetration begins, there will be twice as many units affected by the volume restriction effect, leading to a stronger entropic effect. One cannot say, therefore, that one configuration is better than another. In most practical systems, both configurations will be involved.

10.7.2. Solvent Effects in Steric Stabilization

From the picture of steric stabilization presented, it should be clear that the solvent must play a critical role in determining the effectiveness of a given stabilizer–colloid–solvent system. If the nature of the solvent is changed such that it becomes a better solvent for the monomer units acting as anchors for the stabilizer, then the chains may be more weakly adsorbed, providing for the possible onset of depletion flocculation or related phenomena (see below). Conversely, if the solvent is changed from a "good" to a "poor" solvent for the loops and tails, the thickness of the protective layer will be reduced as the polymer chains collapse in on the particle surface. The result will be reduced stability against flocculation.

Because of the complex thermodynamics of polymer solutions, most polymers exhibit transitions in their solubility in a given solvent as a function of temperature. In a "good" solvent, the polymer chains will be extended in relatively open, random-coil configurations, giving optimum protective layer thickness. As the temperature is changed, the quality of the solvent may decrease, at some point becoming "poor," and the polymer chains will collapse into a more compact configuration. The point at which the transition from good to poor solvent properties occurs is termed the θ point. In terms of colloidal stability, the temperature at which the solvent character changes from "good" to "bad" is the critical flocculation temperature (CFT) of the system.

The θ point can also be attained by the addition of a miscible nonsolvent for the polymer loops and tails. The way in which the interaction potential is affected by the quality of the solvent is illustrated in Figure 10.10. In a poor

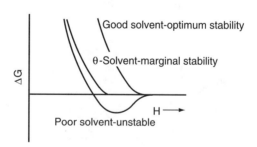

FIGURE 10.10. Because steric stabilization depends to a great extent on the thickness of the adsorbed polymer layer, the quality of the solvent for the adsorbed polymer is important. Curve 1 represents the steric repulsion curve in a "good" solvent (but not so good as to desorb the stabilizing polymer). Curve 2 represents a situation in which the solvent is borderline (a so-called θ solvent). Curve 3 is the general curve for a "poor" solvent that produces little or no stability.

solvent, the adsorbed layers may even add an additional attractive potential to the curve due to van der Waals attraction between the two.

10.7.3. Effects of Polymer Molecular Weight

In general, for a class of protective lyophilic colloids, a higher-molecular-weight material can be expected to provide better protection against flocculation, the reason being, of course, that longer chains imply longer loops and tails and the formation of a thicker protective layer around the particle. Like most physical systems, however, there are certain limits that must be adhered to, because "overkill" in the molecular-weight department may lead to what is termed sensitization and bridging flocculation.

If a very high-molecular-weight polymer with more than one potential point of attachment to the particle surface is added to a colloidal dispersion, there exists the possibility that the various possible points of attachment will encounter two different particles rather than attach to the same particle. That is especially the case where there is a large excess of particles relative to the concentration of polymer. Attachment of the same polymer chain to two particles essentially ties them together and brings them closer, in effect sensitizing the particles to flocculation. The process is illustrated in Figure 10.11.

While sensitization and bridging flocculation can be potential hazards in the formulation of sterically stabilized colloids, they also have their positive aspects. For example, the addition of small amounts of a high-molecular-weight acrylamide polymer to water leads to the flocculation of particulate matter that may be difficult to remove otherwise. Similar applications are found in, for example, the treatment of coal washing effluents and the flocculation of fines from uranium containing calcium phosphate minerals. In most cases of bridging flocculation, the resulting flocs are relatively open and rigid, which means that separation and filtration is relative easy. That fact is also exploited in the use of soil "improvement" polymers, which flocculate the

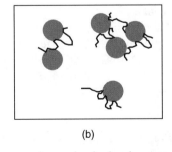

(a)   (b)

FIGURE 10.11. In sterically stabilized systems, a given adsorbed polymer molecule will be associated with one particle (*a*). In systems containing a low polymer concentration and/or very high-molecular-weight polymer, molecules can become adsorbed to two or more particles leading to "bridging" flocculation (*b*).

(a)   (b)

FIGURE 10.12. In a sterically stabilized system containing low-molecular-weight or weakly adsorbed polymer (*a*), as two particles approach, the loosely bound polymer may desorb, leaving "bare" spots on the approaching surfaces, leading to an enhanced flocculation tendency (*b*). That process is referred to as "depletion flocculation."

soil particles forming an open structure that allows for the freer movement of moisture and air throughout the soil.

10.7.4. Depletion Flocculation

If one adds a polymer that is not adsorbed or poorly adsorbed on the particles to a colloidal solution, there may occur another phenomenon, termed depletion flocculation . In depletion flocculation, as two particles approach, polymer chains that are weakly adsorbed, or simply are located between the particles, become squeezed out of the area of closest approach, leaving "bare" surfaces that are attracted in the normal way. However, there may arise an additional attractive force as a result of the removal of polymer from the intervening region (Fig. 10.12).

As polymer is forced out of the area between the approaching particles, the local osmotic balance is displaced; that is, the solution concentration between the particles is less than that in the bulk. Osmosis then forces solvent to flow from between the particles out into the solution. The net effect on the particles is that they are drawn together by the solvent flow (a type of hydrodynamic "suction" effect, if you will), resulting in a loss of stability and flocculation.

The above picture of depletion flocculation is, of course, very schematic and simple-minded, but it should serve to illustrate the concepts involved. A more detailed discussion would involve the introduction of complex theories of polymer adsorption and solution phenomena that are beyond the scope of this book.

10.8. COAGULATION KINETICS

The coagulation of emulsions and dispersions due to random Brownian motion has historically been the topic of most general interest to surface and colloid scientists because of the experimental accessibility of data (with sufficient innovation and diligence on the part of the experimenter) and a reasonably firm theoretical basis. The presence of an intervening fluid in such cases carries with it certain advantages of preparation, manipulation, and interpretation that are not present in an aerosol, for example. The same circumstances can, of course, carry with them certain disadvantages, especially with regard to purification, contamination, specific solvent effects, etc. Fundamentally, the same theoretical considerations apply to both solid dispersions in a liquid medium, aerosols, emulsions, and foams. However, in each case, certain extensions and modifications may be required.

10.8.1. Kinetics of Particle Collisions: Fast Coagulation

When one considers the question of the kinetics of coagulation, the discussion must begin with the question of whether there exists some barrier to coagulation between two approaching particles. Since the overall process, in its simplest form, involves the interaction of two particles or units, it can be thought of in terms of chemical kinetic processes; that is, a bimolecular, second-order process. If no barrier to particle approach, contact, and adherence exists, the particles are said to be noninteracting (except on contact, of course). The chemical equivalent would be a reaction with zero activation energy. If the particles are noninteracting and the primary minimum for the interaction energy is sufficiently deep, every collision between particles will be "sticky;" that is, each will result in the formation of a multiple particle (dimer, trimer, etc.) floc. In such a case, the rate of coagulation will be controlled entirely by diffusion kinetics, analogous to a diffusion-controlled bimolecular reaction.

The first theory relating the rate of coagulation to diffusion was that of Smoluchowski. The complete derivation of the final equation will not be given here. Suffice it to say that the process begins with Fick's first law of diffusion given by

$$J = 4\pi r^2 D \frac{\delta n}{\delta a} \tag{10.18}$$

where J is the "flux" or number of collisions per unit time between particles in the system and some "central" or reference particle (Fig. 10.13), a is the radius of the central particle, and D is the diffusion coefficient. By way of illustration one can visualize the process in the following way: a particle of radius a is placed with its center at the origin of a rectangular coordinate system, where a will be some characteristic radius of the particle in question (e.g., hard sphere, hydrodynamic). Other particles in the system, in the process of random Brownian motion, move through the coordinate system at distances r from the center until one particle center approaches to within a critical distance r_c of the central particle. At that distance, the particles "touch" and

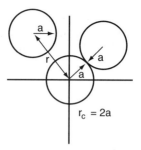

FIGURE 10.13. In a colloidal system, the rate of particle flocculation will depend on the rate of particle collision. That rate, in turn, will depend on the diffusion coefficients of the respective particles and their effective particle diameters (or collision cross sections).

coagulation is assumed to occur. For a system of monodisperse particles, $r_c = 2a$; r_c, then, is the collision diameter of the particles.

At time t = 0, the particle concentration in the system well away from the central particle ($r \to \infty$) is n_0 (particles per unit volume) while at distance r_c, $n = 0$, so that the number of collisions per unit time between the central and a second particle will be

$$J = 4\pi D r_c n_0 \qquad (10.19)$$

The diffusion coefficient D is given by

$$D = \frac{kT}{6\pi\eta a} \qquad (10.20)$$

where η is the viscosity of the dispersion medium. If the constraint of holding the central particle stationary is removed, the value of D to be employed in Equation (10.19) becomes $D = D_1 + D_2$ (for particles of different radii), or for identical particles, $2D$. The kinetic equation for the disappearance of primary particles now becomes

$$\frac{dn}{dt} = -8\pi r_c D n^2 \qquad (10.21)$$

where n is the concentration of primary particles and dn/dt is the rate at which they disappear. From Equation (10.21), the time for the reduction of the number of particles by one half, the half-life, will be

$$t_{1/2} = (8\pi r_c D n_o)^{-1} \qquad (10.22)$$

For water at 25°C

$$t_{1/2} = \frac{3\eta}{4kTn_{\rm o}} = \frac{2 \times 10^{11}}{n_0} \quad \text{(seconds)}$$

so that for a typical dilute dispersion containing 10^{12} particles cm^{-3}, the half-life is in the range of 0.2–1 s. Those represent very short times, shorter than are normally encountered in practice (seconds to minutes, depending on the circumstances) for noninteracting dispersions.

The primary reason for the discrepancy is the fact that as two particles approach one another, it is necessary for solvent molecules between the particles to be moved out of the way. This process is accounted for by the viscosity term in Equation (10.20) for large distances of separation, but at smaller distances, of the order of molecular dimensions, the simple viscosity relationships no longer strictly apply, so that the mutual diffusion coefficient D is no longer equal to $D_1 + D_2$ (see Chapter 4). One could say that the "microscopic" viscosity of the solvent increases so that diffusion is slowed and the particles approach at a reduced velocity. The exact calculation of this hydrodynamic effect represents a difficult problem in fluid dynamics. However, a relatively simple formula for two spheres of equal diameter is

$$D_{\rm eff} = D \, \frac{6s^2 - 20s + 16}{6s^2 - 11s} \tag{10.23}$$

where $s = r/a$. Experimentally, it is found that this hydrodynamic effect results in a collision rate of approximately one half that predicted by Equation (10.23). Therefore the measured half-life will be about twice as long.

The simple theory presented so far considers only collisions between single primary particles, not those involving dimers, trimers, or other particles. When all particles are taken into consideration, the overall kinetics are still second-order overall, but the value of n in the equation is the number of all particles present, regardless of their size and composition, and $t_{1/2}$ refers to the time for disappearance of one half of all particles, not just the primary ones represented by n_0. Because of the complications inherent in trying to define the necessary values of n, r_c, D, and so on for the multiparticle aggregates, the relevant experiments are normally interpreted for the early stages of the coagulation process only, where the majority of particles are primary units. In practice, the experimental process usually involves measuring the value of n versus time using, for example, light scattering, and turbidity, and extrapolating the data back as $t \to 0$ (Fig. 10.14). The regime in which a coagulation process follows the above theory is usually referred to as fast or rapid coagulation.

10.8.2. Slow Coagulation

If some energy barrier to particle contact and adherence exists—that is, if some of the collisions are not "sticky"—the collision process can be seen as

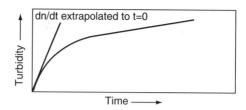

FIGURE 10.14. Experimentally, the rate of particle flocculation can be determined by measuring the change in turbidity of the system as a function of time and extrapolating back to zero time. A typical turbidity curve would have the form shown.

analogous to a bimolecular reaction in which there is a substantial activation energy. In that case, a slow coagulation process occurs that can be described kinetically by the relationship

$$\frac{dn}{dt} = -8\pi r_c Dn^2 \exp \frac{-G_{max}}{kT} \tag{10.24}$$

where G_{max} is the height of the activation energy barrier opposing coagulation. The rates of fast and slow coagulation [Eqs. (10.21) and (10.24)] obviously differ by the exponential factor $\exp(-G_{max}/kT)$. Since the coagulation process is slowed down by that factor, dividing Equation (10.21) by Equation (10.24) gives the stability ratio W:

$$W = exp \frac{G_{max}}{kT} \tag{10.25}$$

Equation 10.26 would be valid if colloidal diffusion processes were exactly analogous to those for individual molecules. However, the interactions between particles in colloidal systems tend to extend over distances much greater than those involved in the formation of atomic or molecular activated complexes (say, 10–100 nm vs. 0.1–1.0 nm). As a result, the effects of those interactions will begin to be felt by the particles well before they approach to the critical distance r_c. Their mutual diffusion rate will therefore be reduced and the collision frequency will drop accordingly. The collision frequency will also be reduced by the hydrostatic effect mentioned above for rapid coagulation.

A more accurate expression for the stability ratio, W, taking into consideration the above retarding effect of the interaction potential on collision frequency, is

$$W = r_c \int_{r=2a}^{r=\infty} \left(\frac{1}{r^2}\right) \exp \left(\frac{G_{(r)}}{kT}\right) dr \tag{10.26}$$

where $G_{(r)}$ is the interaction potential when the two particle centers are separated by a distance r.

For charged particles experiencing electrostatic repulsion, an approximate equation for the stability ratio is

$$W = \left(\frac{1}{\kappa r_c}\right) \exp \frac{G_{max}}{kT} \qquad (10.27)$$

where κ is the thickness of the electrical double layer. A more practically useful form of this equation for purposes of interpretation of data is

$$\ln W = -\ln \kappa r_c + \frac{G_{max}}{kT} \qquad (10.28)$$

where G_{max} is a roughly linear function of the concentration of electrolyte in the system, c_o, so that a plot of $\ln W$ versus $\ln c_o$ is also roughly linear. That approximate relationship is generally born out by experiment, although the hydrodynamic effect, among others, may cause slight bothersome deviations from the ideal curve shape.

10.8.3. Critical Coagulation Concentration

As a practical application of the relationship in Equation (10.28), one may characterize the stability of a charge-stabilized colloidal system by its critical coagulation concentration (ccc), the concentration of electrolyte necessary to bring the system into the regime of rapid coagulation. The process involves the extrapolation of the curve of $\ln W$ versus $\ln c_o$ to $\ln W = 0$, which gives $\ln$ (ccc). However, what is the practical use of the ccc and what does it mean in theoretical terms?

Theoretical and practical interest in the effect of added electrolytes on the stability of colloidal dispersions was reported as early as 1856 by Faraday, although it was almost certainly of great interest to others long before that time. Since the earliest days of theoretical surface and colloid science, the understanding of coagulation phenomena and the formulation of workable theories has been the focus of a major portion of the scientific efforts in the field. Near the turn of the century, the studies by Schultze and Hardy indicated that the primary factor controlling the effect of an electrolyte on a colloid of opposite electrical charge (and for a given concentration) was the valence of the added counterion. It was found that, in general, the valence of ions of the same charge as the colloid was of minor importance.

The studies by Schultze and Hardy led to the formation of the so-called Schultze–Hardy rule, which states that the critical coagulation concentration of a colloid is determined primarily by the valence of the counterions. The early relationship indicated that the ccc varied as the inverse sixth power of the valence, although experimental results obtained under carefully controlled conditions of concentration, temperature, and other factors, give ratios for electrolytes of valence 1, 2, and 3 of $1:0.013:0.0016$. That is, for valence 2, the exponent is -6.27 and for valence 3, -5.85. The fact that exponents are found to be fractional almost certainly stems from the fact that the properties of ions in solution vary slightly according to their ionic radius, hydration radius, among other properties. As a result, their effect on such solution characteristics as surface tension and conductivity, their adsorption at interfaces, the degree of ionization of their salts, their interaction with proteins, and their ccc for a given colloidal system will vary within a given valence. It is found that for monovalent ions, the effectiveness for coagulating negatively charged colloids has the order $Cs^+ > Rb^+ > K^+ > Na^+ > Li^+$, while for divalent cations the order is $Ba^{2+} > Sr^{2+} > Ca^{2+} > Mg^{2+}$.

Clearly, the stability of an electrostatically stabilized colloid, as measured by its ccc, is a function of the concentration and charge of the counterions in the system. The question is, what type of theory can incorporate all of the observed facts about colloidal stability and be able to serve as a workable predictive model for new systems?

10.8.4. The Deryagin–Landau–Verwey–Overbeek (DLVO) Theory

The overall stability of a colloid will depend on the net form of the interaction energy curve for the system—the sum of the attractive and repulsive energy terms as a function of the distance of separation of the particles. For the moment, we will consider only two contributing factors: the attractive van der Waals term and the repulsive double-layer term, leaving aside any consideration of entropic or steric stabilization.

For a one-to-one electrolyte of valence z and bulk concentration c_o, the excess charge density, ρ_e, at a point in the electrical double layer of potential ψ is given by

$$[c^+ - c^-]z = \rho_e = zc_0\left[\exp\left(\frac{-ze\psi}{kT}\right) - \exp\left(\frac{+e\psi}{kT}\right)\right] \qquad (10.29)$$

Poisson's equation of electrostatics relates ρ_e to the variation of ψ with distance to the charged surface, x, in the form

$$\frac{\delta^2\psi}{\delta x^2} = \frac{-\rho_e}{\varepsilon} \qquad (10.30)$$

where ε is the permitivity of the medium. Combination of Equations (10.29) and (10.30) gives the Poisson–Boltzmann equation, which shows that the potential within the EDL will fall off approximately exponentially with distance, for systems with relatively high surface potentials and distances well removed from the surface:

$$\psi = \left(\frac{4kT}{ze}\right) exp\ (-\kappa x) \tag{10.31}$$

where κ is, as previously defined, the thickness of the EDL.

In terms of the interaction of two approaching particles, as the distance of separation decreases, the two electrical double layers begin to overlap. To a first approximation, the two overlapping electrical potentials become additive, resulting in an increased electrical contribution to the total free energy of the system—a repulsive term, in this case, since the process increases the total energy. The form of the EDL overlap potential is also approximately exponential with distance of separation, H. For two parallel surfaces, the free energy contribution of the electrostatic repulsion term is given by

$$\Delta G_{elec} = \left(\frac{64c_o\ kT}{\kappa}\right) exp\ (-\kappa H) \tag{10.32}$$

As previously shown, κ is dependent on the concentration of ions in the system, and the primary concentration effect in Equation (10.32) comes from the exponential term.

If one sums the attractive van der Waals potential and the repulsive electrostatic potential, the total potential-energy expression becomes

$$\Delta G_T = \left[\left(\frac{64c_okT}{\kappa}\right) exp\ (-\kappa H)\right] - \frac{A_H}{(12\pi H^2)} \tag{10.33}$$

where A_H is the Hamaker constant for the colloidal material. When this equation 10.33 is solved for various electrolyte concentrations, a series of curves of the general forms as shown in Figure 10.15 is obtained. The maxima in the curves represent the barriers to coagulation imposed by the electrical double layer. Clearly, as the electrolyte concentration increases, the height of the barrier decreases. As a rule of thumb, one can say that conditions for rapid coagulation (the ccc) will be met when $\Delta G_T = 0$ and $d\Delta G_T/dH = 0$. Mathematically, that is

$$c_o\ (ccc) \propto \frac{1}{(A_H{}^2z^6)} \tag{10.34}$$

which agrees with the Schultze–Hardy rule and predicts ratios of $1:0.016:00014$ for ions of valence 1, 2, and 3, respectively, and agrees generally with experimental observations noted earlier.

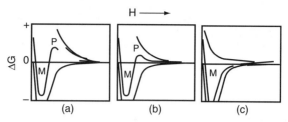

FIGURE 10.15. In an electrostatically stabilized colloid, the concentration of electrolyte will greatly affect the stability of the system: (*a*) with low electrolyte concentration, a relatively high energy barrier will impart stability; (*b*) as electrolyte is added, the stabilizing barrier will be reduced, but reasonable stability may be maintained (depending on the valency of the added electrolyte); (*c*) at a critical electrolyte concentration the energy barrier will effectively disappear and rapid flocculation will occur.

Another consequence of Equation (10.34) is that as the Hamaker constant A_H increases, the ccc values of dispersions with the same electrical characteristics will decrease; that is, the colloidal dispersions will become progressively less stable. As indicated previously, the Hamaker constant in a medium will be a composite of those of the dispersed particle and the medium in vacuum, so that as the characteristics of the medium and dispersed phase become more similar, A_H will decrease, leading to an additional stability for the system, in agreement with observation.

The fact that the DLVO theory predicts (roughly at least) the Schultze–Hardy rule would seem to confirm the validity of the theory. However, several approximations and assumptions are included in the derivation that weaken its claim to complete success. For example, at low surface potentials, the theory predicts that the ccc will be proportional to z^{-2} rather than z^{-6}. A more complete derivation of the theory (i.e., taking into consideration such factors as specific adsorption of ions and hydration effects) can remove some of the discrepancies so far encountered in the simple theory.

An additional important prediction of the DLVO theory is that under certain conditions, a colloid may undergo a form of reversible flocculation brought about by the existence of a so-called secondary minimum in the potential-energy curve. The existence of the secondary minimum has been confirmed experimentally and since it represents a potentially important theoretical and practical aspect of the DLVO theory, it will be discussed briefly below.

The great value of the DLVO theory for the practicing colloid chemist, regardless of the exact nature of the work involved, is that it illustrates dramatically the importance of understanding the electrical properties of a colloid of potential interest, and the importance of understanding the effect of the ionic environment to which that colloid may be exposed.

10.8.5. Reversible Flocculation and the Secondary Minimum

An interesting, and sometimes useful, consequence of the complex interaction between attractive and repulsive terms in the overall equation for colloidal stability is the existence of a so-called secondary minimum in the total energy curve, usually at relatively large inter-particle distances (Fig. 10.16). The primary minimum is, of course, the deep energy well into which the system falls in search of "energetic" stability (the cup in the putting green). Before going over the barrier to enter that well, however, a colloid may encounter a much more shallow energy minimum of the order of a few kT where small, relatively weak aggregates, in this case flocs, may form.

Weak or secondary minimum flocculation, unlike flocculation in the primary minimum, is often found to be subject to significant entropy effects, leading to the realization of a number of interesting phenomena. Because of the shallow nature of the secondary minimum, the flocs formed are held together rather weakly and tend to be unstable; that is, they can be broken up by rather small energy inputs such as gentle stirring. In fact, for very small particles ($r < 100$ nm), the minimum may be so shallow that Brownian motion prevents flocculation altogether. For larger particles the secondary minimum may cause observable effects such as an apparent phase separation into a concentrated, highly ordered dispersion, sometimes exhibiting birefrengence, and a second dilute, isotropic phase. In that case, gentle stirring will regenerate the original homogeneous dispersion. In other cases a dynamic equilibrium may develop between small flocs and individual colloidal particles. Such an equilibrium may be treated theoretically much like the process of molecular aggregate formation in a vaopr (or a solution), in which the concentration of aggregating molecules is below the saturation point, that is, below the

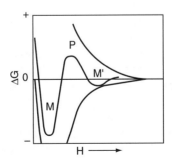

FIGURE 10.16. In many colloidal systems, the interaction energy curve will have a small minimum, the secondary minimum, M', that allows the particles to undergo a lose, reversible flocculation. In some systems of relatively large, monodisperse particles, the secondary minimum may lead to an optical phenomenon called "opalescence" in which a very regular structure is developed (similar to a crystal structure) that produces beautiful and interesting patterns with incident light.

concentration necessary for the formation of significant numbers of aggregates of the critical size necessary for condensation or crystallization to occur.

10.9. THE COMPLETE INTERACTION CURVE

The two stabilization mechanisms discussed above—electrostatic and steric—represent the extremes of the range of practical techniques that allow us to prepare colloidal systems that are at least kinetically stable. The two mechanisms, however, are not mutually exclusive. In fact, most natural and many technologically important colloids involve a combination of both effects (plus possibly others, yet to be fully determined). A total potential-energy curve, then, might contain a number of terms

$$\Delta G_{total} = \Delta G_{elec} + \Delta G_{steric} + \Delta G_{misc} - \Delta G_{att} \tag{10.35}$$

Proteins, for example, are excellent protective colloids. They are polymeric and tend to adsorb on various surfaces, providing good steric stabilization, and may also be charged (depending on the pH), providing electrostatic stabilization as well. Such a dual nature can have certain practical advantages. For example, if a colloid (e.g., silver iodide), is prepared in the presence of gelatin at a pH above the isoelectric point (i.e., a net negative charge is present on the protein molecule), a very stable dispersion will result. If the pH is lowered (or raised, depending on the gelatin type used) to the isoelectric point (zero net charge), the gelatin chains tend to collapse onto the particles, reducing the degree of steric stabilization in addition to that lost by neutralization of the charges on the polymer. The system will tend to flocculate, allowing for the separation of the silver iodide–gelatin complex from most of the water. If the pH is then returned to is initial value, the adsorbed polymer will swell and redisperse the original colloid. The presence of the gelatin prevents the individual particles from making contact so that true flocculation or coagulation cannot occur. Similar processes using both natural and synthetic polymers have found application in several important areas, including photography, microencapsulation, and "carbonless" carbon papers.

Obviously, knowing how colloids can be stabilized provides an invaluable tool for the preparation of many useful systems. It also can provide clues to how an unwanted colloid can be destabilized and removed. The above-mentioned ideas, at times in a slightly different guise, will appear again in the following chapters on emulsions, foams, aerosols, and similar compounds.

PROBLEMS

10.1. If a cube of material is subdivided into 1000 smaller cubes of equal size, the surface area of the smaller cubes will be what fraction of the original?

10.2. The total surface of the cubes in Problem 10.20 will be how many times greater than the original?

10.3. A particle of gold (density $= 19.3$ g cm^{-3}) has a radius of 1 nm. What is the weight of the particle? About how many atoms of gold make up the particle?

10.4. Calculate the reversible, thermodynamic energy input required to transform 1 kg of glass ($\sigma = 245$ mJ m^{-2}) into particles of 1 mm diameter. Assume that the particles are all equivalent in size and surface properties.

10.5. Most comminution processes involve the production of significant amounts of heat. A new laboratory technician given the task of preparing a fine powder of an organic compound of melting point 48°C can accomplish only the production of a gooey, unusable paste. Suggest what the technicians problem may be and how to overcome it.

10.6. Two identical spherical colloids with a surface potential of 25 mV and effective Hamaker constant of 7×10^{-21} J are dispersed in an aqueous medium. Calculate the minimum separation distance, h_0, at which the interaction energy will be zero. Assume $h_0 \ll \kappa^{-1}$.

10.7. A gold sol of 100 nm diameter was found to have the following rapid coagulation characteristics:

t (s)	Number of Particles m^{-3} ($\times 10^{-14}$)
0	20.2
30	14.7
60	11.2
120	7.74
240	4.78
480	2.71

Calculate the collision rate constant and compare it with that expected for a diffusion-controlled process.

10.8. Calculate the expected critical coagulation concentrations (ccc) for a charged colloidal system in water with a Hamaker constant $= 2.0 \times 10^{-20}$ J in the presence of mono-, di- and trivalent counterions.

10.9. Estimate the stability ratio, W, for spherical colloids of 100 nm diameter with a surface potential $= 35$ mV, and Hamaker constant $= 5 \times 10^{-20}$ J in 2 mM KCl at 27°C.

10.10. A ferric hydroxide sol was found to have the following cccs in the presence of electrolyte:

Electrolyte	ccc (mM)
NaCl	9.25
KBr	12.2
$BaCl_2$	4.8
K_2SO_4	0.20
$MgSO_4$	0.22

Estimate the surface potential of the sol from the data. What can you conclude about the sol on the basis of these observations?

10.11. As a rule, the flocculating power of an electrolyte can be increased by adding an organic solute such as alcohol or acetone. Can this be supported by the stability theory discussed in the text? Explain.

10.12. What would you expect to occur when a 1 M KI solution is added to a positively charged AgI sol? What if it were added to a negative sol?

10.13. A dispersion flocculated with a multivalent counter-ion usually cannot be redispersed (repeptized) simply by washing with pure water. Sometimes, however, redispersion will occur if one washes first with a concentrated monovalent electrolyte solution followed by pure water. Explain.

10.14. Arable lands containing high percentages of clay when flooded by seawater will usually be severely damaged in terms of porosity and productivity. Suggest some logical explanation based on colloidal properties and the theories discussed in the text.

10.15. Calculate the bimolecular rate constant for reactions in water that are completely diffusion-controlled. The second-order rate constant is given by

$$\frac{-dc}{dt} = k\,c^2$$

The reaction $OH + I^- \rightarrow OH^- + I$ has a rate constant $k = 1.0 \times 10^{10}$ L mol^{-1} s^{-1}. Is the reaction diffusion-controlled?

10.16. Given the following data for a selenium sol in 40 mM KCl

t (h)	$v \times 10^{-9}$ cm^3	t (h)	$v \times 10^{-9}$ cm^3
0	33.5	263	23.4
23	30.8	407	15.9
71	28.4	1080	10.6
95	24.8	2330	5.4
115	24.7	—	—

where ν = particle concentration at time t, do the data follow the Smoluchowski rate law? Show why or why not.

10.17. According to the Smoluchowski theory, the concentration of primary, secondary, tertiary (etc.) particles, ν_1, ν_2, ν_3 (etc.) are given by the equations

$$\nu_1 = \frac{\nu_o}{1 + (t/T)}; \quad \nu_2 = \frac{\nu_o(t/T)}{[1 + (t/T)]^3} \nu_i = \frac{\nu_o(t/T)^{i-1}}{[1 + (t/T)]^{i+1}}$$

where ν_o = initial concentration of primary particles. Predict how long after the onset of coagulation the concentration of quadruplets (ν_4) reaches its maximum value. How many primary particles will be involved in quadruplet systems at that time?

10.18. The flocculation process is often studied by measuring the light scattering or turbidity of the sol as a function of time. The scattering of a monodisperse system of ν primary particles is given by

$$I_s = I_o K \nu_o M_1^2$$

where K is a constant and M_1 is the mass of the primary particle. Show that the scattered intensity after time t is

$$I_s = I_o K \nu_o M_1^2 \left[1 + \left(\frac{2t}{T} \right) \right]$$

Assume that all particles present are of such a size that no internal interference occurs. Use the equations given in Problem 10.14.

10.19. Colloidal particles may be expected to be very small if formed under the following set of conditions: (a) rapid nucleation and rapid growth; (b) slow nucleation and rapid growth; (c) rapid nucleation and slow growth; (d) slow nucleation and slow growth.

10.20. The degree of hydration of a colloidal particle has a marked effect on its (a) diffusion coefficient; (b) intrinsic viscosity; (c) sedimentation rate; (d) all of these; (e) none of these.

10.21. The addition of a small amount of a reagent to a sol results in flocculation, but the addition of larger amounts has no effect. From these observations, one can conclude that (a) the reagent is an uncharged protective colloid and the sol is close to flocculation; (b) the charge on the sol is reversed after the addition of large amounts of reagent; (c) there is not sufficient information to draw a conclusion.

10.22. The total driving force for the sedimentation of a particle will not change if (a) the density of the continuous phase is decreased; (b) the shape of the particle is changed; (c) the system is transported in the space shuttle; (d) the density of the particle is increased.

10.23. A spherical particle of radius R falls in water with a terminal velocity of 1 cm s^{-1}. One can then say with confidence that a similar particle of radius $10R$ will have a terminal velocity of (a) 10 cm s^{-1}; (b) less than 100 cm s^{-1}; (c) 100 cm s^{-1}; (d) over 100 cm s^{-1}.

11 Emulsions

The preparation, stabilization, and application of emulsions impacts almost every aspect of our lives, from food to pharmaceuticals. Chapter 10 introduced some of the important physicochemical factors in the preparation and stabilization of dispersions of solid particles in a liquid (sols). Most of those same factors are important in emulsions, although their significance and the approaches used to understand them often differ between the two classes of colloids. Although colloidal in character, emulsions are usually systems whose dispersed phase dimensions fall outside the "normal" defined range for colloids. Commonly encountered emulsions will have average drop sizes of at least several microns, with a rather broad size distribution. In addition, most, but not all, sols have a volume fraction of dispersed material that seldom exceeds 50 percent. In emulsions, that quantity is seldom less than 10 percent and sometimes as high as 90 percent. In foams, to be discussed in the following chapter, the dispersed phase content is often even higher. However, as stated in Chapter 10, since emulsions look like colloids and act like colloids—size notwithstanding—they must be colloids.

11.1. FUNDAMENTAL CONCEPTS IN EMULSION SCIENCE AND TECHNOLOGY

As is the case in most discussions of interfacial systems and their applications, definitions and nomenclature can play a significant role in the way the material is presented. The definition of an emulsion to be followed here is that they are heterogeneous mixtures of at least one immiscible liquid dispersed in another in the form of droplets, the diameters of which are, in general, greater than $0.1 \ \mu$m. Such systems possess a minimal stability, generally defined rather arbitrarily by the application of some relevant reference system such as time to phase separation or some related phenomenon. Stability may be, and usually is, enhanced by the inclusion of additives such as surfactants, finely divided solids, and polymers. Such a definition excludes foams and sols from classification as emulsions, although it is possible that systems prepared as emulsions may, at some subsequent time, become dispersions of solid particles or foams.

When discussing emulsions, it is always necessary to specify the role of each of the immiscible phases of the system. Since in almost all cases, at least one liquid will be water or an aqueous solution, it is common practice to describe an emulsion as being either oil-in-water (o/w) or water-in-oil (w/o), where the first phase mentioned represents the dispersed phase and the second the continuous phase.

There is, in principle, no reason why one cannot prepare an oil-in-oil emulsion (o/o). However, the generally high miscibility of most organic liquids is an important limitation. More important, however, is the fact that the nature of interfaces dictates that a system tends to attain a situation of minimum energy, in this case minimum interfacial area, so that some additive must be employed to retard that process. Unfortunately, few materials are sufficiently surface active at such oil–oil interfaces to impart the required minimal stability necessary for the preparation and maintenance of such emulsions. Oil-in-oil emulsions of short persistence can, however, constitute an intermediate step in the preparation of nonaqueous emulsion polymers.

Three major characteristics of an emulsion must be considered:

1. Which of the two liquid phases will be the continuous phase and which the dispersed phase when the emulsion is formed, and what factors can be used to control that result.
2. What factors control the stability of the system; that is, what factors affect the creaming or sedimentation of the dispersed phase, drop coalescence, flocculation, and so on.
3. What factors control the often complex rheology of emulsified systems, and how they can be effectively controlled.

The following discussions will touch on each of these three questions. While the information presented is far from comprehensive, it will hopefully provide a useful introduction to some of the problems and solutions encountered in the practice of emulsion technology.

11.2. EMULSION FORMATION

The preparation of an emulsion requires the formation of a very large amount of interfacial area between two immiscible liquids. If a sample of 10 mL of an oil is emulsified in water to give a droplet diameter of 0.2 μm, the resulting o/w interfacial area will have been increased by a factor of approximately 10^6. The work required to generate one square centimeter of new interface is given by

$$W = \sigma_i \Delta A \qquad (11.1)$$

where σ_i is the interfacial tension between the two liquid phases and ΔA is the change in interfacial area. If the interfacial tension between the oil and water is assumed to be 52 mN m^{-1} (as for a hydrocarbon liquid), the reversible work required to carry out the dispersion process will be on the order of 2 J. Since that amount of work remains in the systems as potential energy, the system is thermodynamically unstable and rapidly undergoes whatever transformations possible to reduce that energy, in this case, by reducing the interfacial area. If some material can be added to the system to reduce the value of σ_i to approximately 1 mN m^{-1}, the magnitude of W will be reduced to 0.3 J—a substantial reduction in W—but the system will still be unstable. Only if the interfacial tension (and therefore W) is zero can a truly stable system be obtained. Obviously, thermodynamics is the constant enemy of the emulsion maker.

Luckily, although thermodynamics will be the factor controlling the ultimate long-term stability of an emulsion, kinetics can play an important role over the short term, and it is through kinetic pathways that most useful emulsions achieve their needed stability. It is clear, then, that while lowering the interfacial tension between phases is an important factor in the formation and stabilization of emulsions, that may not always represent the most important factor in their preparation and ultimate application.

As anyone familiar with the preparation of practical emulsions knows, the process is still almost as much an "art" as a science. The results obtained for a given oil–water system will depend on the dispersing process used, the characteristics and quantities of additives employed, mixing temperature, order of mixing, and other factors. Then, of course, there is always the question of the phase of the moon!

Art and magic aside, there are three principal methods of emulsion preparation which are most often employed. A fairly comprehensive coverage of those methods is presented in the work by Becher et al., cited in the Bibliography. The three methods most often employed include (1) physical emulsification by drop rupture, (2) emulsification by phase inversion, and (3) "spontaneous" emulsification. The latter two methods may be described as "chemically based" processes in that the nature of the final emulsion will be controlled primarily by the chemical makeup of the system (the chemical nature of additives, the ratios of the two phases, temperature, etc.), while in the first it will depend more on the mechanical nature of the process (e.g., amount and form of energy input.), as well as the rheological and chemical properties of the components. Other possibilities exist (see Table 11.1); however, most are of limited practical importance.

11.3. EMULSIONS AND THE LIQUID–LIQUID INTERFACE

In almost all practical emulsions, some additive (an emulsifier) is required to facilitate the formation of drops of the desired size and stability. Normally, one additive, at least, will be a material (defined below) that has the necessary characteristics to facilitate the formation of small droplets and produce the type of emulsion desired (o/w or w/o). The additive, an emulsifier and/or

TABLE 11.1. Some Typical "Mechanical" Methods of Emulsification

Method	Energy Input[a]	Process[b]	Drop Formation[c]
Shaking	L	B	T
Stirring			
Simple	L	B,C	T,V
Rotor–stator	M–H	B,C	T,V
Vibrator	L	B,C	T,V
Scraper	L–M	B,C	V
Pipe flow			
Laminar	L–M	C	V
Turbulent	L–M	C	T
Colloid mill	M–H	C	V
Ball and roller mill	M	B,C	V
Homogenizer	H	B,C	T,V,C
Ultrasonic	M–H	B,C	C,T
Injection	L	B,C	T,V
Electrical	M	B,C	—
Condensation	L–M	B,C	—
Aerosol to liquid	L–M	B,C	—

[a] L = low, M = medium, H = high.
[b] B = batch, C = continuous.
[c] T = turbulence, V = viscous forces in laminar flow, C = cavitation.

stabilizer, may perform two primary functions: (1) lower the energy requirements of drop formation (i.e., lower the interfacial tension) and (2) retard the process of drop reversion to separate bulk phases. In order to function properly, it must adsorb at the L–L interface.

In its second function, the additive must form some type of film or barrier (monomolecular, electrostatic, steric, or liquid crystalline) at the new L–L interface that will prevent or retard droplet flocculation and coalescence. The process of barrier formation or adsorption must be rapid relative to the rate of drop coalescence or a rather coarse emulsion will result. Also, with the formation of more interface, the adsorption of the emulsifier depletes its bulk concentration, so that attention must be paid to the quantity of the material employed relative to the final result desired, as well as its quality as an emulsifier. As will be seen below, the exact role of an emulsifier in emulsion formation can be quite complex, and is not always completely understood. In any case, its (or, in many cases, their) presence will be vital to successful emulsion formation and stability.

11.3.1. Classification of Emulsifiers and Stabilizers

There are four general classes of materials that can, under the proper circumstances, act as emulsifiers and/or stabilizers for emulsions. The list includes common (nonsurfactant according to the definition in Chapter 3) ionic materials, colloidal solids, polymers, and surfactants. Each class varies greatly in its effectiveness in a given role, and in its mode of action (Fig. 11.1).

Adsorbed nonsurfactant ions (Fig. 11.1*a*) will usually do little to affect interfacial tension (except to raise it in some cases) and therefore do little to facilitate emulsification. However, some may, under the proper circumstances, aid in stabilizing the system by imposing a slight electrostatic barrier between approaching drops. Alternatively, they may affect the stability of the system by their action in orienting solvent molecules in the neighborhood of the interface, altering some local physical properties such as dielectric constant, viscosity, and density, thereby producing a small stabilizing effect (solvation effects).

Small colloidal materials (sols), while not directly affecting interfacial tensions, can stabilize an emulsion by forming a physical barrier between drops, thereby retarding or preventing drop coalescence (Fig. 11.1*b*). The action of such materials will depend on several factors, the most important of which

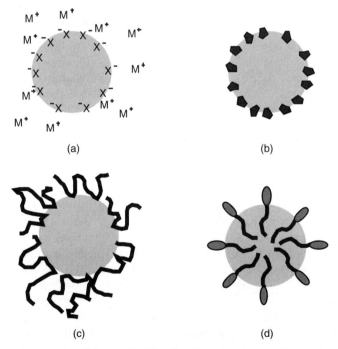

FIGURE 11.1. There are four primary mechanisms for the stabilization of emulsions (plus combinations, of course). Some emulsions may be weakly stabilized by the presence of adsorbed ions and nonsurface-active salts (*a*). The presence of colloidal sols partially wetted by both phases of the emulsion may form a mechanical barrier to drop contact and coalescence (*b*). Many emulsions are stabilized by adsorbed polymer molecules (*c*). Along with polymers, adsorbed surfactant molecules represent the most common stabilization mechanism (*d*).

are the particle size and the specific interfacial interactions between the solid surface and the two liquid phases making up the system. In general, a particle should be partially wetted by both liquid phases, but with a slight preference for the external phase.

Polymeric additives may aid in emulsion formation as a result of surface-active properties but are usually more important as stabilizers. Their action may result from steric or electrostatic interactions, from changes in the interfacial viscosity or elasticity, or from changes in the bulk viscosity of the system. In many if not most cases, the function of polymeric stabilizers is a combination of several actions (Fig.11.1c).

Finally, "normal" monomeric surfactants are usually added in order to decrease the interfacial tension and impart added stability to the system (Fig. 11.1d). The type and quantity of surfactant employed will be determined by the specific properties of the liquid phases, the type of emulsion desired, conditions of use, and other factors.

Of the possible emulsifiers, most are what are considered true surfactants, in that they are effective at lowering significantly the interfacial tension between the two liquid phases. Other additives such as polymers and sols function primarily as stabilizers, rather than emulsifiers. Most polymers are not sufficiently effective at lowering interfacial tensions to act in that regard. In addition, because of their molecular size, the adsorption process for polymers is generally very slow relative to the timescale of the emulsification process. The same applies to stabilizing colloids, in which their action requires the wetting of the particles by the two liquid phases to facilitate their location at the interface. The primary function of polymers and sols in emulsions is in the retardation of droplet flocculation and coalescence.

11.3.2. What Determines Emulsion Type?

The processes of flocculation and coalescence in the context of emulsion stability will be treated in a bit more detail below. At this point it is useful to point out their role in the determination of the final nature of the emulsion. The process leading to emulsion formation usually begins with the production of preliminary large drops, probably of both liquid phases. The continuous phase-to-be will be determined by many factors, to be outlined below. In any case, droplets of that phase must disappear rapidly during the process through flocculation and coalescence. The ultimate dispersed phase, on the other hand, must maintain (or reduce) its droplet size during and after processing.

The emulsification process is so dynamic and complex that an accurate model and theoretical treatment is almost impossible. With certain limitations its is possible to obtain order-of-magnitude estimates of such steps as droplet formation rate and surfactant transport and adsorption rates. However, the work involved is seldom worth the trouble in practice. Flocculation and coagulation rates during preparation are difficult to analyze because of the dynamics of the process and the turbidity of the flow involved. Collision rate theory has been found useful in the analysis of emulsion flocculation and coalescence in "quiet" emulsions and to some extent in idealized "turbulent" systems. In reality, however, those events are beyond prediction in the context of most emulsion preparation processes. There are simply too many events (simultaneous and sequential) occurring in a very short time frame and, in essence, chaos rules. Luckily, if conditions are closely controlled, an adequate degree of reproducibility can be obtained so that useful systems can be formulated and produced with reasonable confidence.

11.4. ADSORPTION AT LIQUID–LIQUID INTERFACES

The relationship between the adsorption of a molecule at an oil–water interface and the resulting interfacial tension is an important one and warrants a brief review here. The Gibbs equation for a system composed of one phase containing a nonionic solute adsorbing at the interface with a second phase is written

$$\Delta\Gamma_i = \frac{-1}{RT}\left(\frac{\partial\sigma}{\partial\ln a}\right)_T \tag{11.2}$$

where a is the activity of the surfactant in solution and the other terms are as defined previously. The equation states that at a liquid–liquid interface, as in the liquid–vapor case, the amount of surfactant adsorbed can be determined from the slope of the σ_i versus ln a curve. In dilute surfactant systems, the concentration, c (mol L^{-1}) can be substituted for activity without serious loss of accuracy. The simple relationship of the Gibbs equation can have significant practical application in the preparation of emulsions, especially in defining the relationship between emulsion droplet size and total surfactant concentration. Using Equation (11.2), the adsorption of surfaceactive molecules at an interface can be calculated from determinations of the interfacial tension. If ionic species are involved, life is complicated a bit by the presence the counterion and a variable surface potential that increases as adsorption proceeds.

It is of interest to try to relate the adsorption characteristics of a surfactant to the stability of an emulsion stabilized solely by an adsorbed monomolecular film. The total number of molecules that can be adsorbed in a given interfacial area will be controlled mainly by the effective "area per molecule" of the adsorbing species. That is, how many of the molecules can fit into the limited space of the interface? For most "normal" surfactant species, the area per molecule is determined primarily by the hydrophilic group and its hydration layer. The relative solubility of the surfactant in the two phases will also affect the result, but that factor is difficult to determine and is most often ignored. A few representative molecular "areas" at the oil–water interface are given in Table 11.2.

From the data in Table 11.2, it can be seen that the "experimental" area occupied by the sulfate and carboxylate groups are relatively large compared

TABLE 11.2. Typical Molecular Areas of Common Surfactants at Aqueous–Oil Interfaces and Saturation Adsorption

Surfactant and Oil Phase	Area (nm^2)
$CH_3(CH_2)_{11}SO_4Na$–n-hexane	0.45
$CH_3(CH_2)_{11}SO_4Na$–n-octane	0.48
$CH_3(CH_2)_{11}SO_4Na$–n-decane	0.49
$CH_3(CH_2)_{11}SO_4Na$–n-heptadecane	0.51
$CH_3(CH_2)_{11}SO_4Na$–benzene	0.65
$CH_3(CH_2)_{11}SO_4Na$–carbon tetrachloride	0.53
$CH_3(CH_2)_{11}SO_4Na$–1-hexene	0.57
Sodium diisoctylsulfosuccinate–n-heptane	1.11
Sodium lauroyl taurate–n-heptane	0.57
$CH_3(CH_2)_{11}SO_3Na$–benzene	0.57
Sodium laurate–n-heptane	0.45

to their "projected" areas calculated from molecular models (0.45 vs. 0.25 nm^2 for carboxyl and 0.48 vs. 0.28 nm^2 for sulfate). Also, while the effective area per molecule is found to vary by only about 6% for normal hydrocarbon oils with from 7 to 17 carbon atoms, in the presence of benzene, unsaturation, and nonhydrocarbon liquids, the head group "size" is often found to increase by as much as 35%.

Several explanations for those observations have been proposed, although the truth probably lies in a mixture of events. The relatively large sizes of the two head groups considered almost certainly result from electrostatic repulsion between the charges, although solvation is undoubtedly involved to some extent. When the ionic strength of the solution is increased, the effective molecular area is seen to decrease due to screening of the charges and possibly because of solvation changes as counterions begin to be more tightly bound at the interface (i.e., ion pairing). In the second case, normal hydrocarbon oils are expected to behave essentially the same with respect to their interaction with adsorbed molecules, regardless of their chain length (within limits, of course). However, when benzene, unsaturated compounds, or a nonhydrocarbon liquid is used, there exists the probability of significantly more interaction between the oil containing π electrons or large, polarizable atoms and the hydrophilic head group. Such specific interactions between aromatic molecules, unsaturated groups, and polar species are well established, as reflected in the interfacial tension for benzene–water (35 mN m^{-1}) vs. that for octane–water (50.8 mN m^{-1}) at 20°C. For the surfactants with bulky hydrophobic tails or "kinks" in their structures, such as the sulfosuccinates and taurates, the packing density at the interface is controlled by the hydrophobe leading to the observed increase in "size."

Since most nonionic surfactants have large, highly solvated polyoxyethylene head groups, it is more difficult to relate the interfacial area occupied by a given molecule to its structure. The problem is further exacerbated by the fact that such surfactant systems are composed not of a single molecular type, but rather an homologous series of varying composition and molecular weight. In such a case one must rely on some more empirical surfactant characterization system such as those discussed below for obtaining some idea of the activity of a given system.

11.5. GENERAL CONSIDERATIONS OF EMULSION FORMATION AND STABILITY

The question of emulsion stability has already been raised in the context of emulsion preparation. However, the preparation process is very dynamic and represents a complicated combination of events that is not easily analyzed. Once prepared, however, and left at rest to "do its own thing," the fates of individual droplets become more readily determined and some semblance of understanding can be extracted from the initial chaos. That is not to say, however, that there exists a good general theory of emulsion stability that one can apply under all, or even most, circumstances.

Even though emulsions as defined have been in use for thousands of years (even longer if natural emulsions are considered), no comprehensive theory of emulsion formation and stabilization has yet been developed that adequately describes, and predicts, the characteristics of many of the complex formulations encountered in practice. Except in very limited and specialized areas, the accurate prediction of such aspects of emulsion technology as droplet size, size distribution, and stability remain more in the realm of art than true science.

When discussing the stability of an emulsified system (like the colloidal sols of Chapter 10), it is important to have a clear idea of the physical condition of the components and the terminology employed. Four terms commonly encountered in emulsion science and technology related to stability are: "breaking," "coalescence," "creaming," and "flocculation." Although they are sometimes found to be used almost interchangeably, those terms are in fact quite distinct in meaning so far as the condition of an emulsion is concerned.

Coalescence , for example, refers to the joining of two (or more) drops to form a single drop of greater volume, but smaller interfacial area (Fig. 11.2a). Such a process is obviously energetically favorable in all cases in which there exists a positive (albeit small) interfacial tension. Although coalescence results in significant microscopic changes in the condition of the dispersed phase (e.g., changes in average particle size and distribution), it may not immediately result in a macroscopically apparent alteration of the system.

The breaking of an emulsion (Fig. 11.2b) refers to a process in which a gross separation of the two phases occurs. The process is a macroscopically

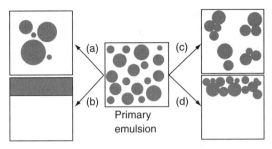

FIGURE 11.2. An emulsion may have several fates depending on the system and conditions. The droplets may coalesce to form larger drops (*a*). All drops may coalesce to form two separated phases (*b*). Droplets may flocculate to form larger aggregates while maintaining their individuality (*c*). Flocculated droplets may produce two apparently separate phases: one rich in dispersed phase, but with individual droplet identity, and one rich in continuous phase with few or no droplets present (*d*).

apparent consequence of the microscopic process of drop coalescence. In such an event, the identity of individual drops is lost, along with the physical and chemical properties of the emulsion. Such a process obviously represents a true loss in the stability of the emulsion.

Between the two clearly distinguishable states of an emulsion lies flocculation, which refers to the mutual attachment of individual emulsion drops to form flocs or loose assemblies of particles in which the identity of each is maintained (Fig. 11.2*c*), a condition that clearly differentiates it from the action of coalescence. Flocculation can be, in many cases, a reversible process, overcome by the input of much less energy than was required in the original emulsification process.

Finally, creaming is a process which is related to flocculation in that it occurs without the loss of individual drop identities (Fig. 11.2*d*). Creaming will occur over time with almost all emulsion systems in which there is a difference in the density of the two phases. The rate of creaming will be dependent on the physical characteristics of the system, especially the viscosity of the continuous phase. It does not necessarily represent a change in the dispersed state of the system, however, and can often be reversed with minimal energy input. If the dispersed phase happens to be the more dense of the two phases, the separation process is termed sedimentation.

Obviously, both flocculation and creaming represent conditions in which drops "touch" but do coalesce. The key to understanding the true stability of emulsions, then, lies on the line separating the processes of flocculation and coalescence.

11.6. SOME MECHANISTIC DETAILS OF STABILIZATION

Even in the infancy of emulsion technology it was recognized that in order to obtain a useful emulsion with any long-term persistence, it was necessary to include a third component, which served some "magical" purpose and imparted the required degree of stability. Such additives, as outlined above, included: simple inorganic electrolytes; natural resins and other macromolecular compounds; finely divided, insoluble solid particles or sols; and surfactants or surface active materials, which were soluble in one or both phases and significantly altered the interfacial characteristics of the system. In practice, one commonly finds that a combination of two or more additives is most effective for producing true, long-term emulsion stability.

Although some brief comments on the actions of emulsifiers and stabilizers have been presented, it is useful to have a little more detailed concept of just how these materials complete their function in the emulsified system. The actions of the most important systems—polymers, sols, and surfactants—will therefore be explained in a bit more detail.

11.6.1. Polymeric Emulsifiers and Stabilizers

In nature as well as in technology, polymeric emulsifiers and stabilizers play a major role in the preparation and stabilization of emulsions. Natural materials such as proteins, starches, gums, cellulosics, and their modifications, as well as synthetic materials such as polyvinyl alcohol, polyacrylic acid, and polyvinylpyrrolidone, have several characteristics that make them extremely useful in emulsion technology. By the proper choice of chemical composition, such materials can be made to adsorb strongly at the interface between the continuous and dispersed phases. By their presence, they can reduce interfacial tension and/or form a barrier (electrostatic and/or steric) between drops. In addition, their solvation properties serve to increase the effective adsorbed layer thickness, increase interfacial viscosity, and introduce other factors that tend to favor the stabilization of the system.

Finally, as can be seen from Equations (10.19) and (10.20), the frequency of particle encounters potentially leading to flocculation and/or coagulation is inversely proportional to the viscosity of the continuous phase. The presence of a polymer in solution will increase the viscosity of the system and impart added kinetic stability to the emulsion.

The effectiveness of polymeric materials at lowering interfacial tensions is usually quite limited. More important to their function is the fact that polymers can form a substantial mechanical and thermodynamic barrier at the interface that retards the approach and coalescence of individual emulsion droplets. The polymeric nature of the materials means that each molecule can be strongly adsorbed at many sites on the interface. As a result, the chance of desorption is greatly reduced or effectively eliminated, and the interfacial layer attains a degree of strength and rigidity not easily found in systems of monomeric materials. In addition, the presence of polymeric materials in the system can retard processes such as creaming by increasing the viscosity of the continuous phase thereby reducing the rate of droplet encounters that could lead to flocculation or coalescence. The concept of steric stabilization of colloids has already been introduce in Chapter 10 and will not be considered further here. It can be assumed that the mechanisms involved apply equally to solid and liquid dispersed systems, although slight variations on the theme are to be expected.

The subject of the mechanisms and degree of polymer adsorption at interfaces is also discussed in more detail in Chapter 14. For now, suffice it to say that macromolecular additives to emulsion systems constitute a major pathway for attaining workable, long-lived practical emulsions. In fact, their use is essential to many important product types, not the least of which are food colloids, inks, pharmaceuticals, and the photographic industry.

11.6.2. Solid Particles

A second class of effective emulsifying agents and stabilizers commonly encountered is finely divided solid particles. It has been known for some time that particles of true colloidal dimensions (e.g., < 100 nm in diameter) that are partially wetted by both aqueous and organic liquids can form stabilizing films and produce both o/w and w/o emulsions with significant stability. Emulsion stabilization by solid particles relies on the specific location of the particles at the interface to produce a strong, rigid barrier that prevents or inhibits the coalescence of drops (Fig. 11.3). It may also impart a degree of electrostatic repulsion, which enhances the overall stabilizing power of the system. There are three keys to the use of particulate solids as emulsion stabilizers: (1) particle size, (2) the state of stabilizer particle dispersion, and (3) the relative wettability of the particles by each liquid component of the emulsion system.

In practice it is found that the stabilizer particles must be small compared to the size of the emulsion droplet and in a state of incipient flocculation; that

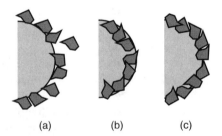

(a)   (b)   (c)

FIGURE 11.3. To effectively stabilize an emulsion, colloidal particles must have a "proper" balance of surface properties. If the particles are preferentially wetted by the continuous phase, they will be poorly adsorbed and easily desorbed from the interface and provide little stability (*a*). If they are preferentially wetted by the dispersed phase they will not be adsorbed at the interface and will again provide little stability (*b*). For optimum effectiveness the particles should be partially wetted by both phases to insure their location at the interface (*c*).

is, they should have limited colloidal stability in both liquids, otherwise their tendency to "locate" at the oil–water interface will not be sufficiently strong for them to "complete their mission." For the third condition, the solid must exhibit a significant contact angle at the three-phase (oil–water–solid) contact line, conventionally as measured through the aqueous phase. For maximum efficiency, the stabilizer usually should be preferentially wetted by the continuous phase (but not excessively so). If the solid particles are too strongly wetted by either of the two liquid phases the required stabilizing action will not result. It is usually necessary, therefore, to control such factors closely by controlling the system pH or by the addition of materials that adsorb onto the particles and impart the required surface characteristics (see Chapter 17).

11.6.3. Surfactants

The last major class of emulsifiers and stabilizers is that of the monomeric surfactants which adsorb at interfaces, lower the interfacial tension, and, hopefully, impose a stabilizing barrier between emulsion drops. Surfactants are the most widely studied and perhaps best understood class of emulsifiers and stabilizers. Perhaps because they are more amenable to both experimental and theoretical analysis, they have been used to probe the finer points of emulsified systems. They will therefore be discussed in more detail than polymers and sols.

Because of their effectiveness at lowering interfacial tensions, they are of vital importance to most practical systems, facilitating the formation of small droplets with a minimum of power input. However, because of their relative mobility into and out of the interface, their practical effectiveness as stabilizers acting alone has been questioned. There is no doubt that their presence significantly prolongs the life of most emulsions; however, the assumed role of the monomolecular adsorbed film is being challenged by the concept of a more complex activity in the guise of liquid crystals or other less ordered structures.

11.6.4. Surfactant Structure and Emulsion Performance

It would be nice if the world of emulsion formulation were such that a simple correlation could be obtained between the chemical structure of a surfactant and its performance in practice. Unfortunately, the complicated nature of typical emulsion formulations (the nature of the oil phase, additives in the liquid phases, specific surfactant interactions, etc.) make correlations between surfactant structure and properties in emulsification processes very empirical.

In the absence of a handy quantitative and absolute method for choosing a surfactant for a given application, it is possible to outline a few rules of thumb that have historically proved useful for narrowing down the possibilities and limiting the amount of experimentation required for the final selection of surfactant(s) for a given application. First and foremost, of course, the surfactant must exhibit sufficient surface activity to ensure significant adsorption at the oil–water interface. That activity must be related to the actual conditions of use and not inferred from its activity in water alone. The presence of materials such as electrolytes and polymers can greatly alter the role of the surfactant in stabilizing an emulsion as well as in controlling the type of emulsion formed.

The surfactant (or surfactants) employed in an emulsion formulation should produce as strong an interfacial film as possible, consistent with their ability to produce the required droplet size under the conditions of emulsification. It is useful, therefore, to choose a surfactant system with maximum lateral interaction among the surfactant molecules concurrent with efficient and effective lowering of the interfacial tension.

At the molecular level, the choice of surfactant for a given application must take into consideration the type of emulsion desired and the nature of the oil phase. As a general rule, oil–soluble surfactants will preferentially produce w/o emulsions while water-soluble surfactants yield o/w systems. Because of the role of the interfacial layer in emulsion stabilization, it is often found that a mixture of surfactants with widely differing solubility properties will produce emulsions with enhanced stability. Finally, it is usually safe to say that the more polar the oil phase, the more polar will be the surfactant required to provide optimum emulsification and stability. Such rules of thumb, while having great practical utility, are less than satisfying on a theoretical level. One would really like to have a neat, quantitative formula for the design of complete emulsion systems. A number of attempts have been made over the years to develop just such a quantitative approach to surfactant selection. Some such approaches are briefly discussed.

11.6.5. Liquid Crystals and Emulsion Stability

The mechanical strength of the interfacial film stabilizing an emulsion can have a significant impact on the overall stability of the system. Liquid crystalline phases occur in solutions of surface-active materials as the concentration is increased from that of a dilute solution to a saturated system in which true crystallization occurs (see also Chapter 15). Such phases possess a degree of order that produces substantial changes in the properties of the system relative to those of molecular or micellar solutions, including a higher degree of rigidity, larger structural units, and less fluctuation in composition. In the present context, such phases, if present at the o/w interface, might be expected to impart an added degree of stability to systems in which they are produced (Fig. 11.4). In a utilitarian sense, then, surfactant liquid crystals at interfaces may be considered to act in a manner similar to colloidal sols or perhaps the mixed interfacial complexes discussed below.

The presence of liquid-crystalline phases at the oil–water interface has been shown to produce improvements in the stability of various emulsions, although the exact mechanism of their action is still subject to some question.

FIGURE 11.4. In a manner similar to stabilization by colloidal particles, surfactant liquid crystals may adsorb at the emulsion interface and provide mechanical, steric, and/or electrostatic stabilization.

Even in the absence of complete understanding, the use of liquid crystals at o/w interfaces has been demonstrated in practical applications.

11.6.6. Mixed Surfactant Systems and Interfacial Complexes

It has been found that the presence of two surfactant species, one water and the other oil soluble, can greatly enhance the stability of an emulsion system. The effect has been explained by invoking two possible mechanisms: (1) to the production of very low interfacial tensions through a synergistic effect that increases the effectiveness of adsorption of the combination and (2) the formation of cooperative surfactant "complexes," which impart greater strength (i.e., a more effective energy barrier to flocculation) to the o/w interface.

An "interfacial" complex may be defined as an association of two or more surface-active molecules at an interface in a relationship that does not exist in either of the bulk phases (Fig. 11.5). Each bulk phase must contain at least one component of the complex, although the presence of both in any one phase is not ruled out. According to le Chatelier's principle, the formation of

FIGURE 11.5. When two surfactants are employed in an emulsion, one more soluble in water and the other in oil, a synergistic effect may be observed that produces a stability better than either material alone. That added stability is sometimes attributed to the formation of an interfacial "complex," although the exact nature of such a complex may not be clear.

an interfacial complex will increase the Gibbs interfacial excess Γ_i for each individual solute involved, and consequently, the interfacial tension of the system will decrease more rapidly with increasing concentration of either component.

The proposed existence of the interfacial complex is distinct from the situation of simple simultaneous adsorption of oil-soluble and water-soluble surfactants. In the case of simultaneous adsorption, each component is competing for available space in the interfacial region and contributes a weighted effect to the overall energetics of the system. The effect of the complex will be synergistic, with the net effect exceeding that of either component or any simple combination of the two.

A possible beneficial effect of interfacial complex formation, in addition to improved surface energetics, is that such structures may possess a greater mechanical strength than a simple mixed interfacial layer. The closer molecular packing density and greater extent of lateral interaction between hydrophobic chains may result in significant decreases in the mobility of molecules at the interface and a decrease in the rate of drop coalescence. Such an effect has often been mentioned in terms of increased interfacial viscosity or elasticity, although the exact role of interfacial rheology in emulsion stabilization is not completely understood.

11.6.7. Emulsion Type

The idea that surfactant molecules preferentially orient at the oil–water interface not only helps clarify the picture of monomolecular film stabilization, but also sheds light on the problem of explaining the emulsion type obtained as a function of the chemical structure of the adsorbed species. It was recognized early that the nature of the surfactant employed in the preparation of an emulsion could influence the type of emulsion formed. For example, while the alkali metal salts of fatty acid soaps normally produce o/w emulsions under a given set of circumstances, the use of di- and trivalent soaps often results in the formation of w/o systems. The invocation of a monolayer mechanism for the stabilization of emulsion droplets requires the formation of a relatively close-packed surfactant film at the interface. It is clear, then, that the geometry of the adsorbed molecules must play an important role in the effect obtained. For efficiency of packing, it can be seen from Figure 11.6 that the formation of w/o systems with polyvalent soaps seems almost inevitable.

The steric requirements of surfactant molecules have historically been referred to in terms of an "oriented wedge" of surfactant molecules at the interface. The concept lead to the "rule of thumb" that if the hydrophilic head of the surfactant was larger than the tail, the result would be an o/w emulsion. If the relationship were reversed, the emulsion would be of the w/o type—a very neat and simple relationship which, due to numerous exceptions and lack of theoretical foundation, fell out of favor for some time. More recently, however, consideration of the critical role of the structure of the

FIGURE 11.6. The geometry of the surfactant at the liquid–liquid interface will generally determine the type of emulsion formed, as illustrated for the case of a divalent metal soap, that will usually produce a w/o (water-in-oil) emulsion.

surfactant has again come into vogue, this time with some theoretical backing. Molecular geometric considerations will be mentioned in a bit more detail below.

A related "rule" concerning surfactant structure and the type of emulsion formed is related to the solubility of the surfactant in the two liquids. The rule states that the liquid in which the surfactant is most soluble will be the continuous phase in the final emulsion. That is, if the surfactant is more soluble in the oil phase, a w/o emulsion will result. A more water-soluble material will produce a o/w system.

That concept was extended from a theoretical standpoint by the postulation that the presence of an absorbed interfacial film requires the existence of two interfacial tensions—one at the oil–monolayer interface and a second at the water–monolayer interface. Since the two tensions will not, except in very unusual circumstances, be equal, the interfacial layer will spontaneously curve, with the direction of curvature determined by the relative magnitudes of the two tensions. Logically, the film will curve in the direction of the higher interfacial tension so that the phase associated with that interface will become the dispersed phase in the system. Unfortunately, this seemingly useful rule also falls on the sword of too many exceptions. Although such simple views of the role of the adsorbed monolayer in determining the nature of the emulsion can be quite useful, the many exceptions make them less than satisfying from a theoretical point of view.

In addition to the molecular nature of the emulsifier employed, the relative amounts of the two phases in the system might be expected to affect strongly the type of emulsion obtained. If one assumes that an emulsion is composed of rigid, spherical droplets of equal size, simple geometry shows that the maximum volume fraction of dispersed phase which can be obtained is 74.02%. One may speculate, then, that any emulsified system in which that level was exceeded must result in phase inversion to an emulsion of the opposite type. It has been shown, however, that it is possible to prepare emulsions of dispersed volume fractions far exceeding that "theoretical" limit. Looking at the reality of the situation more closely, it is possible to identify several reasons to invalidate such a simple geometric approach.

In the first place, emulsion droplets are not, and will never be, perfectly monodisperse; as a result, it is possible for smaller droplets to locate themselves in the spaces between close packed, larger droplets (Fig. 11.7), increasing the total potential packing density of the system. In addition, emulsion droplets are not rigid spheres, but highly deformable so that their shape can be changed from spherical to various elongated or polyhedral shapes to fit the needs of the system. Large excursions from sphericity will be generally unfavorable, of course, since that will require the formation of additional interfacial area for a given dispersed volume fraction. Such an increase in interfacial area may strain the ability of the adsorbed emulsifier or stabilizer film to the point where droplet coalescence occurs.

Even though the early rules of thumb concerning emulsifier type and emulsion type are rife with exceptions, there remains something inherently satisfying in the ideas they contain, aside from their continued practical utility. However, science forever strives to develop the "unified theory of" everything that allows one to predict exactly what will happen in a given set of circumstances, and the area of emulsion science and technology is no exception. The first really practical strides to that goal were taken with the introduction of the so-called hydrophile–lipophile balance (HLB) number.

11.6.8. The Hydrophile–Lipophile Balance (HLB)

It has been a long-term goal of surface science to devise a quantitative way of correlating the chemical structure of surfactant molecules with their surface activity that would facilitate the choice of material for use in a given formulation. The greatest success along these lines has been achieved in the field of emulsions. The first reasonably successful attempt at that goal was the HLB system first developed by Griffin. His work was an attempt to place the choice

(a)   (b)

FIGURE 11.7. Although ideal hexagonal close packing of spheres predicts a maximum dispersed phase volume fraction of about 0.74 (*a*), the inherent polydispersity of most emulsions allows for the preparation of emulsions of much greater internal volume fraction, as smaller droplets are located in the interstitial spaces between larger drops (*b*).

of the optimum nonionic surfactant for the stabilization of a given emulsion on a quantitative, somewhat theoretical basis. In this system, Griffin proposed to calculate the HLB number of a surfactant from its chemical structure and match that number with the HLB of the oil phase which was to be dispersed. The system employed certain empirical formulas to calculate the HLB number for molecular structures, producing numbers between 0 and 20 on an arbitrary dimensionless scale.

At the high end of the scale (8–18) lie hydrophilic surfactants, which possess high water solubility and generally act as good aqueous solubilizing agents, detergents, and stabilizers for o/w emulsions; at the low end (3–6) are surfactants with low water solubility, which act as solublizers of water in oils and good w/o emulsion stabilizers. In the middle are materials that are very surface-active, in terms of lowering surface and interfacial tensions, but generally perform poorly as emulsion stabilizers, possibly because of their balanced solubility characteristics in the two phases. The effectiveness of a given surfactant in stabilizing a particular emulsion system would then depend on the balance between the HLB of the surfactant and the oil phase involved.

For nonionic surfactants with polyoxyethylene solubilizing groups, the HLB may be calculated from the formula

$$\text{HLB} = \frac{\text{mol\% hydrophilic group}}{5} \tag{11.3}$$

In such a calculation, an unsubstituted polyoxyethylene glycol would have an HLB of 20. HLB values for some typical nonionic surfactants are given in Table 11.3. Surfactants based upon polyhydric alcohol fatty acid esters such as glycerol monostearate can be handled by the relationship

TABLE 11.3. HLB Values for Typical Nonionic Surfactant Structures

Surfactant	Typical Commercial Name	HLB
Sorbitan trioleate	Span 85	1.8
Sorbitan tristearate	Span 65	2.1
Propylene glycol monostearate	"Pure"	3.4
Glycerol monostearate	Atmul 67	3.8
Sorbitan monooleate	Span 80	4.3
Sorbitan monostearate	Span 60	4.7
Diethylene glycol monolaurate	Glaurin	6.1
Sorbitan monolaurate	Span 20	8.6
Glycerol monostearate	Aldo 28	11
Polyoxyethylene(2) cetyl ether	Brij 52	5.3
Polyoxyethylene(10) cetyl ether	Brij 56	12.9
Polyoxyethylene(2) cetyl ether	Brij 58	15.7
Polyoxyethylene(6) tridecyl ether	Renex 36	11.4
Polyoxyethylene(12) tridecyl ether	Renex 30	14.5
Polyoxyethylene(15) tridecyl ether	Renex 31	15.4

$$HLB = 20 \left(\frac{1-S}{A}\right) \qquad (11.4)$$

where S is the saponification number of the ester and A is the acid number of the acid. A typical surfactant of this type, polyoxyethylene–20–sorbitan monolaurate (Tween 20), with $S = 45.5$ and $A = 276$, would have an HLB of 16.7. For materials that cannot be completely saponified, an empirical formula of the form

$$HLB = \frac{E + P}{5} \qquad (11.5)$$

can be employed, where E is the weight percent of polyoxyethylene chain and P is the weight percent of polyhydric alcohol (glycerol, sorbitan, etc.) in the molecule.

Although the system proposed by Griffin proved to be very useful from a formulation chemists point of view, its empirical nature did not satisfy the desire of many for a more sound theoretical basis for surfactant characterization. Davies and Rideal suggested that HLB numbers could be calculated based upon group contributions according to the formula

$$HLB = 7 + \Sigma(\text{hydrophilic group numbers}) - \qquad (11.6)$$
$$\Sigma(\text{hydrophobic group numbers})$$

Some typical group numbers as listed by Davies and Rideal, as well as other investigators, are listed in Table 11.4.

The use of the HLB system for choosing the best emulsifier for a given application originally required the performance of a number of experiments

TABLE 11.4. Group Numbers for the Calculation of HLBs According to Davies and Rideal

Group	HLB Number	Group	HLB Number
Hydrophilic		Hydrophobic	
$-SO_4NA$	38.7	$-CH-$	-0.475
$-COOK$	21.1	$-CH_2-$	-0.475
$-COONa$	19.1	$-CH_3$	-0.475
$-N$ (tertiary amine)	9.4	$=CH-$	-0.475
Ester (sorbitan)	6.8	$-CF_2-$	-0.87
Ester (free)	2.4	$-CF_3$	-0.87
$-COOH$	2.1	Miscellaneous	
$-OH$ (free)	1.9	$-(CH_2CH_2O)-$	0.33
$-O-$	1.3	$-(CH_2CH_2CH_2O)-$	-0.15
$-OH$ (sorbitan)	0.5		

in which surfactants or surfactant mixtures with a range of HLB numbers are employed to prepare emulsions of the oil in question, and the stability of the resulting emulsions evaluated by measuring the amount of creaming that occurred with time. The use of surfactant mixtures can become complicated by the fact that such mixtures often produce more stable emulsions than a single surfactant with the same nominal HLB number. The HLB of a mixture is usually assumed to be an algebraic mean of those of the components

$$HLB_{mix} = f_A \times HLB_A + (1-f_A) \times HLB_B \qquad (11.7)$$

where f_A is the weight fraction of surfactant A in the mixture. While strict adherence to this equation not always found, the concept remains a useful tool for surfactant formulation purposes, especially in the absence of a better option.

While it goes a long way toward simplifying the choice of surfactants for the preparation of a given o/w emulsion, the HLB system does not always provide a clearcut answer for a given system. It does not, for example, take into consideration the effects of a surfactant on the physical properties of the continuous phase, especially its rheological characteristics. As noted previously, the viscosity of the continuous phase may significantly affect the rate of creaming, as will alterations in the relative densities of the two phases. As a result, it is possible to prepare very stable emulsions with surfactants whose HLB numbers lie well away from the "optimal" which would be predicted by the strict application of the HLB approach. Regardless of its faults, the HLB system as originally proposed by Griffin and subsequently expanded by others has found extensive practical use. A number of theoretical approaches to "explain" the success of the HLB approach have been developed. Perhaps one of the most interesting is that based on cohesive energy density or the overall attractive forces between molecules in a condensed phase. That approach, discussed briefly below, goes a long way toward justifying (in a theoretical sense) the empiricism of the original Griffin method and turning it into a more "satisfying" theory.

11.6.9. Cohesive Energies and the Solubility Parameter

Interactions among atoms and molecules, as we have seen, are a result of various forces stemming from their atomic or molecular structure, including electrostatic or charge interactions, steric or entropic phenomena, and the ever-present van der Waals forces. Of these, electrostatic and steric interactions may be repulsive in that they act to force the interacting units apart or at least reduce the net attraction between units. The van der Waals forces, on the other hand, are usually (but not always) attractive. When one discusses the use of a surfactant as an emulsion stabilizer, as in the above sections, the concept of the function of the surfactant is that it have a strong tendency to locate or adsorb at the liquid–liquid interface, there to better form a barrier to retard drop contact, flocculation, and coagulation. Logically, then, one would expect that the more effective surfactant, having by definition two distinctly different molecular parts, would be that which can more closely approximate the chemical natures of the two phases with each end of the molecule (Fig. 11.8).

On the macro scale, the colligative properties (e.g., boiling point, melting point, viscosity) of a material will depend on the total of the intermolecular interactions among the molecules or atoms. While the concept of the mutual attraction or cohesion among units in gases, liquids, and solids has been around for many years (It was recognized by van der Waals in his studies on why the volume of gases decreased more under pressure than predicted by the ideal gas law), it has been only since the 1950s that we have been in a position to really quantify the concept to the point where it can serve as a tool in improving our understanding of physical, chemical, and biological systems. Practical areas of interest include problems such as adhesion (or lack thereof), viscosity, stress fracture and material strength, protein (e.g., enzyme) conformation and function, transport through biological membranes, RNA and DNA structure and replication, emulsions and foams, and so on. Our ability to understand, quantify, modify, and use these interactions promises much in many areas of chemistry, physics, materials science, medicine, pharmacology, biotechnology, microelectronics, food science, and many more.

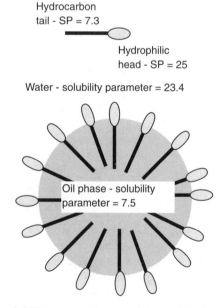

FIGURE 11.8. The solubility parameter concept predicts that the optimum interactions among surfactant, oil phase, and aqueous phase will occur when the two portions of the surfactant molecule closely match the parameters of each phase.

The first firm steps in the quantification of material interactions in terms of their molecular cohesive (stick together) properties came in 1950 with the studies of Hildebrand on the solubility of nonelectrolytes. Hildebrand characterized the cohesive energy density of a material as an intensive property he called the "solubility parameter," usually given the symbol δ, measured in $(J\ cm^{-3})^{1/2}$. The reference to solubility stems from the fact that the original studies were based on the solubility of the materials of interest in various solvents and the correlations between the chemical structures of the two. The square root was chosen because it was found to allow one to calculate an average value of δ for mixtures of materials and to estimate values for materials based on their atomic and functional group composition.

Conceptually, the phenomenon can be visualized (at the molecular level) as the propensity for neighboring molecules to mutually attract and "stick" together. The higher the cohesive energy density, the more "sticky" the interaction, as reflected by such bulk properties as higher boiling point, higher viscosity, and so on. For example, water (MW = 18) is found to have a cohesive energy density of $\delta = 47.9$ $(J\ cm^{-3})^{1/2}$ with a boiling point of 100°C and viscosity of 1.0 cP (centipoise) at 20°C, compared to diethyl ether (MW = 74, $\delta = 15.1$) having a boiling point of 35°C and viscosity of 0.23 cP at 20°C. Obviously, water molecules "stick" to each other more strongly than ether molecules.

While Hildebrand's original work was carried out on what are considered "nonpolar" molecules (hydrocarbons, etc.), it has since been extended to include weakly polar compounds (esters, ethers, amides, etc.), and strongly hydrogen bonding materials such as water, alcohols, and amines. The extension of the solubility parameter concept to polar and hydrogen bonding compounds greatly expands the range of potentially useful materials that can be treated. We know from earlier chapters that the forces responsible for intermolecular interactions (excluding ionic interactions) include three components: (1) the universal London dispersion force, (2) the normal dipole and induced dipole interactions, and (3) the stronger hydrogen bonding interactions found in water, carboxylic acids, etc. From the values given in Table 11.5, it is clear that materials classed as nonpolar have low values for δ (i.e., low cohesive energy density), polar materials have intermediate values, and hydrogen bonding materials quite high values.

When one applies the solubility parameter idea to mixtures, the same basic concepts allow one to visualize the molecular situation in terms of the interactions among the various molecular species present. As a first approximation, one can estimate the cohesive interaction between two unlike molecules, $E_{c(ab)}$, as the product of the two solubility parameters

$$E_{c(ab)} = \delta_a \times \delta_b \tag{11.8}$$

where subscripts a and b refer to the two components of the mixture.

TABLE 11.5. Solubility Parameters (Cohesive Energy Densities) of Some Commonly Encountered Materials

Material	Solubility Parameter [δ, (J cm^{-3})$^{1/2}$]	Material	Solubility Parameter [δ, (J cm^{-3})$^{1/2}$]
Alkanes		Ethers	
n-Heptane	15.3	Diethyl ether	15.8
n-Decane	15.8	Dibenzyl ether	19.3
n-Hexadecane	16.4	Methoxy benzene	19.5
Cyclohexane	16.8	Aromatics	
Halogenated compounds		Benzene	18.6
Chloroform	19.0	Toluene	18.2
Trichloro ethylene	19.0	Naphthalene	20.3
Chlorobenzene	19.6	Styrene	19.0
Aldehydes and ketones		Nitrogen-containing compounds	
Acetaldehyde	21.1	Acetonitrile	24.4
Benzaldehyde	21.5	Benzonitrile	19.9
2-Butanone	19.0	Nitrobenzene	22.2
Acetophenone	21.8	Ethanol amine	31.5
Furfural	24.4	Pyridine	21.8
Esters		Diethyl amine	16.8
Ethyl acetate	18.1	Formamide	36.6
n-Butyl acetate	17.4	Miscellaneous	
Di-n-butylphthalate	20.21	Dimethyl sulfide	19.2
Alcohols		Dimethyl carbonate	20.3
Ethanol	26.5	Thiophene	20.1
1-Propanol	24.5	Tricresyl phosphate	17.2
1-Butanol	23.1	Triphenyl	
Benzyl alcohol	23.8	phosphate	17.6
1-Dodecanol	20.0	Dimethylsulfoxide	24.5
Glycerol	33.7	Dimethyl siloxanes	10–12
		Water	47.9

Source: A. F. M. Barton, *CRC Handbook of Solubility Parameters and Other Cohesion Parameters,"* CRC Press, Boca Raton, FL, 1983, Table 2, pp. 142–149.

Take, for example, a mixture of n-hexane ($\delta = 14.9$) and water ($\delta = 47.9$). As illustrated in Figure 11.9a, the water molecules have a very strong (sticky) mutual attraction ($E_c = 2294$ J cm^{-3}) while the attraction for hexane [$E_{c(ab)} = 714$] is relatively small. The mutual attraction among hexane molecules is also small ($E_c = 222$), but they have little choice in the matter. The result—the mutual attraction of water molecules overwhelms that between water and hexane and hexane is left "out in the cold"; in other words, oil and water don't mix.

If one looks at two more closely related materials (in terms of the polarity of the molecule) such as water and glycerol ($\delta = 33.7$), one expects, and finds,

(a)

(b)

FIGURE 11.9. The relative cohesive interactions between the molecules of a mixture of two liquids determines the extent of their miscibility.

a much stronger interaction [$E_{c(ab)}$ = 1614]. That result is reflected in the infinite miscibility of those two materials. The most important aspect of the solubility parameter in the present context is that it reflects the chemical character of a molecule and allows one to use the "like dissolves like" rule of thumb as a guide for determining the affinity of one material (e.g., a solute) for another (the solvent).

While the original Hildebrand concept was developed for essentially nonpolar materials, its has since been expanded to take into consideration weakly polar and strong (e.g., hydrogen bonding) interactions. An in-depth discussion of solubility parameter theory and application is, of course, beyond the scope of this book. An extensive compendium of theory and practice is referenced in the Bibliography. Our purpose here is to introduce the concept within the context of surfactants and their application in emulsions. Their extension to other areas of surface and colloid science should be fairly obvious and will be mentioned in the appropriate place.

11.7. SOLUBILITY PARAMETERS, SURFACTANTS, AND EMULSIONS

Surfactants are, by definition, molecules that possess a "split personality" in that two or more sections of the molecule exhibit very different chemical (i.e., polarity or solubility) characteristics. Since the early studies of Hildebrand were carried out on relatively simple (chemically speaking) materials, it was not immediately obvious that the concepts he developed would be applicable to highly unsymmetric systems exemplified by normal surfactant molecules. Luckily, the cohesive energy density of a molecule can be relatively accurately broken down into the individual contribution of its component parts, which usually allows one to calculate a "good" ($\pm 10\%$) value for a given chemical structure using tabulated group values, some of which are given in Table 11.6.

The basic calculation for estimating solubility parameters from group contributions at 25°C is

TABLE 11.6. Molar Attraction Constants (Group Numbers) for Calculation of Solubility Parameters at 298 K

Group	F_g	Group	F_g
$-CH_3$	438	Conjugation	40–60
$-CH_2-$ (single bonded)	272	$-H$ (variable)	160–205
$-CH<$	57	$-O-$ (ether)	143
$>C<$	190	$>CO$ (ketones)	563
$CH_2=$	389	$-COO$ (esters)	634
$-CH=$ (double bonded)	227	$-CN$	839
$>C=$	39	$-Cl$ (single)	552
$-CH\equiv C-$	583	$-Cl$ ($>CCl_2$)	532
$-C\equiv C-$	454	$-Cl$ ($>CCl_3$)	511
Phenyl	1503	$-Cl$ ($>C(Cl)-C(Cl)<$	497
Phenylene (o,m,p)	1346	$-Br$ (single)	340
Naphthyl	2344	$-I$ (single)	425
5-Membered ring	215–235	$-CF_2$ (n-fluorocarbons)	150
6-Membered ring	195–215	$-CF_3$ (n-fluorocarbons)	274
$-OH$ (single)	348	$-PO_4$ (organic phosphate)	1020
$-S-$ (sulfides)	460	$-ONO_2$ (nitrate)	900
$-SH$ (thiol)	644	$-NO_2$ (aliphatic nitro)	900

Source: After P. A. Small, *J. Appl. Chem.* **3,** 71 (1953).

$$\delta_d = \frac{\Sigma F_g}{V} \tag{11.9}$$

where F_g is the group contribution for each simple unit and V is the molecular volume. The subscript d on the solubility parameter indicates that the calculation is for the dispersion component of the energy density. However, a reasonable first estimate for polar and even hydrogen bonding materials can be obtained with the formula. More accurate results for highly polar materials can be obtained using the methods described in the bibliographic reference.

To effectively use calculations according to Eq. (11.9), one must keep in mind that many materials, especially surfactants, are not molecularly pure and that small amounts of impurities can affect properties, including cohesive energy density to an extent well beyond what one might expect. In addition, some materials containing long, flexible chains can fold or coil, replacing intermolecular interactions with intramolecular ones and thereby altering the "character" of the molecule in terms of its apparent solubility parameter. Because of the complex nature of interactions involving ionic surfactants, the calculation of their solubility parameters can become quite tedious. Values of (and the corresponding HLB for several common surfactants are given in Table 11.7.

In general, one can expect a high degree of interaction between materials that have solubility parameters differing by $\pm$ 4 $(\mathrm{J\ cm}^{-3})^{1/2}$, although it is sometimes necessary to extend that range to 8 or 10 units. For example, a material (or group) with $\delta = 39.3$ $(\mathrm{J\ cm}^{-3})^{1/2}$ (formamide) would be expected

TABLE 11.7. Comparison of Solubility Parameters and HLB for Some Common Simple Surfactants

Surfactant	$\delta/(\mathrm{J\ cm}^{-3})^{1/2}$	HLB
Oleic acid	16.8	1.0
Glycerol monostearate	17.0	3.8
Sorbitan monolaurate	17.6	8.6
Polyoxyethylene (10) oleyl ether	18.2	12.4
Polyoxyethylene (20) cetyl ether	18.6	15.7
Sodium octadecanoate (stearate)	19.0	18.0
Sodium hexadecanoate (palmitate)	19.2	19.0
Sodium dodecanoate (laurate)	19.6	20.9
Sodium dinonylnaphthalene sulfonate	21.5	28.5
Dioctyl sodium sulfosuccinate	25.0	32.0
Sodium dodecyl sulfate	28.8	40.0
Sodium decyl sulfate	30.0	40.0
Sodium octyl sulfate	32.3	41.9

Source: From A. F. M. Barton, *CRC Handbook of Solubility Parameters and Other Cohesion Parameters*, CRC Press, Boca Raton, FL, 1983, Table 7, p. 445.

to have a greater interaction with water than a material with $\delta = 29.9$ (ethylene glycol), although both are water-miscible.

In the context of emulsion stabilization, it is important that a surfactant molecule have a strong interaction with both the water and oil phases. If water interactions are too dominant (i.e., $E_{c(ab)}$ is too large), the molecule will tend to be too soluble in that phase and will lose effectiveness at the oil–water interface. If $E_{c(ab)}$ is too small, the opposite effect will result (Fig. 11.10). The goal, then is to balance the cohesive interaction of the surfactant tail with that of the oil phase and that of the head with the aqueous phase. That end can (in theory) be accomplished in one of two ways: (1) by adjusting the structure of the surfactant or (2) by adjusting the composition (e.g., the polarity) of one or both phases. In many critical applications (e.g., pharmaceuticals, cosmetics, foods) the choice of surfactant is limited by law and/or activity, so that it may be more feasible to adjust the characteristics of one or both liquid phases.

As an example of the modification of the oil phase, one can consider an emulsion of mineral oil ($\delta = 14.4$) in water. A typical surfactant tail group [e.g., lauryl, $CH_3(CH_2)_{11}-$] has a "partial" solubility parameter of 16.7. In order to improve the "match" between surfactant tail and oil phase, one might add a more polar solute such as cetyl alcohol [$CH_2(CH_2)_{16}OH$, $\delta \approx 20$], thereby increasing the polarity of the oil phase. Alternatively, one can decrease the cohesion between the aqueous phase and the surfactant head group by the addition of a less polar solute such as propylene glycol ($\delta = 30.7$).

As a first approximation, and over a limited range of δ values, one can estimate the solubility parameter of a mixture based on the mole fraction of each component as

$$\delta_{mix} = X_a(\delta_a) + (1-X_a)(\delta_b) \tag{11.10}$$

| (a) | (b) | (c) |

FIGURE 11.10. The optimum functioning of a surfactant as an emulsion stabilizer depends on the correct balance of its solubility parameter with those of the two phases. The proper balance gives optimum adsorption at the interface (*a*). If its value is too close to that of the external phase, the surfactant will be too soluble in that phase and will not adequately adsorb (*b*). If it is too close to that of the internal phase, the same will occur, but in the internal phase (*c*).

TABLE 11.8. Solubility Parameters for Some Common Oil Phases

Oil	$\delta/(\text{J cm}^{-3})^{1/2}$	Oil	$\delta/(\text{J cm}^{-3})^{1/2}$
Mineral oil (white refined)	14.5	Butyl stearate	15.3
Pine oil	14.9	Dibutyl phthalate	20.1
Linseed oil (white refined)	14.9	Dioctyl phthalate	18.2
Soy oil	15.1	1,1,1-Trichloroethane	17.5
Castor oil	18.2	Cyclohexanone	20.3
Lanolin	18.1	1-Dodecanol	20.0
Carnuba wax	18.1	Trichloroethylene	19.0
Beeswax	17.7	o-Dichlorobenzene	20.5

From A. F. M. Barton, *CRC Handbook of Solubility Parameters and Other Cohesion Parameters,* CRC Press, Boca Raton, FL, 1983, Table 9, p. 167 and Eq. 11.11.

where X_a is the mole fraction of component a and δ_a and δ_b are the respective solubility parameters. The solubility parameters of selected oils and solvents are given in Table 11.8.

11.8. THE RELATIONSHIP BETWEEN HLB AND SOLUBILITY PARAMETER

It should not be surprising to find some quantitative relationship between the Griffin HLB number and the Hildebrand δ, given that both concepts are related to the balance of interactions between hydrophobic tails and oil phases, and head groups and aqueous phases. For example, a study of various surfactant–solvent systems by Little led to the following relationship between δ and HLB:

$$\delta = \frac{243}{(54 - \text{HLB})} + 12.3 \qquad (11.11)$$

Alternatively, a polynomial regression analysis of the data in Table 11.7 produces the following expression

$$\delta = 0.0003 \times \text{HLB}^3 - 0.0096 \times \text{HLB} + 16.404 \qquad (R^2 = 0.9831) \quad (11.12)$$

Equations (11.11) and (11.12) are, of course, highly empirical relationships, and alternative methods of relating HLB and solubility parameter have been proposed. In the absence of better information, such relationships can be useful as tools for estimating the solubility parameter or HLB of a particular surfactant structure and to help in its evaluation as an emulsifier candidate.

11.9. THE GEOMETRICAL APPROACH

As an additional tool for approaching the problem of relating surfactant structure to emulsion formulation (to supplement, but not replace the "classic" approaches) Israelachvilli, Mitchell, and Ninham considered the geometrical constraints imposed by the particular molecular characteristics of a surfactant molecule that control the formation of aggregates (e.g., micellization) and other interfacial interactions.

In analyzing the relationships between the aggregation characteristics of a surface active material (aggregate size, shape, curvature, etc.) and molecular structure, the authors defined a geometric factor F by the equation

$$F = \frac{v}{a_o l_c} \tag{11.13}$$

where v is the molecular volume of the hydrophobic group, a_o is the head group area, and l_c is the critical length of the hydrophobe. The factor F can be viewed as a type of HLB number, based on volume fraction instead of weight fraction of hydrophobe, and the geometry of the hydrophobic chain.

The "geometry" of the molecules expected at the oil–water interface predicted by Equation (11.13) is illustrated in Figure 11.11. By use of the geometric considerations (see also Chapter 15), it can be seen that the value of F determined from molecular geometry should predict the type of emulsion (o/w or

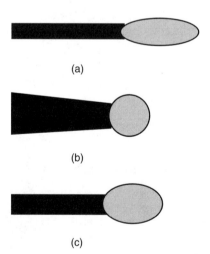

(a)

(b)

(c)

FIGURE 11.11. The type of emulsion formed by a given surfactant may also depend to a great extent on the geometry of the molecule as given by its geometric factor, F, in Equation (11.13.) For $F < 1$, the tendency will be for o/w emulsion formation (a). For $F > 1$ w/o, emulsions will probably be formed (b). For $F = 1$, other factors (procedure, luck, magic, etc.) may be the determining factors (c).

w/o) formed by a particular surfactant. For instance, if $F < 1$, the curvature of the oil–water interface should be concave toward the oil phase, leading to an o/w emulsion. For $F > 1$, the reverse would be expected. At $F = 1$, a critical condition is expected where phase inversion would occur, or multiple emulsion formation will be favored.

11.9.1. Phase Inversion Temperature (PIT)

An important class of surfactants for use as emulsifiers and stabilizers is that of the nonionic polyoxyethylene (POE) adducts. This class of materials is solubilized in water through hydrogen bonding with the POE chain. As mentioned in Chapter 4, hydrogen bonding is a temperature sensitive interaction and decreases as the temperature increases. Nonionic materials therefore often exhibit an inverse temperature–solubility relationship leading to the appearance of a cloud point for many examples of the class. The cloud point has already been discussed in general terms, but its existence has interesting ramifications when viewed in the context of emulsions.

Since the cloud point of a surfactant is a structure related phenomenon, it should also be related to HLB, solubility parameter, cmc, and other parameters, as is found to be the case. Clearly, temperature can play an important role in determining surfactant effectiveness where hydration (or hydrogen bonding) is the principal mechanism of solubilization. Because of the temperature sensitivity of such materials, their activity as emulsifiers and stabilizers also becomes temperature sensitive. In particular, their ability to form and stabilize o/w and w/o emulsions may change dramatically over a very narrow temperature range. In fact, an emulsion may "invert" to produce the opposite emulsion type as a result of temperature changes. Such a process is termed "phase inversion," and the temperature at which it occurs for a given system is its phase inversion temperature (PIT).

The potential importance of the temperature effect on surfactant properties has been recognized for some time and led to the concept of using the PIT as a quantitative tool for the evaluation of surfactants in emulsion systems. As a general procedure, emulsions of oil, aqueous phase, and approximately 5% surfactant were prepared by shaking at various temperatures. The temperature at which the emulsion was found to be inverted from o/w to w/o (or vice versa) was then defined as the PIT of the system. Since the effect of temperature on the solubility of nonionic surfactants is reasonably well understood, the physical principles underlying the PIT phenomenon follow directly.

It is generally found that the same circumstances that affect the solution characteristics of nonionic surfactants (their cmc, micelle size, cloud point, etc.) will also affect the PIT of emulsions prepared with the same materials. For typical polyoxyethylene nonionic surfactants, increasing the length of the POE chain will result in a higher PIT for a given oil–aqueous phase combination (Fig. 11.12), as will a broadening of the POE chain length distribution. The use of phase inversion temperatures, therefore, represents a very useful

FIGURE 11.12. The phase inversion temperature (PIT) of an emulsion will depend on the balance of properties of the surfactant and the oil phase. For a series of nonionic polyoxyethylene surfactants and oil phases, the PIT is found to increase with the oil phase carbon number and the POE content of the surfactant.

tool for the comparative evaluation of emulsion stability. Although the PIT approach to surfactant evaluation is considerably newer than the HLB number, the effects of variables on the relationships among PIT, surfactant structure, and emulsion stability show an almost linear correlation between the HLB of a surfactant under a given set of conditions and its PIT under the same circumstances. In essence, the higher the HLB of the surfactant system, the higher will be its PIT.

The PIT system of surfactant evaluation theoretically applies only to nonionic materials. However, it is often found that for a given oil–water system, a combination of two or more surfactants (e.g., a nonionic and an ionic) will produce better results than either surfactant alone, at the same (or less) total surfactant concentration. Ionic surfactants usually have the "normal" temperature–solubility relationship—higher temperature means greater solubility—and in mixtures can often swamp out the phase inversion effect of a nonionic material. However, if the ionic/nonionic mixture is used with an aqueous phase of relatively high ionic strength, the HLB/δ/F value of the molecule will be reduced and the phase inversion effect may reappear and become a useful tool again.

11.9.2. Application of HLB and PIT in Emulsion Formulation

The choice of a particular emulsifier system for an application will depend on several factors, one of which will be chemically related (optimum HLB,

δ, F, PIT, etc.); but others will be driven by the three "es"—economics, environment, and esthetics. The relative importance of the latter factors will depend mostly on price and value-added considerations, legal and functional restrictions, and subjective appeal for each individual system. Here we are concerned with the technical aspects of emulsion formation and stabilization, so other factors will be ignored.

In most general applications, the HLB system has been found most useful in guiding the formulator to a choice of surfactant most suited to his needs. Table 11.9 lists the ranges of HLB numbers that have been found to be most useful for various applications. Obviously, the ranges in which surfactants of various HLB numbers can be employed are quite broad. Specific requirements for many systems have been tabulated in the work of Becher and other cited references. While such tabulations can be very useful to the formulations chemist, it must be kept in mind that there is nothing really magic about a given HLB number. Many surfactants or mixtures may possess the same HLB, yet subtle differences in their chemical structures or interfacial properties (e.g., δ or F) may result in significant differences in performance. Particularly important may be the formation of interfacial complexes, liquid–crystalline phases, and similar. Even though the additive nature of surfactant mixtures has not been found to be linear over a wide range of compositions, over the short range of one or two HLB units usually encountered in formulation work linearity can usually be assumed with little risk. It is therefore possible to fine-tune a surfactant mixture with a minimum of actual experimental effort as indicated earlier [Eq. (11.7)].

One approach to the application of surfactant HLB to formulation is to match that of the surfactant to the oil phase being employed. The HLB (or δ) of the oil can be determined empirically or calculated using the data in Table 11.6. It is usually found that the principle of additivity will hold for mixtures of oils in a way similar to that for surfactants. Therefore, in formulating an emulsion, it is possible to determine the HLB of the oil phase and vary the surfactant or mixture HLB to achieve the optimum performance. The HLB numbers of some commonly used oil phases are given in Table 11.8. It should be pointed out that the values listed in the table are for the formulation

TABLE 11.9. HLB Ranges and Their General Areas of Application

Range	Application
3–6	w/o emulsions
7–9	Wetting
8–18	o/w emulsions
3–15	Detergency
15–18	Solubilization

of o/w emulsions. When w/o emulsions are required, the appropriate HLB value will usually be smaller.

It will be noted that HLB numbers are most often used in connection with nonionic surfactants. While ionic surfactants have been included in the HLB system, the more complex nature of the solution properties of the ionic materials makes them less suitable for the normal approaches to HLB classification. In cases where an electrical charge is desirable for reasons of stability, it is often found that surfactants that have limited water solubility and whose hydrophobic structure is such as to inhibit efficient packing into micellar structures should be most effective emulsifiers. Surfactants such as the sodium trialkylnaphthalene sulfonates and dialkylsulfo-succinates, which do not readily form large micelles in aqueous solution, have found some use in that context, usually providing advantages in droplet size and stability over simpler materials such as sodium dodecyl sulfate.

Clearly, the process of selecting the best surfactant or surfactants for the preparation of an emulsion has been greatly simplified by the development of the more or less empirical but theoretically based approaches exemplified by the HLB, solubility parameter, and PIT methods. Unfortunately, each method has its significant limitations and cannot eliminate the need for some amount of trial-and-error experimentation. As our fundamental understanding of the complex phenomena occurring at oil–water interfaces, and of the effects of additives and environmental factors on those phenomena, improves it may become possible for a single, comprehensive theory of emulsion formation and stabilization to lead to a single, quantitative scheme for the selection of the proper surfactant system.

11.9.3. Some Other Factors Affecting Emulsion Stability

A discussion of emulsion stability should include not only possible mechanisms of stabilization but also some comments concerning the timeframe of the stability requirements, because there exist external and internal factors unrelated to interfacial and colloidal phenomena that work unrelentingly to destroy the most "stable" system. The rates of colloidal degradation of emulsions vary immensely, so that it is not possible to define a single number that can be used as a measure of acceptable or unacceptable persistence. In any emulsion, especially one that is only very poorly stabilized, the breaking process will involve the coalescence of droplets brought together by the action of Brownian motion, convection currents, and other random disturbances. Their stability may be measured on the order of seconds or minutes.

Emulsions that contain more effective stabilizing additives such as one of those described above may be stable for hours, days, months, or even years. In such systems the action of random or induced motion and droplet collision will continue, but the rheological properties of the continuous phase will slow down such processes and/or interfacial layers will posses sufficient strength and rigidity so that coalescence will occur on a relatively long timescale.

In addition to the mechanical actions and interfacial energy considerations that will act to reduce the degree of dispersion of an emulsion, there are other considerations that act to limit the stability of emulsions. One such factor is the phenomenon, commonly termed "Ostwald ripening," in which large drops are found to grow at the expense of smaller ones. Such growth, whether in a crystal or an emulsion, results from differences in the chemical potential (and therefore solubility) of molecules in small particles relative to those in larger ones. Such differences arise from the fact that the pressure (or chemical potential), Δp, of a material inside a drop is inversely proportional to the drop radius, r, as given by that old friend, the Kelvin equation in which the effect of radius on solubility is given by

$$\ln (S_1/S_2) = \sigma_i V/RT \left(\frac{1}{r_1} - \frac{1}{r_2} \right) \tag{11.14}$$

where S_1 and S_2 are the solubilities of the particles of principal radii r_1 and r_2 and V is the molar volume of the phase inside the drops or crystals. The effect of the Kelvin relationship is often readily apparent in foam systems where the solubility of gases in the liquid phase can be substantial. In emulsion systems, on the other hand, the solubility of the dispersed phase may be so low that diffusion from small to large droplets will be exceedingly slow. Even in such circumstances, the process will occur, but at such a rate that it will not be apparent for long periods. In the presence of excess surfactant, however, micellar solubilization (Chapter 16) may aggravate the situation.

In the present context, it is often possible to greatly reduce the rate of droplet growth due to Ostwald ripening by employing emulsifiers and stabilizers that form a barrier to the passage of dispersed phase molecules into the continuous phase. This process can be especially important in multiple emulsion systems discussed later.

Other external factors affecting the stability of emulsions include the actions of bacteria and other microorganisms, physical abuse (e.g., shearing or rapid agitation), and freezing, especially in o/w emulsions. During the process of freezing, for example, the formation of ice crystals in the continuous phase forces the emulsion droplets together under significant pressure, often resulting in the rupture of the interfacial film and drop coalescence. It is obvious, then, that stability to such action will require an interfacial film of considerable strength. Even though the protection of an emulsion from breaking due to freezing action is of considerable economic importance, there has been relatively little fundamental research published in the area.

Bacterial action can be of importance in areas such as food, pharmaceutical, and cosmetic emulsions, or other systems which contain components subject to biological degradation such as proteins and natural gums. Such systems are obviously of great economic importance, so that a great deal of research has been devoted to the problem. In cases where biological stability is important, some advantages can be gained by the proper choice of surfactant in the stabilizing formulation, since many such materials show significant microbiocidal activity. Alternatively, bactericides and antioxidants, for instance, can be added as extra protection to prolong the life of the emulsion.

In addition to direct degradation, bacterial action affecting emulsion stability may be of a second-hand nature. The breaking or curdling of milk emulsions, for example, involves changes in the pH of the system. As the bacteria in the system propagate, they produce acidic waste products. The lowering of the pH of the system by those products decreases the degree of ionization of the milk protein (casein) stabilizing the o/w emulsion so that at a certain pH, the casein is no longer able to function as an efficient emulsion stabilizer and the emulsion breaks. It is sometimes observed that milk or cream that appears perfectly good in the bottle "breaks" when poured into a cup of hot coffee. In that case the coffee, which is acidic, may simply finish the destabilizing job begun by the bacteria or a limited stability provided by the neutralized casein (steric stabilization) is removed in a PIT-like effect.

11.10. MULTIPLE EMULSIONS

While a great deal of information has been published over the years on the theoretical and practical aspects of emulsion formation and stabilization, until recently little has been said about more complex systems generally referred to as "multiple emulsions." Multiple emulsions, as the name implies, are composed of droplets of one liquid dispersed in larger droplets of a second liquid, which is then dispersed in a final continuous phase. Typically, the internal droplet phase will be miscible with or identical to the final continuous phase. Such systems may be w/o/w emulsions as indicated in Figure 11.13, where the internal and external phases are aqueous; or o/w/o, which have the reverse composition. Although known for almost a century, such systems have only recently become of practical interest for possible use in cosmetics,

Aqueous phase   Oil phase

FIGURE 11.13. A multiple emulsion may be either o/w/o or w/o/w, the result depending on the surfactant(s) used for each emulsification step, the process employed, the natures of the respective phases, etc.

pharmaceuticals, controlled drug delivery, wastewater treatment, and separations technology. Other useful applications will no doubt become evident as our understanding of the physical chemistry of such systems improves.

Because they involve various phases and interfaces, multiple emulsions must be inherently unstable, even more so than conventional "simple" emulsions. Their surfactant requirements are such that two stabilizing systems must be employed: one for each oil–water interface. Each surfactant or mixture must be optimized for the type of emulsion being prepared but must not interfere with the companion system designed for the opposite interface. Long-term stability, therefore, requires careful consideration of the characteristics of the various phases and surfactant solubilities.

11.10.1. Nomenclature for Multiple Emulsions

For systems as potentially complex as multiple emulsions, it is very important that a clear and consistent system of nomenclature be employed. For a w/o/w system, for example, in which the final continuous phase is aqueous, the primary emulsion will be a w/o emulsion, which is then emulsified into the final aqueous phase. The surfactant or emulsifier system used to prepare the primary emulsion is denoted as the primary surfactant. To avoid further ambiguity as to components or their locations in the system, subscripts may be used. For example, in a w/o/w system the aqueous phase of the primary emulsion would be denoted as w_1 and the primary emulsion as w_1/o. After the primary emulsion is further dispersed in the second aqueous phase w_2, the complete system may be denoted $w_1/o/w_2$. In the case of an o/w/o multiple emulsion in which the oil phases are different, the notation becomes $o_1/w/o_2$. Additional refinements to fit even more complex systems, including the "order" of multiple emulsions, have been suggested.

11.10.2. Preparation and Stability of Multiple Emulsions

In principle, multiple emulsions can be prepared by any of the many methods for the preparation of conventional emulsion systems, including sonication, agitation, and phase inversion. Great care must be exercised in the preparation of the final system, however, because vigorous treatments normally employed for the preparation of primary emulsions will often break that system if used in secondary emulsion formation, resulting in loss of the identity of the primary phase.

Multiple emulsions reportedly have been prepared conveniently by the phase inversion technique mentioned earlier; however, such systems will generally have a limited persistence. It requires a very judicious choice of surfactant or surfactant combinations to produce a system that has useful characteristics of formation and stability. A general procedure for the preparation of a $w_1/o/w_2$ multiple emulsion, illustrated in Figure 11.14, may involve the formation of a primary emulsion of water-in-oil using a surfactant suitable for the

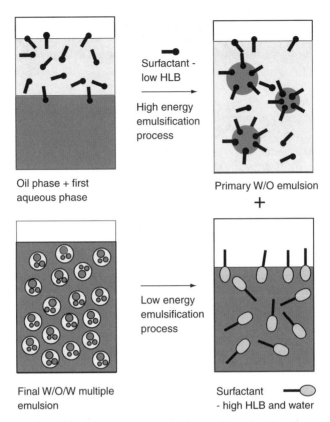

FIGURE 11.14. The preparation of a multiple emulsion involves two steps: (1) the formation of the primary emulsion with a surfactant more soluble in the first continuous phase and a high-energy emulsification process; and (2) the use of a second surfactant more soluble in the second continuous phase and a low energy emulsification process.

stabilization of such w_1/o systems. Generally, that involves the use of an oil soluble surfactant with a low HLB (2–8). The primary emulsion will then be emulsified in a second aqueous solution containing a second surfactant system appropriate for the stabilization of the secondary o/w_2 emulsion (HLB 6–16).

As noted above, because of the possible instability of the primary emulsion, great care must be taken in the choice of the secondary dispersion method. Excessive mechanical agitation such as in colloid mills, high-speed mixers, and sonication could result in coalescence of the primary emulsion and the production of a "simple" emulsion. The evaluation of the yield of filled secondary emulsion drops, therefore, is very important in assessing the value of different preparation methods and surfactant combinations.

The nature of the droplets in a multiple emulsion will depend on the size and stability of the primary emulsion. Three main classes of droplets have been suggested for w/o/w emulsions, based on the nature of the oil–phase

FIGURE 11.15. The primary emulsion may have one (or several) characteristic drop compositions, depending on the system and processes employed. Type A emulsions contain (on average) one primary emulsion (PE) drop. Type B PEs contain several smaller drops, but a relatively low volume fraction of dispersed phase. Type C PEs have a higher dispersed phase volume fraction with relatively small average drop sizes.

droplets: *type A* systems (Fig. 11.15*a*), which are characterized as having one large internal drop essentially encapsulated by the oil phase; *type B* (Fig. 11.15*b*), which contain several small, well-separated internal drops; and *type C* (Fig. 11.15*c*), which contain many small internal drops in close proximity. It is understood that any given system will in all probability contain all three classes of drops, but one will be found to dominate depending primarily on the surfactant system employed.

11.10.3. Primary Emulsion Breakdown

There are several possible pathways for the breakdown of multiple emulsions. A few are shown schematically in Figure 11.16. Although all possible mecha-

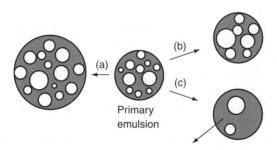

FIGURE 11.16. Multiple emulsion degradation can take place by several mechanisms. Important pathways include (*a*) secondary emulsion coalescence with little change in drop size in the PE, (*b*) PE drop coalescence with little change in secondary emulsion characteristics, and (*c*) loss of PE internal phase to the final external phase due to diffusion or solubilization.

nisms for droplet coalescence cannot be illustrated conveniently in a single figure, a consideration of just a few possibilities can help to clarify the reasons for instability in a given system. Even though there may be a number of factors involved, one of the primary driving forces will be, as always, a reduction in the free energy of the system through a decrease in the total interfacial area. As has been noted previously, a major role of surfactants at any interface is to reduce the interfacial energy through adsorption. In a typical multiple emulsion system, the primary mechanism for short-term instability will usually be droplet coalescence in the primary emulsion. It becomes important, then, to select as the primary emulsifier a surfactant or combination of surfactants that provides maximum stability for that system, whether w_1/o or o_1/w.

A second important pathway for the loss of "filled" emulsion droplets is the loss of internal drops by the rupture of the oil layer separating the small drops from the continuous phase. Such an expulsion mechanism would be expected to account for the loss of larger internal droplets. Unless the two phases are totally immiscible (in fact, a rare situation), there will always exist the possibility that osmotic pressure differences between the internal and continuous portions of the system will cause material transfer to the bulk phase. The high pressures in the smaller droplets would be expected to provide a driving force for the loss of material from smaller drops in favor of larger neighbors (Ostwald ripening), as well as to the continuous phase. Alternatively, osmotic pressure may cause solvent from the external phase (w_2) to migrate to the internal phase (w_1) swelling the droplets and rupturing the stabilizing layer (e.g., by surfactant depletion), leading to loss of the primary emulsion.

Finally, the presence of surfactant always raises the possibility of micelle formation in the primary continuous phase and the subsequent solubilization of the primary dispersed phase. Solubilization, therefore, represents a convenient mechanism for the transport of primary emulsion components to the secondary continuous phase. Such a solubilization process also represents a convenient mechanism for the transport of material. In the context of a critical application such as controlled drug delivery, in which the mechanism of delivery is diffusion-controlled, such breakdown mechanisms would be very detrimental to the action of the system since they could result in a rapid release of active solute with possibly dangerous effects.

These natural or spontaneous mechanisms of emulsion breakdown, as well as others, must be addressed in the formulation stage in order to understand and control a particular multiple emulsion system of interest. In all cases, the final stability of the system will depend on the nature of the oil phase of interest, the characteristics of the primary and secondary emulsifier systems, and the relationship between the internal and continuous phases.

As with simple emulsions, multiple emulsions will also be sensitive to breaking due to physical abuse. Shear and shock sensitivity must always be considered. In some applications, however, such sensitivity may be advantageous. In a cosmetic skin cream, for example, the shear imposed on the system during the process of application may serve as a convenient mechanism for release of components encapsulated in the primary emulsion.

11.10.4. The Surfactants and Phase Components

Choice of surfactant(s) for the preparation of multiple emulsions can, in principle, be made from any of the four classes of surfactants discussed in Chapter 3, although nonionics tend to be materials of choice because they are more easily "tailored" to meet the needs of the system. The choice will be determined by the characteristics of the final emulsion type desired, such as the natures of the various phases, additives, and solubilities. In many applications (e.g., foods, drugs, cosmetics), the choice may be further influenced by such questions as toxicity, interaction with other addenda, and biological degradation. In a given system, several different surfactants may perform adequately in terms of stability, but produce different types of multiple emulsions (A, B, or C in Figure 11.15), so that the choice will depend on application as well as function.

Clearly, multiple emulsions represent a fertile field of research in both applied and academic surface science. Although there are an ever-increasing number of publications appearing on the subject, the area remains somewhat empirical in that each system is highly specific. As yet there are few general rules that can guide the interested formulator in the selection of the optimum surfactants for his application. A great deal remains to be done to gain a better understanding of the colloidal stability of such complex systems, and the effects of the various components in each phase on the overall process of preparation and stabilization. A sound understanding of the role of surfactant and other addenda in simple emulsions and an intuitive feel for the effect of the multiple interfaces present can serve as a good starting guide.

PROBLEMS

11.1. Using the viscosity of pure water, estimate the creaming rate for a fat droplet of density 0.925 g cm^{-3} and diameter 5×10^{-6} m.

11.2. Tetradecane was emulsified at 25°C in two 0.5% (w/w) surfactant solutions: (a) $C_{12}H_{25}-(OCH_2CH_2)_5OH$; (b) $C_{12}H_{25}OSO_3Na$. What class of emulsion would you expect in each case? What would you expect to be the natures of the two emulsions when heated to 50°C?

11.3. Calculate the initial emulsifier concentration needed to cover emulsion droplets of oil to the extent of $\Gamma = 0.95\Gamma_o$. Assume a droplet radius of 500 nm and 1 L of o/w emulsion with a 50% (vol/vol) dispersed phase. Assume also that $\Gamma = (\Gamma_o(c/a)]/[1 = (c/a)]$, where $\Gamma_o = 10^{-10}$ mol cm^{-2} and $a = 10^{-7}$ mol cm^{-3}.

11.4. One of the processes by which an emulsion, like a foam, destroys itself is by Ostwald ripening: the diffusion of liquid from small to large droplets. Calculate the time required for a benzene droplet to disappear when it is positioned near much larger droplets at a distance comparable to its radius. Assume droplet radii of 100 and 1000 nm. The solubility of benzene in water may be taken as 0.2% (vol/vol); the diffusion constant of benzene in water $D = 10^{-5}$ cm^2 s^{-1}; the interfacial tension of water–benzene s = 25 mN m^{-1}; and the molar volume of benzene Vm = 100 cm^3.

11.5. Calculate (to the nearest whole number) the maximum possible value for the dispersed phase fraction, ϕ, in an emulsion consisting of uniform spherical particles.

11.6. A simple geometric theory for the stabilization of emulsions is that of the oriented wedge, in which the adsorbed surfactant molecules are assumed to form a uniform structure of wedges around the emulsion droplet. If an emulsion of 1000-nm-diameter droplets is stabilized by a surfactant whose head group occupies a surface area of 0.45 nm^2, what must be the cross-sectional area of the hydrophobic tail for maximum effectiveness?

11.7. A mixture of 70% Tween 60 and 30% Span 65 was found to give optimum stability to a particular emulsion system. What composition of a mixture of sodium dodecyl sulfate and cetyl alcohol should be expected to produce the same result?

11.8. A surfactant mixture with an HLB of 8 is expected to produce a stable emulsion with lanolin. A new chemist in the firm is given the job of formulating a suitable emulsion, with the requirement that the mixture contain 10% cetyl alcohol. Suggest at least two alternative surfactant compositions that meet the stated requirements.

11.9. An emulsion is to prepared of white petroleum oil in 0.001 M KCl using sodium dodecylsulfate as emulsifier. If the surfactant adsorbs at the oil–water interface occupying 0.5 nm^2 at equilibrium concentrations above 0.001 M, how much is needed per liter of emulsion if the average drop size is to be 50 nm in diameter?

12 Foams

It was noted previously that emulsions and foams are related by the fact that each represents a physical state in which one fluid phase is finely dispersed in a second phase, and that the state of dispersion and the long-term stability (persistence) normally is dependent on the composition of the system. In emulsions, each phase is a liquid so that such factors as mutual solubility and the solubility of additives in each phase must be considered. In foams, the dispersed phase is a gas so that problems of solubility are less critical, although as will be seen, the transfer of the dispersed gas from one bubble to another or out into the adjacent atmosphere is important.

Because of the forces involved in their formation and stabilization, foams will have a definite structure. Early investigators proposed the classification of foams into two morphological classes: (1) the "kugelschaum" or spherical foams consisting of widely separated spherical bubbles, also called gas emulsions; and (2) "polyederschaum" or polyhedral foams consisting of bubbles that are nearly polyhedral in shape, having narrow lamellar films of very low curvature separating the dispersed phase. The two types of foam are illustrated schematically in Figure 12.1. The morphological classification of foams is useful because each type of foam will undergo distinct changes with time, leading to collapse or a final "persistent" configuration.

The second morphological class of foams, usually considered the "true" foams, wherever three bubbles meet, they will form three equal angles of 120°. The three interfacial tensions involved, σ, must be equal so that the interfacial forces are equal where they meet. Mechanical equilibrium therefore requires that the three angles between them be equal. In an ideal foam, with all bubbles of the same size, the foam would assume the shape of pentagonal dodecahedrons. In almost all foams, however, there will be a variety of different bubble volumes present and their shapes will be far from the ideal.

12.1. THE IMPORTANCE OF FOAMS

The presence of foam in an industrial product or process may or may not be desirable. Foams have wide technical importance, as such, in the fields of fire fighting, polymeric foamed insulation, foam rubbers, and foamed structural materials such as concrete, whipped cream, shaving cream, and many areas of the baking industry. They also have certain esthetic utility in many detergent

FIGURE 12.1. There are two general classes of foams, the "dilute" foams (*a*) in which the gas bubbles are well separated and spherical, and the "concentrated" foams (*b*) in which the bubbles are closely packed so that their shape becomes distorted from the spherical.

and personal care products, although their presence may not add much to the overall effectiveness of the process. Foams also serve useful purposes in industrial processes such as mineral separation (foam or froth flotation), in the secondary recovery of petroleum by fluid displacement, and for environmental reasons in some electroplating operations. In the latter case, the presence of a foam blanket over the electroplating solution helps prevent solution splattering and the loss of volatile materials, therefore reducing the costs of maintaining an acceptable working environment. Unwanted foams, on the other hand, may be a significant problem in many technical processes, including sewage treatment, coatings applications, surfactant manufacture, extraction processes, and crude-oil processing. Some of these uses of foams will be discussed in more detail under specific subject headings.

By understanding the basic laws governing foam formation and the physical and chemical characteristics of materials that produce and sustain foams, or prevent and destroy them, the investigator or operator is well equipped to maximize (or minimize) the desired foaming effect. In the following sections, some of the basic physical principles of foam formation and stabilization will be covered along with some practical approaches to problems of foam characterization and control.

12.2. FOAM FORMATION

Like other colloidal systems, foams may be formed either by dispersion or condensation processes. In the former process, the incipient dispersed gas phase is present as a bulk or condensed phase. Small volumes of the future dispersed phase are introduced into the liquid by agitation or converted into gas by some mechanism such as heating, or pressure reduction. In the case of condensation, the gas phase is introduced at the molecular level and allowed to "condense" within the liquid to form bubbles.

The formation of the "head" on a glass of beer is a classic example of foam formation by condensation. In such a system, when the can, bottle, or tap is opened, carbon dioxide produced by fermentation in the container and solubilized under pressure is liberated. The solution becomes supersaturated, and the excess gas forms a dispersed phase which rises to the top and forms the head. Many industrial processes for the formation of solid foams employ a similar process in which a "blowing agent" is added to the polymerizing system creating the foam.

The simplest way to form a nearly ideal foam is to introduce the gas into the liquid through a capillary tube. In that way individual bubbles of equal (almost) size will break off from the capillary tip under the action of surface tension. The process, however, must be slow in order to ensure that interfacial equilibrium is achieved for each bubble; otherwise a monodisperse foam will not be produced. A much more rapid, but less controllable, procedure is to bubble gas into the system through a porous plug. In that process a highly polydisperse foam will result since many small bubbles will have the opportunity to coalesce while still attached to the plug. Even less consistent results will be obtained for foams produced by agitation.

In all the methods for the formation of foams, the initial bubbles will be separated by relatively thick layers of the continuous phase to produce spherical foams. However, in most cases, gravity will transform them into the polyhedral structure, with the foam at the top of the container and a reservoir of liquid accumulated at the bottom. As the bubbles rise, the external hydrodynamic pressure will decrease and the bubble volume increase, reducing the internal pressure of each bubble, although that internal pressure will still be greater than that externally. There will therefore be a mechanical driving force impelling the bubble to release the excess pressure by rupture. The fact that in many cases a reasonably long-lived foam is established indicates that some mechanism acting within the narrow lamellar films separating the bubbles is sufficient to withstand that mechanical pressure.

12.3. BASIC PROPERTIES OF FOAMS

The continuous phase of foams may consist entirely of liquid components or a mixture of various liquids and solutes. The long-term stability of a given foam will depend on that composition, but in the presence of special additives, even a completely liquid lamellar film will have some degree of rigidity. When a "stable" polyhedral foam structure is formed, it represents a transient minimum in the surface energy of the system—a metastable configuration that could, in theory, remain for a significant period of time. It would require a "push" from some external source to cause it to increase its surface area and "break." Under normal circumstances, such "pushes" are common enough in the form of dust particles, air movement, convection due to temperature differentials, and other variables. Therefore the weaker foams will rapidly collapse from the metastable state and revert to a phase separated state. If the liquid phase is "fortified" by the addition of various components that can increase that lamellar rigidity, enhanced stability will result. Details of such action will be given below.

A primary characteristic of foams is that they have very low densities. An aqueous foam with bubble diameters of about 1 cm and lamellar thickness of 10^{-3} cm will have a density of approximately 0.003 g cm^{-3}. That low density makes foams very useful in a number of applications, including firefighting and various separation techniques. In the former case, the ingredients for producing the foam are easily transported, can be produced rapidly to cover a large area, and consist primarily in water, which is relatively inexpensive and easy to obtain, and has the added advantage that it serves as an efficient mechanism for the removal of heat from the system. The foam blanket performs the role of any blanket in relation to the fire—it prevents contact between the air and the combustible material below. For firefighting, the use of spherical rather than polyhedral foams may have some advantages since the higher water content would aid in the removal of heat from the system.

Related to the low density of foams is the characteristic that they will have a large surface area for a given weight of foam. For the example above, the foam will have a surface area of about 2000 cm^2 g^{-1}. Under properly controlled conditions, specific finely powdered mineral ores will attach to foam bubbles and be carried to the surface of a solution where they can be skimmed off and the desired mineral significantly enriched. Unwanted materials such as rock and dirt will be left to sink to the bottom of the container. The procedure, known generally as "flotation," is a valuable tool in many mineral purification processes. Similar events have been postulated to occur in many detergency processes. Techniques based on the flotation principle have been proposed for the purification or removal of surface active soluble materials from solution, although they seem to have found little large-scale use.

12.4. FOAM STABILITY OR PERSISTENCE

Foams, like emulsions, are inherently unstable systems. Because they are encountered in so many technological areas, they have been the subject of a significant amount of investigation and discussion in the literature. A number of reviews have been published over the years that cover most aspects of foam formation and stabilization (see Bibliography). While the theoretical aspects of stabilization are reasonably well worked out, a great deal remains to be understood concerning the practical details of foam formation, persistence, and prevention.

12.4.1. Thermodynamic Conditions for Stability

Like almost all systems containing two or more immiscible phases, foams involve thermodynamic conditions in which the primary driving force is to reduce the total interfacial area between the phases—that is, they are thermodynamically unstable. In spite of their ultimate tendency to collapse, however, foams can be prepared that have a lifetime (persistence) of minutes, days, or even months.

There are three fundamental physical mechanisms for the collapse of a foam: (1) the diffusion of gas from one bubble (small, high internal pressure) to another (larger, lower internal pressure) or into the bulk gas phase surrounding the foam, (2) bubble coalescence due to capillary flow leading to rupture of the lamellar film between the adjacent bubbles (usually slower than 1 and occurring even in stabilized systems), and (3) rapid hydrodynamic drainage of liquid between bubbles leading to rapid collapse (in the absence of any of the stabilizing mechanisms discussed below). Each mechanism can be important for a given foam system and will be addressed in turn. In most nonrigid systems, however, all three mechanisms will be operative to some extent during some phase of the collapsing process so that an analysis of the system and its stability may not be a simple undertaking.

The first mechanism, gas diffusion, is due to the difference in gas pressure inside the bubble as a result of differences in curvature of the lamellar films. As a simple example, consider the system of two contacting bubbles shown in Figure 12.2. The Laplace equation states that the pressure difference, Δp, on either side of a curved interface will be given by

$$\Delta p = \sigma \left(\frac{1}{r_1} + \frac{1}{r_2} \right) \tag{12.1}$$

where r_1 and r_2 are the major radii of curvature of the system and σ the interfacial tension. For the two bubble system, let R_1 be the radius of the larger and R_2, that of the smaller bubble. The common interfaces or partition will have a radius R_0. From Equation (12.1), the pressure in the larger bubble will be atmospheric pressure, $p_a + 4\sigma/R_1$. The factor of 4 arises here because the bubble wall involves two interfaces, each of which will contribute to the total. The small bubble will likewise have a total pressure of $p_a + 4\sigma/R_2$.

At equilibrium, the common interface (the septum) must be concave toward the smaller bubble with a radius of curvature given by

$$\frac{4\sigma}{R_0} = \frac{4\sigma}{R_2} - \frac{4\sigma}{R_1} \tag{12.2}$$

FIGURE 12.2. When two gas bubbles are in contact, but do not rupture and coalesce, the radius of curvature of the septum between bubbles will be concave toward the smaller bubble and have a radius r_0 equal to $r_1 r_2/(r_1 - r_2)$.

or $R_o = R_1R_2/(R_1 - R_2)$. If the septum were completely impermeable to the gas, such a situation would be mechanically stable. In reality, such impermeability does not exist, and gas will spontaneously diffuse from the region of high to that of low pressure. As a result, the smaller bubble will shrink while the larger will grow. If the difference in the two radii R_1 and R_2 is initially small, diffusion will be slow. With time, however, the difference will increase, as will the rate, meaning that in a foam the average bubble size and dispersity will increase with aging.

The second major reason for film drainage and ultimate rupture is capillary flow. When several bubbles are in contact, especially when the foam has reached a more-or-less stable polyhedral structure, the liquid region of multiple bubble contact will have a much greater curvature (i.e., smaller radius of curvature) than the lamellar films, which may be almost planar. Those regions of high curvature toward the gas phase, referred to as "plateau borders" (Fig. 12.3), act as a capillary "pump" to evacuate the region between bubbles. Since the gas pressure within the bubble must be the same throughout, the liquid pressure within the plateau borders will be lower than in the more parallel areas. That pressure difference will drive the liquid into the plateau borders, thinning the film and advancing the process of ultimate film rupture. Liquid will also be drained from the lamellae due to gravitational forces; as a result the lamellae will become thinner and thinner until a critical thickness may be reached, at which time the system can no longer sustain the pressure and collapse occurs.

Of the three mechanisms, hydrodynamic drainage due to gravity is usually the most rapid and, if the foam is particularly unstable, leads to total collapse before other mechanisms can become important. In those cases, once the loss of liquid from the lamellar layer produces a critical thickness of 5–15 nm, the liquid film can no longer support the pressure of the gas in the bubble, and film rupture occurs. As a model for gravity drainage, a film may be treated as a vertical slit of thickness δ (not to be confused with the solubility parameter

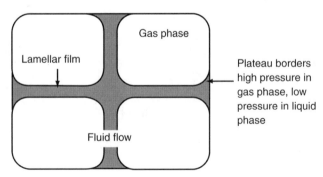

FIGURE 12.3. Low capillary pressure in the plateau border regions of the liquid phases causes a net flow into those areas, draining the lamellar regions between bubbles and leading, in most cases, to ultimate rupture and foam breakage.

introduced in Chapter 11) and width w between two parallel walls (Fig. 12.4). Assuming a constant thickness of the lamellar film between bubbles, the rate of film drainage will be

$$dV/dt = \frac{g\rho w\delta^3}{12\eta} \tag{12.3}$$

where g is the gravity constant, ρ is the density difference between the liquid and the gas around the foam, and η the viscosity of the liquid. In reality, of course, the slit or film will not be of constant size; δ will decrease as drainage occurs, so that Equation 12.3 is not quantitatively valid. Qualitatively, however, the relationship provides a clue as to how one may influence the drainage process. Also, the lamellar film in a foam is not really a rigid structure, so that the mechanical characteristics of the wall may affect the rate of drainage, especially in thinner films.

The overall question of foam stability and bubble coalescence requires the consideration of both the static and dynamic aspects of bubble interactions. In the initial stages of film drainage, where relatively thick lamellar films exist between gas bubbles, gravity can make a significant contribution to the drainage of liquid from between foam bubbles. Once the films have thinned to a thickness of a few hundred nanometers, however, gravity effects become negligible and interfacial interactions begin to predominate. When the two sides of the lamellar film are in sufficiently close proximity, interactions can occur involving the interfacial forces discussed in previous chapters. Such forces (per unit area), acting normal to (across) the lamellar film, are collectively exhibited as the so-called disjoining pressure of the system $\pi(\delta)$. The net interaction energy between bubbles as a function of distance of separation

FIGURE 12.4. Film drainage in foams can be modeled as a thin film of liquid of thickness, δ, between two parallel vertical plates of width, w. The rate of hydraulic (gravity-driven) drainage will depend on the viscosity of the liquid as well as other factors, as is found in real foams [Eq. (12.3)].

will have a form similar to that in Figure 12.5, where the minima will correspond to metastable states in which $\pi = 0$ and the films have some degree of equilibrium stability. In the plane parallel regions of the lamellar film, the Laplace (or capillary) pressure given by eq. 1 will be zero. In the plateau border regions, however, that will not be the case, and mechanical equilibrium requires that

$$\Delta p = -\pi(\delta) \tag{12.4}$$

Thus, the internal pressure of the bubbles is just balanced by the interfacial forces acting across the lamellar film. The most important interfacial interactions contributing to $\pi(\delta)$ are electrostatic repulsion between charged interfaces and steric interactions due to adsorbed species. Those topics have already been discussed in the context of colloidal stability and will not be treated further here.

Internal thermodynamic and hydrodynamic factors aside for the moment, it should be remembered that foams are sensitive to a number of external environmental stresses, which act to bring about bubble coalescence and foam collapse. Those include vibration, the presence of solid particles, organic contaminants, and temperature differentials. It will therefore be important to take such factors into consideration when carrying out foaming studies or formulating a foam systems.

12.4.2. Stabilization Mechanisms

Practical mechanisms for extending the persistence of foams can include one or several of the following conditions: (1) a high viscosity in the liquid phase, which retards hydrodynamic drainage, as well as providing a cushion effect

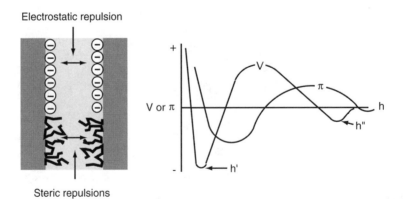

FIGURE 12.5. Thin lamellar films in foams will develop a disjoining pressure, $\pi(h)$, between the opposing interfaces when electrostatic or steric stabilization mechanisms are operative. The stability of the film (and the foam) will depend on the value of π, among other things.

to absorb shocks resulting from random or induced motion; (2) a high surface viscosity, which also retards liquid loss from between interfaces and dampens film deformation prior to bubble collapse; (3) surface effects such as the Gibbs and Marangoni effects, which act to "heal" areas of film thinning due to liquid loss; (4) electrostatic and steric repulsion between adjacent interfaces due the adsorption of ionic and nonionic surfactants, polymers, and other agents, which can oppose drainage through the effects of the disjoining pressure; and (5) retardation of gas diffusion from smaller to larger bubbles.

The addition of surfactants and/or polymers to a foaming system can alter any or all of the above-mentioned system characteristics and therefore enhance the stability of the foam. They may also have the effect of lowering the surface tension of the system, thereby reducing the work required for the initial formation of the foam, as well as producing smaller, more uniform bubbles.

12.5. PRACTICAL CONTROL OF FOAMABILITY AND PERSISTENCE

For a liquid to produce a foam of any degree of utility, it must be able to (1) expand its surface area so as to form a membrane around gas bubbles, (2) possess the correct rheological and surface properties to retard the thinning of the lamellae leading to bubble coalescence, (3) and/or retard the diffusion of trapped gas from small to large bubbles or to the surrounding atmosphere. Foaming does not occur in pure liquids because such a system can lack mechanisms for the completion of any of those three tasks. When surface active molecules or polymers are present, however, rheological effects and adsorption at the gas–liquid interface serve to retard the loss of liquid from the lamellae and, in some instances, to produce a more mechanically stable system.

Theories related to such film formation and persistence, especially film elasticity, derive from a number of experimental observations about the surface tension of liquids. First, as is well known from the Gibbs adsorption equation, the surface tension of a liquid will decrease as the concentration of the surface-active material in solution increases (assuming positive adsorption) up to the point of surface saturation. Second, the instantaneous (dynamic) surface tension at a newly formed surface is always higher than the equilibrium value; that is, there is a finite time requirement during which the surface-active molecules in the solution (or bulk molecules, if no solute is present) must diffuse to the interface in order to lower the surface tension. The time lag in reaching the equilibrium surface tension due to diffusion is generally known as the Marangoni effect. The two surface tension effects due to adsorption and diffusion are usually complementary, and are often discussed as the combined Gibbs–Marangoni effect.

12.5.1. Monomeric Surfactant Stabilization

The fundamental impact of surfactant concentration and diffusion rate in lamellar films can be viewed roughly as follows (Fig. 12.6): as the lamellar

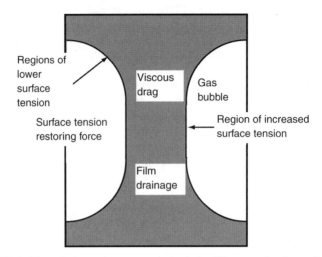

FIGURE 12.6. The microscopic anatomy of a foam. The complex interplay of interfacial and hydraulic forces makes the understanding and control of foams a challenging proposition. The effective control of those interactions with surfactants or other additives determines the persistence of a given system.

film between adjacent bubbles is stretched as a result of gravity, agitation, drainage, or other mechanical action, new surface will be formed having a lower transient surfactant concentration, and a local surface tension increase will occur. A surface tension gradient along the film will be produced, causing liquid to flow from regions of low σ toward the new stretched surface, thereby opposing film thinning. Additional stabilizing action is thought to result from the fact that the diffusion of new surfactant molecules to the surface must also involve the transport of associated solvent into the surface area, again countering the thinning effect of liquid drainage. The mechanism can be characterized as producing a "healing" effect at the site of thinning.

Even though the Gibbs and Marangoni effects are complementary, they are generally important in different surfactant concentration regimes. The Marangoni effect is usually of importance in fairly dilute surfactant solutions and over a relatively narrow concentration range. In the absence of external agitation, the amount of surfactant adsorbed at a new interface can be estimated by

$$n = 2\left(\frac{D}{\pi}\right)^{1/2} ct^{1/2} \frac{N}{1000} \tag{12.5}$$

where n is the number of molecules per square centimeter, D is the bulk diffusion constant (cm^2 s^{-1}), c is the bulk concentration of the surfactant (mol L^{-1}), t is the time in seconds, and N is Avogadro's number. Using Equation (12.5), it is possible to estimate the time required for the adsorption of a given amount of surfactant at a new interface compared to the rate of generation of that interface. If the surfactant solution is too dilute, the surface tension of the solution will not differ sufficiently from that of the pure solvent for the restoring force to counteract the effects of casual thermal and mechanical agitation. As a result, the foam produced will be very transient. In line with the Marangoni theory, there should be an optimum surfactant concentration for producing the maximum amount of foam in a given system, under fixed conditions. Such effects have been verified experimentally in terms of the Ross–Miles test (Table 12.1). It can be seen from the data that the optimum concentration will usually be within a factor of 2 of the critical micelle concentration.

In the case of the Gibbs effect, it is proposed that the rise in surface tension occurring as the film is stretched results from a depletion of the surfactant concentration in the bulk phase just below the newly formed interface. Obviously, in systems such as foams, where the available bulk phase in the narrow lamellae may be small compared to the amount of interface being formed, the effect will be enhanced. As with the Marangoni effect, if the surfactant concentration in the bulk phase is too low, a surface tension gradient of sufficient size to produce the necessary "healing" action will not be produced. Conversely, if the concentration is too large, well above the cmc, the available surfactant will be such that no gradient is formed.

Quantitatively, the Gibbs effect can be described in terms of a coefficient of surface elasticity E, which Gibbs defined as the ratio of the surface stress to the strain per unit area

$$E = 2A \frac{\delta \sigma}{\delta A} \tag{12.6}$$

TABLE 12.1. Typical Surfactant Concentrations Required to Attain Maximum Foam Height (MFH)[a]

Surfactant	Cmc (mM)	Concentration for MFH (mM)
$C_{12}H_{25}SO_3Na^+$	11	13
$C_{12}H_{25}SO_4Na^+$	9	5
$C_{14}H_{29}SO_3K^+$	3	3
$C_{14}H_{29}SO_4Na^+$	2.3	3
$C_{16}H_{33}SO_3K^+$	0.9	0.8
$C_{16}H_{33}SO_4Na^+$	0.7	0.8
$p\text{-}C_8H_{17}C_6H_4SO_3Na^+$	16	13
$p\text{-}C_{10}H_{21}C_6H_4SO_3Na^+$	3	4.5
$p\text{-}C_{12}H_{25}C_6H_4SO_3Na^+$	1.2	4
$o\text{-}C_{12}H_{25}C_6H_4SO_3Na^+$	3	4
$(C_8H_{17})_2CHSO_4Na^+$	2.3	4

[a] Ross–Miles Method, at 60°C. Ross, J; Miles, G. D., *Amer. Soc. for Testing and Materials*, Method D. 1173–53. ASTM, Philadelphia, 1953.

Since the elasticity is the resistance of the film to deformation, the larger the value of E, the greater will be the ability of the film to sustain shocks (its resiliency, so to speak) without rupture. As mentioned earlier, when a film of a pure liquid is stretched, no significant change in surface tension will occur, and the elasticity as defined by Equation (12.6) will be zero. This is the theoretical basis for the observation that pure liquids will not foam. The relationship between surface elasticity and surface transport is important since, if a film has a significant value of $E > 0$, stretching the film will produce an increase in the local surface tension and induce flow of subsurface liquid into the stretched area, again acting to restore the original thickness of the lamellae.

Two surfactant-related processes, then, must be considered in conjunction with these foam-stabilizing mechanisms. One is the rate of surface diffusion of surfactant molecules from regions of low to high surface tension. The second is the rate of adsorption of surfactant from the underlying bulk phase into the surface. In each case, a too rapid arrival of surfactant molecules at the new surface will destroy the surface tension gradient and prevent the restoring action of the Gibbs–Marangoni "healing" process. Conversely, a very low bulk concentration will result in equally ineffective action.

12.5.2. Polymers and Foam Stabilization

Very stable foams can be prepared if polymers are included in the formulation. When polymers (and proteins in particular) are adsorbed at the liquid–air interface, they will assume configurations significantly different from their equilibrium situation in the bulk solution; in the case of proteins, they will become partially denatured. The relatively dense, somewhat structured adsorbed polymer layer will impart a significant degree of rigidity or mechanical strength to the lamellar walls, producing an increase in the stability of the final foam. The presence of polymer will also aid stability in the initial stages after foam formation since the liquid viscosity increase that will result from its presence will slow the process of film drainage. Polymers will not generally be effective in the context of the Gibbs–Marangoni effect since their diffusion rates will be much slower than that of low-molecular weight-surfactants.

The presence of polymers in some systems can cause difficult problems where foam stability is not desirable. Such is particularly the case in waste treatment facilities, where the presence of proteins can cause extreme problems. If polyvalent ions such as Ca^{2+} and Al^{3+} are present, the problem is exacerbated still more. Proteins will bond strongly with such ions and form an essentially cross-linked surface film so rigid that it may approach the strength of a solid foam. Obviously, such a situation will be detrimental to the overall treatment process. In other situations, such as breads and other baked products, the formation of rigid foam walls can be particularly advantageous.

12.6. FOAM FORMATION AND SURFACTANT STRUCTURE

The relationship between the foaming power of a surfactant and its chemical structure can be quite complex. The correlation is further complicated by the fact that there is not necessarily a direct relationship between the ability of a given structure to produce foam and its ability to stabilize that foam. One usually finds that the amount of foam produced by a surfactant under a given set of circumstances will increase with its bulk concentration up to a maximum, which occurs somewhere near the critical micelle concentration (cmc; see also Chapter 15). It would appear, then, that surfactant cmc could be used as a guide in predicting the initial foaming ability of a material, but not necessarily the persistence of the resulting foam. Any structural modification that leads to a lowering of the cmc of a class of surfactants, such as increasing the chain length of an alkyl sulfate, can be expected to increase its efficiency as a foaming agent. Conversely, branching of the hydrophobic chain or moving the hydrophilic group to an internal position, all of which increase the cmc, will result in a lower foaming efficiency. Typical foaming characteristics for several anionic and nonionic surfactants are given in Table 12.2, where foaming efficiency and persistence were determined according to the Ross–Miles procedure.

TABLE 12.2. Foaming Characteristics of Typical Anionic and Nonionic Surfactants in Distilled Water[a]

Surfactant	Concentration	Foam Height (mm)	
		Initial	After (min)
$C_{12}H_{25}SO_3^-Na^+$	0.25	—	205(1)
$C_{12}H_{25}SO_4^-Na^+$	0.25	220	175(5)
$C_{14}H_{29}SO_3^-Na^+$	0.11	—	214(1)
$C_{14}H_{29}SO_4^-Na^+$	0.25	231	184(5)
$C_{16}H_{33}SO_3^-K^+$	0.033	—	233(1)
$C_{16}H_{33}SO_3^-Na^+$	0.25	245	240(5)
$C_{18}H_{37}SO_4^-Na^+$	0.25	227	227(5)
$o\text{-}C_8H_{17}C_6H_4SO_3^-Na^+$	0.15	148	—
$p\text{-}C_8H_{17}C_6H_4SO_3^-Na^+$	0.15	134	—
$o\text{-}C_{12}H_{25}C_6H_4SO_3^-Na^+$	0.25	208	—
$p\text{-}C_{12}H_{25}C_6H_4SO_3^-Na^+$	0.15	201	—
$t\text{-}C_9H_{19}C_6H_4O(CH_2CH_2O)_8H$	0.10	55	45(5)
$t\text{-}C_9H_{19}C_6H_4O(CH_2CH_2O)_9H$	0.10	80	60(5)
$t\text{-}C_9H_{19}C_6H_4O(CH_2CH_2O)_9H$	0.10	110	80(5)
$t\text{-}C_9H_{19}C_6H_4O(CH_2CH_2O)_{13}H$	0.10	130	110(5)
$t\text{-}C_9H_{19}C_6H_4O(CH_2CH_2O)_{20}H$	0.10	120	110(5)

[a] Ross–Miles method, at 60°C.

The ability of a surfactant to perform as a foaming agent is dependent primarily on its effectiveness at reducing the surface tension of the solution, its diffusion characteristics, its properties with regard to disjoining pressures in thin films, and the elastic properties it imparts to interfaces. The amount of foam that can be produced in a solution under given conditions (i.e., for a set amount of work input) will be related to the product of the surface tension and the new surface area generated during the foaming process. Obviously, the lower the surface tension of the solution, the greater will be the surface area that can be expected to be developed by the input of a given amount of work. Maintenance of the foam, however, may be as important as original formation.

It is often observed that the amount of foam produced by the members of an homologous series of surfactants will go through a maximum as the chain length of the hydrophobic group increases. This is probably due to the conflicting effects of the structural changes. In one case, a longer-chain hydrophobe will result in a more rapid lowering of surface tension and a lower cmc. However, if the chain length grows too long, low solubility and slow diffusion and adsorption may become problems.

It has been found in many instances that surfactants with branched hydrophobic groups will lower the surface tension of a solution more rapidly than a straight-chain material of equal carbon number. However, since the branching of the chain increases the cmc and reduces the amount of lateral chain interaction, the cohesive strength of the adsorbed layer, the film elasticity, will be reduced, yielding a system with higher initial foam height but reduced foam stability. Similarly, if the hydrophilic group is moved from a terminal to an internal position along the chain, higher foam heights, but lower persistence, can be expected. In all such cases, comparison of foaming abilities must be compared at concentrations above their cmc.

Ionic surfactants can contribute to foam formation and stabilization as a result of the presence at the interface of the electrical double layer that can interact with the opposing interface in the form of the disjoining pressure. Additional stabilizing effects may be gained from the fact that the ionic group requires a significant degree of solvation, with the associated solvent molecules adding to the steric (or entropic) contribution to the disjoining pressure $\pi(\delta)$. Not surprisingly, it is found that the effectiveness of such surfactants as foaming agents can be related to the nature of the counterion associated with the adsorbed surfactant molecules. The effectiveness of dodecyl sulfate surfactants as foam stabilizers, for example, decreases in the order $NH_4^+ > (CH_3)_4N^+ > (C_2H_5)_4N^+ > (C_4H_9)_4N^+$. Such a series may reflect a change in the solvation state of the surfactant from the essentially totally dissociated ammonium counterion, producing a maximum disjoining pressure and requiring significant solvation, to the more tightly ion-paired tetrabutylammonium counterion with greatly reduced $\pi(\delta)$ and different solvation requirements.

Nonionic surfactants generally produce less initial foam and less stable foams than do ionics in aqueous solution. Because such materials must by nature have rather large surface areas per molecule, it becomes difficult for the adsorbed molecules to interact laterally to a significant degree, resulting in a lower interfacial elasticity. In addition, the bulky, highly solvated nonionic groups will generally result in lower diffusion rates and less efficient "healing" via the Gibbs–Marangoni effect. Polyoxyethylene nonionic surfactants in particular exhibit a strong sensitivity of foaming ability to the length of the POE chain. At short chain lengths, the material may not have sufficient water solubility to lower the surface tension and produce foam. A chain that is too long, on the other hand, will greatly expand the surface area required to accommodate the adsorbed molecules and will also reduce the interfacial elasticity. This characteristic of POE nonionic surfactants has made it possible to design highly surface-active, yet low-foaming surfactant formulations. Even more dramatic effects can be obtained by the use of "double-ended" surfactants in which both ends of the POE chain are substituted. In many cases only a single methyl group on the end of a surfactant chain will significantly reduce foaming where such a result is desired.

Foaming ability of surfactants can also be correlated with the respective solubility parameter, as discussed for emulsions in Chapter 11. In this case, the solubility of the surfactant must be properly balanced—that is, be soluble enough to attain a significant concentration in solution, but not so soluble that significant adsorption does not occur.

If the solubility of a surfactant is highly temperature-dependent, as is the case for many nonionic polyoxyethylene surfactants and long chain fatty acid soaps such as sodium stearate, it will be found that foaming ability will increase in the same direction as its solubility. Nonionic POE surfactants, for example, exhibit a decrease in foam production as the temperature is increased and the cloud point is approached (solubility decreases). Long-chain carboxylate salts, on the other hand, which may have limited solubility and poor foaming properties in water at room temperature, will be more soluble and will foam more as the temperature increases.

12.7. LIQUID CRYSTALS AND FOAM STABILITY

As we have seen, the stability of foams depends on a wide variety of factors involving several aspects of surface science. The potential importance of liquid crystal (LC) formation to emulsion stability was pointed out in the previous chapter. Not surprisingly, an equally important role for such structures has been identified in foaming applications. Although the phenomenon of LC stabilization of aqueous foams has been recognized for some time, their role in nonaqueous foaming systems has been less well documented. Recently, it has been shown that the presence of a liquid crystalline phase can also serve as a sufficient condition for the production of stable foams in organic systems.

The role of the liquid crystal in stabilizing a foam can be related to its effect on several mechanisms involved in foam loss, including hydrodynamic drainage, the mechanical strength of the liquid film, and the diffusion rate of

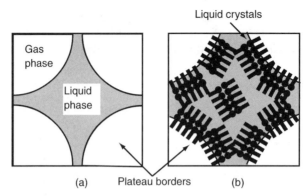

FIGURE 12.7. The presence of surfactant liquid crystals may add stability to a "normal" foam (*a*) by forming a semirigid structure in the plateau border regions (*b*) and/or thick lamellar films that provide mechanical as well as colloidal resistance to drainage.

entrapped gas. The effect of the LC phase on film drainage can be considered to be twofold. In the first place, the more ordered, multilayer nature of the phase imparts a much higher viscosity to the film than a normal surfactant monolayer, thereby preventing or slowing the process of liquid drainage. In addition, it has been found that liquid crystalline phases tend to accumulate in the plateau border areas, where their presence results in an increase in the size of the areas, a larger radius of curvature (Fig. 12.7), and thus a smaller Laplace pressure forcing film drainage. The second stabilizing function for the liquid crystal can be related to the Gibbs–Marangoni effects, in that the presence of a large quantity of surfactant at the Plateau borders allows them to act as a reservoir for surfactant molecules needed to maintain the high surface pressures useful for ensuring foam stability.

The production of a LC phase can not only add to the stability of the foam from a surface chemical standpoint but also significantly enhance the mechanical strength of the system. When thinning reaches the point at which bubble rupture can become important, the mechanical strength and rigidity of such structures can help the system withstand the thermal and mechanical agitation that might otherwise result in film rupture and foam collapse.

Finally, because the LC structure is more highly ordered and, potentially, more dense than a normal fluid, the diffusion rate of gas molecules between bubbles may—theoretically, at least—be expected to be slowed significantly.

12.8. THE EFFECTS OF ADDITIVES ON SURFACTANT FOAMING PROPERTIES

As we have seen, the foaming properties of a surfactant can be related to its solution properties through the cmc. It is not surprising, then, that additives in a formulation can affect foaming properties in much the same way that they affect other surfactant solution properties. The presence of additives can affect the stability of a foam by influencing any of the mechanisms already discussed for foam stabilization. It may, for example, increase the viscosity of the liquid phase or the interfacial layer, or it may alter the interfacial interactions related to Gibbs–Marangoni effects or electrostatic repulsion. By the proper choice of additive, a high-foaming surfactant can be transformed into one exhibiting little or no foam formation. Conversely, a low-foaming material may produce large amounts of foam in the presence of small amounts of another surface-active material, which itself has few if any useful surfactant properties. It is theoretically possible, then, to custom-build a formulation to achieve the best desirable combination of foaming action to suit the individual needs of the system. The addition of small amounts of such additives has become the primary way of adjusting the foaming characteristics of a formulation in many, if not most, practical surfactant applications.

Additives that alter the foaming properties of a surfactant through changes in its micellization characteristics can be divided into three main classes: (1) inorganic electrolytes, which are most effective with ionic surfactants; (2) polar organic additives, which can affect all types of surfactants; and (3) macromolecular materials. The latter materials can affect the foaming properties of a system in many ways, some unrelated to the surface properties of the surfactant itself. Electrolyte additives can act to increase foaming ability by reducing the cmc of ionic foaming agent. On the other hand, an excessive amount of electrolyte may, and probably will, greatly reduce foam persistence by reducing the electrostatically induced disjoining pressure.

From a practical point of view, the most important class of additives is that of the polar organic materials, which have received a great deal of attention both academically and industrially because of the relative ease of application and control of the additive. Some of the earliest uses of polar organic materials as profoaming additives was in the area of foam stabilizers for heavy-duty laundry formulations. As a general rule, it was found that additives that lower the cmc of a surfactant could stabilize foams in the presence of materials that were normally detrimental to foam formation and persistence. The ability of additives to increase foam formation and foam stability by lowering the cmc of the primary surfactant could be related to the extent of such lowering. Straight-chain hydrocarbon additives whose chain length is approximately the same as that of the surfactant are generally the most effective at lowering the cmc and increasing initial foam height. Bulky chains on the additives produce much smaller effects on foaming properties. The effectiveness of polar additives of various types as foam stabilizers is found to be in the approximate order: primary alcohols < glyceryl ethers < sulfonyl ethers < amides < N-substituted amides. This is essentially the same order found for the effects of such materials on the cmc of surfactants. The effects on cmc and foam stability of the addition of polar additives to sodium dodecylbenzene sulfonate are given in Table 12.3.

TABLE 12.3. Effect of Structure of Organic Additives on CMCs and Foaming Characteristics of Sodium 2-n-Dodecylbenzene Sulfonate Solutions

Additive	Cmc (g/liter)	Δcmc (%)	Foam Volume (mL at 2 min)
None	0.59	—	18
Lauryl glycerol ether	0.29	−51	32
Laurylethanolamide	0.31	−48	50
n-Decyl glycerol ether	0.33	−44	34
Laurylsulfolanylamide	0.35	−41	0
n-Octylglycerol ether	0.36	−39	32
n-Decyl alcohol	0.41	−31	26
Caprylamide	0.50	−15	17
Tetradecanol	0.60	0	12

Not only does foam stabilization by additives seem to go hand-in-hand with the effect of the additive on the cmc of the surfactant, but there is also a correlation with the relative amount of additive that is located in the interfacial film. The greater the mole fraction of additive adsorbed at the interface, the more stable is the resulting foam. Many of the most stable foaming systems were found to have surface layers composed of as much as 60–90 mol% additive.

12.9. FOAM INHIBITION

Although the presence of certain additives can enhance the foaming ability and persistence of a surfactant system, chemically similar materials may also significantly reduce foam formation or persistence. The same material can, in fact, function as both a promoter or inhibitor under different circumstances. Materials that reduce the amount of foam formed are termed "foam inhibitors," which act to prevent the formation of foam, or "foam breakers" or "defoamers," which increase the rate of foam collapse. A foam inhibitor may function by interfering with the adsorption of surfactant at the air–solution interface or by reducing the effectiveness of adsorbed surfactant as a stabilizer.

Foam breakers may include inorganic ions such as calcium, which counteract the effects of electrostatic stabilization or reduce the solubility of many ionic surfactants, organic or silicone materials that act by spreading on the interface and displacing the stabilizing surfactant species, or materials that directly interfere with micelle formation.

A foam breaker that acts by spreading may do so as a monolayer or as a lens (Fig. 12.8). In either case, it is assumed that the spreading foam breaker sweeps away the stabilizing layer, leading to rapid bubble collapse from the outside of the foam. The rate of spreading of the defoamer will, of course,

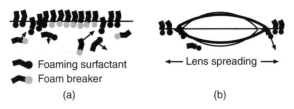

Foaming surfactant

Foam breaker

(a)

◄— Lens spreading —►

(b)

FIGURE 12.8. A foam breaker may act by one or both of two mechanisms: (*a*) the foam breaker may displace stabilizing surfactant molecule by molecule leading to breakdown; or (*b*) the breaker may displace the stabilizing structure by spreading as a lens at the interface.

depend on the nature of the adsorbed layer present initially. If the foaming agent can be desorbed rapidly, the defoamer will spread rapidly, resulting in fast foam collapse. If the foaming agent does not desorb rapidly, on the other hand, spreading will be retarded, or even halted. Foam collapse will then be a much slower process, relying on the thinning of the lamellae by other drainage mechanisms.

In some cases it is found that the action of defoaming agents may depend on the concentration of the surfactant present. If the surfactant concentration is below the cmc, the defoamer will usually be most effective if it spreads as a lens on the surface rather than as a monolayer film. Above the cmc, however, where the defoamer may be solubilized, the micelles may act as a reservoir for extended defoaming action by adsorption as a surface monolayer. If the solubilization limit is exceeded, initial defoaming effect may be due to the lens spreading mechanism with residual action deriving from solubilized material.

So far the discussion of foams and defoaming has centered on aqueous systems. The action of defoamers in organic systems is essentially the same as that in aqueous phases. Unfortunately, the choices of possible candidates is much more restricted, limited principally to silicones and fluorocarbon materials. In potentially critical systems such as lubricating oils, some of the few materials having the required characteristics of limited solubility and adequate surface tension lowering properties, act as foaming agents below their solubility limit, but inhibit foam formation when that limit is exceeded. In other words, they act as foaming agents by adsorption and surface tension lowering until their solubility is exceeded, at which time they became foam breakers, acting via the lens mechanism, assuming that micelle formation does not occur.

Materials that are effective as defoaming agents can be classified into eight general chemical classifications, with the best choice of material depending on such factors as cost, the nature of the liquid phase, the nature of the foaming agent present, and the nature of the environment to which it may be subjected. One of the most common classes of antifoaming agents consists of the polar organic materials composed of highly branched aliphatic alcohols. As noted earlier, linear alcohols in conjunction with surfactants can result in increased foam production and stability, due to mixed monolayer formation and enhanced film strength. The branched materials, on the other hand, reduce the lateral cohesive strength of the interfacial film, which increases the rate of bubble collapse. The higher alcohols also have limited water solubility and are strongly adsorbed at the air-water interface, displacing surfactant molecules in the process.

Fatty acids and esters with limited water solubility are also often used as foam inhibitors. Their mode of action is similar to that of the analogous alcohols. In addition, their generally low toxicity often makes them attractive for use in food applications. Organic compounds with multiple polar groups are, in general, found to be effective foam inhibitors. The presence of several polar groups generally acts to increase the surface area per molecule of the adsorbed foam breaker and results in a loss of stabilization.

Metallic soaps of carboxylic acids, especially the water-insoluble polyvalent salts such as calcium, magnesium, and aluminum, can be effective as defoamers in both aqueous and nonaqueous systems. In water, they are usually employed as solutions in an organic solvent, or as a fine dispersion in the aqueous phase. Water-insoluble organic compounds containing one or more amide groups are found to be effective antifoaming agents in a number of applications, especially for use in boiler systems. It is generally found that greater effectiveness is obtained with materials containing at least 36 carbon atoms (e.g., distearoylethylenediamine, $C_{38}H_{78}N_2O_2$) compared to simple fatty acid amides.

Alkyl phosphate esters are found to possess good antifoaming characteristics in many systems due to their low water solubility and large spreading coefficient. They also find wide application in nonaqueous systems such as inks and adhesives. Organic silicone compounds are also usually found to be outstanding antifoaming agents in both aqueous and organic systems. Because of their inherently low surface energy and limited solubility in many organic solvents, the silicone materials constitute one of the two types of materials that are available to modify the surface properties of most organic liquids.

The final class of materials that have found some application as antifoaming agents are the fluorinated alcohols and acids. Because of their very low surface energies, they are active in liquids where hydrocarbon materials have no effect. They are, in general, expensive, but their activity at very low levels and in very harsh environments may overcome their initial cost barrier.

PROBLEMS

12.1. An important practical application of bulk foaming systems is in firefighting, especially those involving liquid inflammables such as gasoline or jet fuel. Propose some mechanisms by which an aqueous foam system might assist in controlling a fire.

12.2. The drainage of a liquid between two stationary soap films may take several hours. Assuming a liquid viscosity $\eta = 10^{-2}$ P (c.g.s.), a film thickness $d = 10^{-4}$ cm, a liquid density $\rho = 1$ g cm^{-3}, and g (gravity constant) = 1000 cm s^{-2}, show that such a time frame is not unreasonable.

12.3. A vertical soap film can be in mechanical equilibrium only if the force of gravity acting on each film element is balanced by a gradient in the surface tension on both surfaces. Calculate the necessary surface tension gradient for a film with a thickness of 100 nm, given a soap solution density of 1 g cm^{-3} ($g = 980$ cm s^{-2}).

12.4. Calculate the time required for a single gas bubble to disappear completely by shrinkage through diffusion, assuming that the original bubble has a radius of 1 mm and is separated from the atmosphere by a film of 1000 nm thickness. Given: $\sigma_{LV} = 30$ mN m^{-1}; diffusion coefficient of the gas in the solution = 10^{-5} cm^2 s^{-1}; gas solubility = 0.03 vol. gas at STP per vol. liquid at 1 atm.

12.5. The elasticity of a film element, E, was defined by Gibbs as

$$E = 2\left(\frac{d\sigma}{d \ln A_s}\right)$$

where A_s is the surface area of one film surface element. (a) Calculate E for a film element that is covered by an ideal monolayer with a surface pressure $\pi = \Gamma RT$. (b) Show that E has the following form for a solution containing a single nonionic surfactant, assuming ideal solution behavior

$$E = \frac{4RT \, \Gamma^2}{c}\left[\frac{1}{\left(h + 2\dfrac{d\Gamma}{dc}\right)}\right]$$

where c and Γ are the volume and surface concentrations of the surfactant in the film and h is the film thickness.

12.6. The concentration of a surface active impurity in an aqueous solution is 10^{-5} M; the surface concentration at the liquid–air interface is 10^{-10} mol cm^{-2}. To remove the impurity, air is blown into the solution, causing the formation of a foam that contains bubbles with volumes of about 0.1 cm^3 and thickness of about 10^{-4} cm. (a) Estimate the amount of impurity contained in 1 cm^2 of foam film and the approximate ratio of the amount adsorbed to that remaining in the bulk solution. (b) How many square centimeters (m^2) of film should be produced and removed to reduce the impurity content of 1 L of solution by 50%? (c) What will be the volume of foam produced to attain that goal?

12.7. Explain why the addition of a small amount of octyl alcohol to a soap foam breaks it, but does not prevent its formation if added to the solution before foam formation.

12.8. Many modern washing machines and dishwashers recommend the use of low-foaming detergents for optimum efficiency. Suggest, in general terms, molecular structures and/or characteristics for surfactants that might be expected to combine the requirements for good detergency with little foam formation.

13 Aerosols

The previous chapters have introduced several classes of colloids and some of the important surface aspects of their formation, stabilization, and destruction. Emulsions, foams, and dispersions are the most commonly treated and intensely studied examples of colloidal systems. They constitute the majority of practical and ideal systems one encounters. There exists one other class of true, lyophobic colloids—the aerosols—which, although seemingly less important in a theoretical or applied sense, are of great practical importance.

Aerosols are systems in which there exists a condensed phase of one material (solid or liquid) that is dispersed in a gaseous phase and that has dimensions that fall into the colloidal range. There are two subclasses of aerosols depending on whether the dispersed phase is a liquid or a solid. There cannot be, of course, a dispersion of one gas in another. Where the dispersed phase is a liquid, the system is commonly referred to as a "mist" or a "fog." For solid aerosols, one may commonly refer to a "dust" or "smoke." Each class of aerosol has its own characteristics of formation and stabilization and will be discussed briefly below.

13.1. THE IMPORTANCE OF AEROSOLS

Aerosols, both liquid mists and solid smokes, have a great deal of technological and natural importance. Technically they are usefully employed in coating operations, firefighting, medical treatments (allergy and asthma sprays), chemical production processes, spray drying, and other procedures. On the opposite side of the ledger, of course, we have the smoke from industrial smokestacks, smog and haze from industry and automobiles, forest fires, high-flying jet contrails, chemical and biological weapons, and so on. However, the real impact of aerosols (in purely massive terms) comes from natural sources: clouds, smokes, and similar natural airborne particles.

A cloud (natural) is a large collection of water droplets or ice crystals moving through the atmosphere and held together (loosely) by a variety of forces to be discussed below. Other natural aerosols include: airborne pollen; dust and sand (if high enough in the atmosphere causing beautiful red sunsets); volcanic clouds of water, sulfur oxides (producing acids), and other solid and liquid materials; natural "chemical fogs" produced by plant metabolism and decomposition in dense forest areas (e.g., the Smoky Mountains in the southeastern United States); and many more.

Today, probably the most visible aerosols (to the general consciousness) are those resulting from air pollution. Composed of an infernal mixture of water, solid particulate materials, and liquid droplets, pollution aerosols can literally represent the devils own chemical workshop. Under the influence of the suns ultraviolet gaze, and aided by the effects of catalytic processes discussed in Chapter 9, the complex soups we produce daily can undergo continuous chemical changes often leading to disastrous results for ourselves and our environment. The effects of pollution on human health, on vegetation, on materials and structures, and on the atmosphere itself are more apparent and frightening ever day. To pretend that we, the human race, can simply take a step back out of the industrial revolution and return to "better" days is pure fantasy (and folly). Instead, we must improve our technology to reduce the level of pollutants we produce and to control that which is unavoidable so that it never reaches the open light of day. Today, using some of the simple principles described below (and others more complex, of course), we have the technological capability to greatly improve our situation. What are lacking are the economic and political force of will to implement what we know and continue developing new ways to control these inhabitants of the twilight zone we call "aerosols."

13.2. COLLOIDAL PROPERTIES OF AEROSOLS

While aerosols are typical colloids in that they respond to the same forces already introduced—that is, electrostatic and van der Waals interactions—the special conditions that prevail in terms of the intervening gaseous medium results in an apparent qualitative difference from colloids in liquid media. The preceding chapters illustrated the importance of the intervening medium to the character and interactions of colloidal particles due to the screening effect of the continuous phase on particle–particle interactions. In aerosols, although the fundamental rules remain the same, the screening effect of the gaseous medium becomes relatively insignificant so that a number of adjustments in thinking must be made in order to reconcile the apparent differences between aerosols and emulsions, sols, and other colloidal systems.

In a first analysis, we can identify at least four basic differences between aerosols and other colloids related to the dispersion medium: (1) buoyancy effects, (2) the effects of movement of the dispersing medium, (3) particle mobility in undisturbed conditions (i.e., free fall), and (4) modification of interactions by the intervening medium. In emulsions, foams, and sols we have seen that buoyancy can be important in determining the stability of a system (i.e., matching the densities of dispersed and continuous phases can retard creaming or sedimentation). In aerosols, where the density of the continuous phase will always be significantly less than that of the dispersed particle, such effects are practically nonexistent—the colloid is essentially left to its own devices; the usual interactions found for all colloids, the "constant" pull of gravity (assuming that we are not aboard the space shuttle or MIR), and the whims of the winds.

13.2.1. Dynamics of the Aerosol Movement

Study of the dynamics of fluid flow is concerned with the forces acting on the bodies in the fluid. In the earlier chapters on solid dispersions, emulsions, and foams, fluid dynamics was largely ignored in favor of the "true" colloidal interactions. In aerosols, the nature of the continuous medium makes the subject of fluid dynamics much more important to the understanding of the system, so that the following discussion will introduce a few basic relationships that can be important in the study of aerosols.

"Winds," in the form of convection currents or other movements of the medium, are generally more important in gases than liquids. Small temperature differences or mechanical movements that would be damped out quickly in a more viscous liquid may be translated over large distances in gases and produce a much greater effect in aerosols. (Remember the famous Chinese butterfly that can change the weather in Kansas according to chaos-based theories of weather development?)

In a static system of relatively high viscosity (relative to that of gases), inertial forces due to particle movement are seldom significant; specifically, viscous forces dominate. In gases, the forces resulting from particle movement become more important and must be considered in a dynamic analysis of the system. In dynamic fluid flow analysis, the ratio of inertial forces (related to particle mass, velocity, size, etc.) to viscous forces (a characteristic of the medium and not the particles) in a system is a dimensionless number termed the Reynolds number, Re, and is used to define the type of flow occurring in the system (i.e., laminar or turbulent). For spherical particles of radius R and density ρ moving with a velocity v in a medium of viscosity η, the Reynolds number is given by

$$\text{Re} = \frac{2vR\rho}{\eta} \tag{13.1}$$

When $\text{Re} < 1$ the system is said to be in laminar flow (Fig. 13.1a) and the Stokes equation [Eq. (10.20)] is found to apply. When $\text{Re} > 10^3$, the system is in fully turbulent flow (Fig. 13.1b) and flow resistance is controlled by drag forces due to the medium given by

$$F_\text{d} \cong 0.2\pi\rho_\text{m}R^2v^2 \tag{13.2}$$

In the region $1 < \text{Re} < 10^3$, a transition occurs from laminar ($F_\text{d} \propto v$) to turbulent flow ($F_\text{d} \propto v^2$) and the relationship between F_d and v becomes more complex. Also, since drag forces actually apply only to the relative velocity

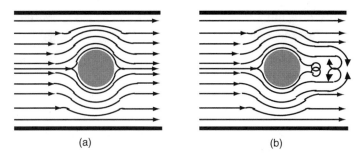

(a)   (b)

FIGURE 13.1. In the movement of aerosol particles, the type of flow in the gas phase will significantly affect the fate of the particles. For Reynolds number, Re, <1, laminar flow will prevail (*a*). However, since gases are usually of very low viscosity compared to liquids, it is more common to encounter the situation where Re > 1000. In that case, turbulent flow is common and particle dynamics is much more difficult to model.

of the particle to the medium, the effects of drag or viscous resistance to flow for a dispersed particle must be adjusted to take into consideration the flow of the medium. Raindrops, for example, generally fall under turbulent flow conditions, so analysis of their behavior should include extrapolation from Equation (13.2).

Even under ideal conditions, the dynamic flow behavior of aerosols in contrast to other colloids can be markedly different. In still air, the average distance a particle will travel before colliding with another particle, the mean free path, λ, is given by

$$\lambda = [(\sqrt{8})\pi\rho_N R^2]^{-1} \tag{13.3}$$

where ρ_N is the particle number density. For an aerosol containing 10^8 particles cm^{-3} and radius 10^{-4} cm, $\lambda = 0.11$ cm. Thus, a particle in random motion would travel an average of 0.11 cm before colliding with a neighboring particle. Such collisions may result in changes in the characteristics of the system— momentum changes in the case of elastic collisions and possibly size changes for inelastic ("sticky") collisions. The potential importance of sticky collisions will be discussed below.

For aerosols of small radius ($< 10^{-4}$ cm) and Re $> 10^3$, Equation (13.2) should be adjusted to take into consideration the effects of particle collisions. A correction factor, C_c (the Cunningham correction factor) can be incorporated into the Stokes equation to give

$$F_d = \frac{6\pi\eta R v}{C_c} \tag{13.4}$$

where

$$C_c = \frac{1 + \lambda}{R} \left[1.26 + 0.4 \exp\left(\frac{-1.1R}{\lambda}\right) \right] \qquad (13.5)$$

Obviously, the correction given in this equation becomes more important for aerosols of smaller particle radius, or in conditions of lower gas pressures (e.g., high altitudes).

According to the Stokes equation, the velocity of free fall of a particle in an undisturbed gravitational field, v_f, is given by

$$v_f = \frac{m_a g}{6\pi R \eta} = \frac{2R^2 g \rho}{9\eta} \qquad (13.6)$$

For simplicity, it is assumed that the density of the gas phase is small compared to that of the particle. For more accurate results, the density difference between particle and gas ($\rho = \rho_p - \rho_g$) should be employed. At 20°C and atmospheric pressure, the viscosity of air is 1.83×10^{-4} cP (centipoise or g cm^{-1} s^{-1}), so that for an aerosol particle of $R = 10^{-4}$ cm and $\rho = 3.0$ g cm^{-3} (e.g., volcanic ash), the rate of fall will be approximately 0.04 cm s^{-1}. Particles from a plume of ash thrown to an altitude of 10,000 m would (theoretically and neglecting all complicating factors mentioned above) take about 290 days to reach the ground! If the particle size grows to 10^{-3} cm radius by flocculation, its rate of fall increases to 3.6 cm s^{-1}, and the same trip will take about 3.2 days. It is easy to understand, then, why volcanic eruptions and other natural (and unnatural) events that produce high-altitude aerosols can affect not only the color of our sunsets but also other more vital global atmospheric interactions.

In water, with a viscosity approximately 50 times that of air, mineral particles similar to those above would have sedimentation rates on the order of 1.6 m day^{-1} and 23 m min^{-1}, respectively. Such calculations (estimations, really) are important for modeling problems of sediment accumulation in dammed reservoirs, for example.

13.2.2. Colloidal Interactions in Aerosols

Although the rules are the same, particle–particle (colloidal) interactions in aerosols can seem to have significantly different quantitative and qualitative characteristics than in liquid media. A gaseous medium, because of its very different unit density, dielectric constant, and other properties, is very ineffective at screening the forces acting between colloidal particles. For that reason, Hamaker constants in aerosols are large, usually falling in the range of 5–20×10^{-20} J (see Table 4.5), resulting in strong attractive interactions between particles and between particles and surfaces regardless of the materials involved. If we use as a measure of the kinetic energy of an aerosol particle the value of kT (Boltzmann's constant × absolute temperature), at ambient temperature, that energy will be about 4×10^{-21} J. The Hamaker constant

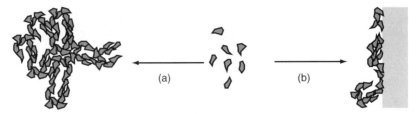

FIGURE 13.2. Because of the low dielectric constant of air, the attractive interactions among aerosol particles is orders of magnitude greater than that prevailing in a liquid medium. An obvious result of that is the appearance of household "fuzz balls" (*a*) and rapid dust accumulation on surfaces (*b*). In technology requiring ultra-clean environments, the resulting problems require million-dollar solutions.

in aerosols will, therefore, be at least an order of magnitude greater, indicating that the attractive force between particles will almost always overwhelm the kinetic energy of the particles and inelastic or sticky collisions will almost always occur.

Some practical results of that strong interaction include household fuzz balls, where dust particles flocculate and form fractal-like structures (homogeneous flocculation; Fig. 13.2*a*) and the adhesion of aerosol particles to extended surfaces (heterogeneous flocculation, Fig. 13.2*b*).

Such attractive interactions can be particularly important in situations where the presence of even a few extraneous particles on a surface can be highly detrimental, as in the production of microchips for the electronics industry (Fig. 13.3). The presence of a single dust particle on the surface of a silicon wafer before coating with the photoresist resin that will be used to engrave the final circuit will, in all probability, result in a defective product in that area. When one considers that modern chips may have circuit line spacings of less than 10^{-4} cm, a particle of that diameter or even smaller will represent a veritable monkey wrench in the works. For that reason, extreme measures must be taken to ensure that aerosol particles are absent (to the extent technologically possible) in production areas.

FIGURE 13.3. The absence of aerosol particles (dust or liquid) is especially important in the microelectronics industry. The presence of dust or liquid contaminants on the surface of virgin semiconductor (*a*) will lead to coating defects in the preparation of the microcircuits (*b*) and defects in the final product (*c*).

When liquid aerosol droplets coalesce, as in rain formation in clouds, the identity of the individual drops is lost and a single large drop is formed. As shown in the preceding section, the radius of the aerosol drop will greatly affect the free-fall rate. If the drop forms under relatively calm conditions (e.g., little vertical convection to retard the drops fall) small, gentle rainfall will probably result. In cloud formations with rapid vertical convection currents (as in thunderstorms), the growing drops will be buoyed up by the air currents, allowing more time for drop growth and resulting in larger and more forceful rain. When the cloud formation reaches a sufficiently high altitude and the drops are maintained suspended by strong internal convection currents, the drops may freeze to produce sleet or hail.

As a practical matter, almost all aerosol particles will rapidly acquire an electric charge leading to electrostatic interactions. The mechanisms for acquiring charge in aerosols are basically the same as those described in Chapter 5, although direct ionization by dissociation will be of minor importance because of the lack of ionizing solvent. Perhaps most important are charge acquisition due to friction (as in walking across a rug on a dry winter day and touching a door knob), electron gain or loss due to collision with ionizing radiation, and adsorption of ions from the air.

It is estimated that a cubic meter of "normal" city air will contain 10^8 ions (both positive and negative). When an uncharged aerosol particle is formed, it will at some time collide with an ion. Because of the more stable energetic situation, the ion will usually stick to the particle producing a charged surface. The larger the aerosol particle, the more charges it can accumulate. One might expect that the presence of surface charge would lead to stabilization of the aerosol and less flocculation. In fact, the coulombic interaction between charged aerosol particles is significantly less that kT and has little influence on particle–particle interactions at distances much greater than those at surface contact; that is, it has little effect on flocculation kinetics (Chapter 10).

The presence of charge on aerosol particles does have its important implications, however. As will be discussed in later sections, many practical applications of aerosols depend on the presence of charge, as does one of the most important process for destroying unwanted aerosols.

13.3. LIQUID AEROSOLS: MISTS AND FOGS

Mists and fogs are colloidal dispersions of a liquid in a gas. They may therefore be thought of as being roughly the inverse of a foam system. The interactions controlling their stability, however, are not in general the same as those involved in foam stabilization, because mists and fogs do not normally possess the thin lamellar stabilizing films encountered in most foams, whether of the spherical or the polyhedral type.

Liquid aerosols may be formed by one of two processes, depending on whether the dispersed system begins as a liquid or undergoes a phase change from vapor to liquid during the formation process. In the first case, since the dispersed material does not change phases, the aerosol is formed by some process that changes the dispersity or unit size of the liquid. To this class belong the spray mists such as those formed at the bottom of a waterfall or by ocean waves, mists produced by vigorous agitation, and those formed by some direct spraying or atomization process. The term "atomization" is a somewhat unfortunate choice because it has nothing to do with the nature of the process being described; however, the term entered the aerosol field many years ago and is still encountered. A more apt term is "nebulazation" (cloud formation). Liquid aerosols can also be formed directly by the application of high electrical potentials to the liquid.

The second class of mists or fogs is that produced by some process in which the incipient liquid phase is introduced as a vapor and forms droplets as a result of some equilibrium condensation process, or the liquid is produced as a result of some chemical reaction. The former mechanism includes, of course, cloud and fog formations, while the latter corresponds to some "chemical" fogs and mists.

13.3.1. Spraying and Related Mechanisms of Mist and Fog Formation

Liquid aerosol formation by spraying is a very important industrial process, even though some of the fundamental details of the process are still not very well understood. Major applications include paint application; fuel injection in diesel, gasoline, and jet engines; spray drying of milk, eggs, and other foods; the production of metal and plastic powders (spray cooling); medicinal nose and throat sprays; the application of pesticides to crops; and many more. In all of those applications, it is vitally important that the characteristics of the aerosol produced be optimized to produce the desired end result. Theories related to the formation of drops in spray systems can be very helpful in approximating the conditions necessary to produce an aerosol of defined characteristics. However, because of the nature of the process and the incidence of hard to control external factors, it is usually necessary to arrive at the optimum spraying system by trial-and-error techniques based on previous experience in the field.

Aerosol spays are usually formed by one of four basic processes as illustrated in Figure 13.4:

1. Directing a jet of liquid against a solid surface, thereby breaking the liquid up into fine droplets.
2. Ejecting a jet of liquid from an orifice into a stream of air or gas.
3. Ejecting a stream of liquid from a small orifice under high pressure.
4. Dropping liquid onto a solid rotating surface from which small droplets are ejected by centrifugal force.

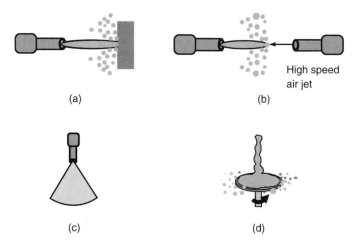

FIGURE 13.4. Of many possible mechanisms for liquid aerosol formation, four of the most common include (*a*) surface impact of a high-pressure liquid stream, (*b*) the collision of high-velocity liquid and gas streams, (*c*) high-pressure spray nozzles, and (*d*) spinning-disk centrifugal atomizers.

Other systems exist but are of much less significance in practice. Of the four, the most important industrially are using the high-pressure orifice and the rotating disk. For that reason they will be discussed briefly below.

13.3.2. Inertial Processes for Drop Formation

Notwithstanding the practical importance of aerosol production by spraying, the mechanics of the processes are still not very well understood. Numerous attempts have been and are being made to quantify and understand the phenomena involved in order to get a better practical handle on the matter. Most of those treatments are quite complex and beyond the scope of this book. However, it may be instructive to work through two relatively simple approaches in order to see how surface tension forces come into play.

Spray production by methods involving high-speed ejection of a liquid through an orifice (nozzle atomization) and ejection from a spinning disk by centrifugal force (rotary atomization) are the simplest and most important situations because they require knowledge of only one material velocity—that of the liquid. Spray production by the action of an incident air stream on a jet of liquid involves, of course, the velocity of both the liquid and the air.

13.3.3. Nozzle Atomization

If a liquid is forced through an orifice (nozzle) under a pressure, the velocity of the liquid in the channel of the orifice becomes so high that turbulent flow is encountered. As a result, the liquid will not flow smoothly in lines parallel to the walls of the orifice but will flow in complex patterns with eddies, swirls, and vortices. When the liquid leaves the orifice in this turbulent—or, to use a more fashionable term, chaotic—state, the angular forces in the vortices will act against the surface tension of the liquid to strip off units of liquid to form droplets (Fig. 13.4c).

For a simple classic analysis of the situation, assume that as the liquid leaves the orifice it has not only a linear velocity due to the pressure forcing it through the system but also some angular velocity w resulting from its chaotic flow pattern. Liquid will therefore rotate within the jet with a period of $2\pi/\omega$. The rotation creates a local centrifugal force. For a column of exiting liquid of radius r and height dz, that force is given by

$$F_\omega = \tfrac{2}{3}\pi \rho r^3 \omega^2 \, dz \tag{13.7}$$

where ρ is the density of the liquid. The pressure disrupting the jet will be given by

$$P = \frac{\tfrac{2}{3}\pi \rho r^3 \omega^2 \, dz}{2\pi r \, dz} = \tfrac{1}{3}\rho r^2 \omega^2 \tag{13.8}$$

The surface tension forces keeping the jet together will be σ/r. The second radius of curvature for the jet being infinitely large. The critical radius at which a continuous jet of liquid becomes unstable and breaks up to form droplets will be

$$r_c = \left(\frac{3\sigma}{r\omega^2}\right)^{1/3} \tag{13.9}$$

It is difficult, of course, to determine the value of ω in a flowing system; thus experimental verification of such an analysis is not a trivial matter. However, if one assumes that (is proportional to the injection pressure, the product of the pressure and r^3 should be constant. In practice, the agreement is not quite exact. If one were to use an excess pressure—that is, the pressure in excess of that at which chaotic flow begins—the agreement might logically be expected to improve.

Since theories for predicting the drop size of a spray based on the characteristics of the liquid and the apparatus are complex and sometimes unsatisfactory, it is usually necessary to measure sizes for each given situation. In general, however, the following rules hold for most fluid ejection systems:

1. Increasing the surface tension of the liquid will decrease the drop size.
2. Increasing the viscosity of the liquid will increase the average radius.

3. Increasing jet pressure decreases the average drop radius.
4. Increasing the nozzle diameter increases drop size.

Variations on the nozzle atomization process include pneumatic nozzles, in which a second fluid (air or other gas) is ejected with the liquid under pressure, rotating pneumatic cups atomizers, and sonic atomizers. The choice of the best system for a given application will be based on equipment costs, materials characteristics, and the characteristics required in the final product.

13.3.4. Rotary Atomization

In rotary atomization, a liquid is fed centrally onto a spinning part and accelerated to high velocity before being ejected into a gaseous atmosphere (Fig. 13.5). Under ideal circumstances, the liquid is extended over the entire surface of the spinning element in a thin film. When it reaches the edge, the liquid can suffer one of three fates: (1) droplets may be formed directly at the edge (Fig. 13.5a), (2) the liquid may leave the surface in filaments that subsequently break up into droplets (Fig. 13.5b), or the liquid may be detached as a sheet which later breaks up to form droplets (Fig.13.5c).

The mechanism of droplet formation will be controlled by

1. The viscosity and surface tension of the liquid
2. The inertia (kinetic energy) of the liquid at the edge

(a)   (b)   (c)

FIGURE 13.5. There are three principle mechanisms for drop formation in disk atomizers: (a) direct formation at the disk edge; (b) drop formation from strings of liquid away from the edge; (c) drop formation from sheets away from the edge. The locus and mechanism will vary according to the conditions of use: the characteristics of the liquid (surface tension, viscosity, density, feed rate, etc.); the kinetic energy of the liquid at the disk edge (speed of rotation, disk diameter, etc.); the effects of the surrounding gas at the disk edge; and the physical configuration of the disk (vanes, curvature, etc.).

3. Frictional effects between the liquid and the air it encounters at the edge
4. Shear stresses present in the liquid as it leaves the spinning edge

At relatively slow spin speeds and low liquid feed rates, viscosity and surface tension forces predominate. In that case the mechanism of drop formation is usually direct formation at the edge. The drops usually consist of a primary drop (relatively large) and several smaller satellite drops. Higher spin speeds and feed rates lead to drop formation by mechanisms 2 and 3, in which inertial and frictional forces dominate.

If the spinning element is a smooth, flat surface, the spreading liquid will tend to "slip" over the surface and not attain the maximum theoretical tangential velocity expected based on the spin speed. That phenomenon is an example of wetting failure common to many high-speed coating operations and is discussed in more detail in Chapter 17. In most spraying operations the spinning disk is not smooth but has a series of vanes that "force" the liquid onto the surface so that more speed is attained before the liquid separates from the edge. That results in a smaller average drop diameter for the same spin speed. Cup-shaped elements are also employed in situations where very small particles are not required. The effect of changes in various conditions on the average particle size to be expected are given in Table 13.1.

The proceding brief treatment of aerosol drop formation by ejection processes illustrates that theoretical analysis can be used in predicting an approximate result based on a given set of circumstances. However, much more complex analyses are necessary to obtain more than a ballpark figure, and even then the results may not justify the effort. In liquid aerosol formation, as in many such areas, experience is often the best guide.

13.3.5. Aerosol Formation by Condensation

A "chemical" method for the production of aerosol involves the direct condensation of drops or particles in the air or other gaseous environment. In order

TABLE 13.1. A Summary of Some Basic Rotary Atomizer and Feed Liquid Characteristics and Their Expected Effects on Average Drop Size of the Aerosol Produced

Characteristic	Expected Effect of Increase on Drop Size
Atomizer	
Disk diameter	Decrease
Disk speed	Decrease
Feed liquid	
Mass liquid feed rate	Increase
Density	Increase
Viscosity	Increase
Surface tension	Increase

for a vapor to condense under conditions far from its critical point, certain conditions must be fulfilled. If the vapor contains no foreign substances that may act as nucleation sites for condensation, the formation of aerosol drops will be controlled by the degree of saturation of the vapor, analogous to the situation for homogeneous crystal formation (Chapter 7).

It will be remembered that the formation of a new phase by homogeneous nucleation involves first the formation of small clusters of molecules, which then may disperse or grow in size by accretion until some critical size is reached, at which point the cluster becomes recognizable as a liquid drop. The drop may then continue to grow by accretion or by coalescence with other drops to produce the final aerosol. Normally, extensive drop formation is not observed unless the vapor pressure of the incipient liquid is considerably higher than its saturation value; that is, unless the vapor is supersaturated.

The barrier to the condensation of the liquid drop is related to the high surface energy possessed by a small drop relative to its total free energy. Thermodynamically, a simple argument can be given to illustrate the process. If one considers the condensation process as being

$$n\,A\;(\text{gas},\,P) \leftrightarrow A_n \quad \text{(liquid drop)}$$

where n denotes the number of molecules of gas A at pressure, P, involved in the process, then in the absence of surface tension effects, the free energy change of the process will be given by

$$\Delta G = -nkT \ln\left(\frac{P}{P_o}\right) \tag{13.10}$$

where P is the pressure or activity of A in the vapor phase and P_o is that in the liquid phase. The ratio P/P_o is often referred to as the degree of supersaturation of the system. A liquid drop of radius r will have a surface energy equal to $4\pi r^2\sigma$, so that the actual free energy change on drop formation will be

$$\Delta G = -nkT \ln\left(\frac{P}{P_o}\right) + 4\pi r^2\sigma \tag{13.11}$$

where both elements to the right can be written in terms of the drop radius, r. If ρ is the density of the liquid and M its molecular weight, the equation becomes

$$\Delta G = -\tfrac{4}{3}\pi r^3 \frac{\rho}{M\,RT} \ln\left(\frac{P}{P_o}\right) + 4\pi r^2\sigma \tag{13.12}$$

where the two terms are of opposite sign and have a different dependence on r. A plot of ΔG versus r exhibits a maximum as illustrated in Figure 13.6 for a hypothetical material with a density of one, molar volume of 20, and pressure or activity ratio of 4 at a given temperature. The radius at which the plot is a maximum may be defined as the critical radius, r_c, which can be determined from Equation (13.12) by setting $(\Delta G)/dr = 0$. That transformation gives the old faithful Kelvin equation, which on rearrangement leads to

$$r_c = \frac{2\sigma V_m}{RT} \ln \left(\frac{P}{P_o} \right) \tag{13.13}$$

where V_m is the molar volume of the liquid. For water at 25°C and supersaturation (P/P_o) of 6, this equation predicts a critical radius of 0.58 nm, corresponding to a cluster size of about 28 water molecules. It is difficult to say whether a drop of that small size actually has the same properties as the bulk liquid. It is probable, in fact, that the relatively high surface:volume ratio in such an assembly will result in an actual surface tension greater than the "true" bulk value of 72 mN m^{-1}. If a larger value for σ is used, the value of r_c decreases. The same occurs as the degree of supersaturation increases. The uncritical quantitative use of Equation (13.13) can be misleading in that it predicts critical cluster sizes for homogeneous nucleation that are unlikely to occur with much frequency if left to the chance of random fluctuation processes. Qualitatively, however, the equation is useful in explaining the difficulty of forming liquid aerosols by direct condensation in highly purified systems.

If one combines Equation (13.12) and (13.13), one may obtain a value for the free energy of formation of a cluster of the critical radius for drop formation, ΔG_{max}:

$$\Delta G_{max} = \frac{4\pi r_c^2 \sigma}{3} = \frac{16\pi\sigma^3 M^2}{[3\rho^2(RT \ln (P/P_o))^2]} \tag{13.14}$$

$$= 16\pi\sigma^3 V_m^{2/3} \left(RT \ln \left(\frac{P}{P_o} \right) \right)^2$$

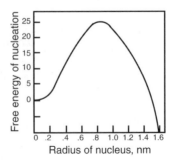

FIGURE 13.6. In nucleation processes there will exist a critical particle radius, r_c, below which free energy considerations will drive the incipient aerosol particle to disappear. Above r_c particle growth should occur.

Conceptually, one can think of the nucleation process in the following terms. If the pressure or activity of the vapor, P, is small relative to P_o, then ΔG for a given cluster of molecules will increase with each added molecule. In other words, the tendency will be for clusters smaller than r_c to return to the vapor phase.

Statistically, one might expect to encounter clusters of all sizes due to random fluctuation processes; however, all except the smallest would be very uncommon. There would therefore be little likelihood of obtaining the critical radius necessary for drop formation to occur. However, as the degree of supersaturation increases, r_c decreases and random fluctuations begin to result in more clusters with that radius. Once that point is reached, the clusters begin to grow spontaneously to form drops. When a specific supersaturation pressure is exceeded, there will develop a steady parade of clusters of the required critical dimensions, resulting in the formation of a visible mist or fog.

Condensation aerosols can also be prepared by chemical reactions involving two gaseous reactants

$$A(g) + B(g) \rightarrow C(l \text{ or } s)$$

In such a process the particle size will be controlled primarily by the concentrations of the reactants in the gas phase.

13.4. SOLID AEROSOLS: DUST AND SMOKE

Aerosols composed of solid particles suspended in a gas are commonly referred to as "dust or smoke," the exact terminology usually depending on the size and sedimentation rate of the particles, or the method of aerosol formation. In some situations, aerosols formed through dispersion processes are termed "dusts" while those arising from condensation processes are called "smokes." Alternatively, some prefer to label as dusts aerosols of sufficient particle size to have relatively rapid (e.g., noticeable over a short time span) sedimentation rates in air, while smokes would be of smaller, lighter particles. Regardless of the terminology employed, it is clear that solid aerosols constitute a very important, and usually undesirable, component of many modern processes.

The majority of industrial aerosols (both wanted and unwanted) are produced by processes of dispersion in which small particles are formed from larger solid masses. Smoke from burning wood or coal, dust in sugar refineries and grain mills, mining dusts, and some exhausts from internal combustion engines all fall into this general category of aerosol formation. Because of the rather random nature of their formation, dusts produced by dispersion techniques do not lend themselves to quantitative analysis of factors controlling particle size, size distribution, or other factors of potential interest in most colloidal systems. Condensation aerosols, on the other hand, are less common but include those formed by the chemical reaction of one or more gaseous materials (e.g., ammonia) or by oxidation or burning magnesium. The conditions of formation by condensation are somewhat more accessible from an experimental standpoint (e.g., control of reactant flow rates, temperature, pressure) and therefore more amenable to theoretical analysis. In most cases, however, experience and intuition are the important factors where control of final aerosol characteristics is important.

Because of the practical importance and potentially harmful impact of solid aerosols produced by dispersion processes, they have received a great deal of attention in terms of how their production can be eliminated or reduced and their dissipation suppressed. In such work, the emphasis is usually placed on process changes that will produce fewer or larger (and therefore more rapidly sedimenting) particles, or, as in the case of coal-burning furnaces, retarding the upward movement of carbon aerosol particles so that they can be more completely burned before leaving the stack. The suppression and destruction of smokes and dusts will be discussed further later.

Solid aerosols produced by condensation processes, while not generally as important from a practical standpoint, are often of more theoretical interest because it is easier to control their nucleation and growth rate, particle size and size distribution, and rate of disappearance. They can therefore be used more readily to study the various theories of aerosol formation and destruction. The condensation methods normally employed in such studies can be divided into two classes: chemical and physical condensation processes. A typical physical method may involve the heating of a material of relatively low volatility (e.g., stearic acid) sufficiently to produce a high degree of supersaturation and passing the vapor into a stream of cold gas, rapidly condensing the vapor into a solid aerosol. Many of the elements of tobacco smoke are also of such a "physical" origin. A representative chemical method would be illustrated by the gas-phase reaction of ammonia and hydrogen chloride to produce a snow of ammonium chloride or by the photochemical oxidation of iron pentacarbonyl $Fe(CO)_5$ in air to produce a smoke of ferric oxide.

While dusts and smokes are generally a nuisance, they do have their uses. Particularly important are pesticide dusts (and sprays) that may be applied to wide areas by airplane or surface dispersal techniques. In addition, some industrial catalysts, whose activity depends on their having a large surface : volume ratio, are effectively employed in the form of an aerosol dust or smoke. Finely dispersed solid materials are often found to exhibit combustion properties quite distinct from those observed for a large solid block of the same material. Dust produced from a low grade of coal, which would not normally be suitable for certain fuel applications, may perform appreciably like a more expensive liquid fuel. On the other hand, normally easily handled materials such as sugar or flour, not normally considered significant fire hazards, when encountered as a fine dust (as in a silo) may constitute a grave explosion hazard that must always be controlled.

13.5 THE DESTRUCTION OF AEROSOLS

Aerosols, like foams, emulsions, and dispersions, may be either advantageous or detrimental, depending on the situation. The previous discussion introduced some of the fundamental aspects of aerosol formation. Of equal or perhaps greater practical importance is the question of the suppression of aerosol formation, the destruction of unavoidable aerosols, or the controlled deposition of aerosols onto surfaces. Perhaps the best approach to solving such problems is through an understanding of some of the general principles involved in their stabilization and destruction. In that context, some of the mechanisms of destruction involved will be essentially the same as those for other colloidal systems: flocculation and coalescence.

The methods of destroying aerosols, whether liquid or solid, are numerous. Some of the more common are illustrated in Figure 13.7. One of the most important from a practical standpoint is the use of a spray (usually water) to "wash" the aerosol from the gas phase. As already mentioned, for aerosols, almost every collision between aerosol particles, collisions with container walls, or collision with a water droplet will be "sticky." For two aerosol particles the result is homocoagulation; for the other cases the process is heterocoagulation. In each case the result will be an increase in the size of

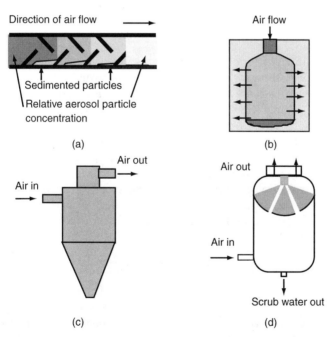

FIGURE 13.7. Although there are a number of methods to control aerosol discharges, some of the most common include (*a*) baffle filters, (*b*) bag filters; (*c*) cyclones, and (*d*) spray scrubbers.

the average particle (except, of course, where it sticks to a wall or other surface), leading to improvements such as a more rapid sedimentation due to gravity, more efficient collection in cyclonic precipitators, and easier filterability.

According to Equation (13.6), the rate of fall of an aerosol particle will be proportional to the radius squared and the density of the particle (actually the difference in density between particle and gas), and inversely proportional to the viscosity of the gas. For very small aerosols the correction given in Equation (13.5) can be applied. Obviously, any process that will increase the particle size (e.g., through flocculation) will help to facilitate the removal of the aerosol by gravitational means. The same applies to centrifugal or other processes based on the inertia of the aerosol particle.

The preceding discussion applies only in still air or other gas. If there exists a net movement of the gas phase, as in an incinerator smoke stack, the velocity of that movement must be incorporated into the equation in order to model the dynamics of particle flow. Equations (13.5) and (13.6) also assume that the viscosity of the aerosol (a liquid in this case) is much greater than that of the gas phase. Since the viscosity of gases increases with temperature (0.05 cP for air at 1000°C vs. 0.018 cP at 20°C) critical calculations may need correction. In that case, the following equation can be applied

$$v_f = \frac{2gr^2(\rho - \rho_1)}{9\eta} \frac{3\eta_1 + 3\eta}{3\eta_1 + 2\eta} \tag{13.15}$$

where η_1 is the viscosity of the liquid aerosol drop.

A common method for removing an aerosol (usually solids) from the air is the cyclone collector. A typical cyclone design is illustrated in Figure 13.8. The principle of operation is based on a double spiral vortex motion in which centrifugal force acting on the particles forces them away from the vortex axis and into the wall of the collector. Aerosol and air spiral downward in the outer spiral. "Clean" air and any entrained particles then spiral up the center vortex and exit the collector. The radial movement of the particle is the result of two opposing forces: viscous forces of the air trying to carry the particle to the vortex axis (and out of the collector) and centrifugal forces carrying the particle away from the axis and toward the wall.

The centrifugal force, F_c, is given by

$$F_c = \frac{mv_t^2}{d} \tag{13.16}$$

where m is the mass of the particle, v_t is its tangential velocity, and d is its distance from the vortex axis. Although centrifugal forces will dominate the force balance, time is required for the particle to reach the wall, so that a certain residence time in the cyclone will be required and the design dimensions

FIGURE 13.8. One of the most used methods of solid aerosol or powder collection is the cyclone collector. The powder-laden air enters the collector tangentially and the more dense powder is forced to the walls by centrifugal force. The "cleaned" air then exits the collector by a central axial duct.

must take that into account. In addition, care must be taken to insure that precipitated particles are not reintroduced into the air stream by the vortex and particles carried out with the "clean" air. Although the operation of the typical cyclone collector is described in terms of a double spiral flow, the actual flow situation is extremely complex, and a great deal remains to be learned.

The cyclone collector is extremely common and can be found with many design variations incorporated to increase efficiency or to meet specific needs. Almost all spray drying processes employ cyclones for primary or secondary product collection. Other industrial processes employ them to control the discharge of unwanted aerosols into the environment.

One mechanism for the destruction of a mist or fog of a volatile liquid is by changing the degree of "saturation" of the surrounding gas. If the gas contains less than the saturation amount of vapor of the liquid phase, the drop will evaporate. The rate of evaporation of a drop of liquid of radius r and mass m is given by

$$\frac{-dm}{dt} = 4\pi Dcr \left[1 + k \left(\frac{2\rho_1 v_t r}{\eta} \right)^{1/2} \right] \tag{13.17}$$

where k is a constant depending on the vapor diffusion coefficient D (cm^2 s^{-1}), c is the concentration of the vapor next to the drop (g cm^{-3}), v_t is the rate of fall as given by Equation (13.6), and the other terms are as previously defined. For water at 20°C, the value of k is 0.229. A hypothetical drop of radius 10^{-4} cm that falls at 0.015 cm s^{-1}, in dry air, will evaporate well before it could fall one meter. In a practical sense, then, a mist or fog can be dissipated by increasing the temperature of the surrounding gas, thereby lowering the degree of vapor saturation and increasing the rate of evaporation.

As mentioned above, most aerosol particles rapidly acquire an electrical charge. If, in a given situation, charge acquisition does not occur spontaneously, it can usually be induced by some mechanism such as ionization by electron bombardment, corona discharge, and ionizing radiation. The rate of ionization of the particle and its final surface charge will be proportional to the square of the radius of the drop. A consequence of the presence of a charge on the particle is that it will move in an electric field. Such migration capacity makes possible a practical and relatively efficient means for the removal of aerosol particles for the air: electrostatic precipitation. The principle, in its simplest form, involves passing air containing aerosol between two electrode surfaces on which an electrical potential is applied (Fig. 13.9).

The force acting on a particle in an electric field is the charge, ze, times the field strength, E, and its velocity of migration, v_m (in the direction of the field), will be

$$v_m = \frac{zeE}{6\pi r\eta} \tag{13.18}$$

Take, for example, a particle with charge $= 2e$, $r = 10^{-4}$ cm, in air at 20°C passing between two vertical electrodes 5 cm apart with an applied potential of 10^6 V m^{-1}. According to Equation (13.18), the velocity of migration in the

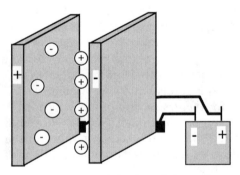

FIGURE 13.9. In the electrostatic precipitation of charged aerosols, the air is made to pass between two oppositely charged electrodes so that the particles will be attracted to and adhere to the electrode surfaces. If the configuration of the electrodes is properly designed, very efficient aerosol removal can be achieved.

electric field will be 0.093 cm s^{-1}. If the particle has a density of 3 g cm^{-3}, its rate of free fall will be 0.04 cm s^{-1} [Eq. (13.6)]. Assuming that the particle enters the electric field equidistant from each electrode, it will reach the corresponding electrode surface before leaving the field so long as the height of the electrodes is greater than 1.08 cm.

The movement of charged aerosol particles in electric fields has many important technological applications, including electrodeposition (especially for painting metallic surfaces such as automobiles) and electrophotography. It is also very important for the purification of air in critical situations.

Additional practical methods of aerosol removal include filtration and centrifugation. Centrifugation, of course, is basically the same as sedimentation except that the force of gravity is replaced by artificial forces of greater strength. Equation (13.6) continues to apply in that case, although the value of g must be multiplied by the appropriate factor.

Filtration methods, while usually relying on separation due to size, can also involve some degree of a centrifugation effect. For example, if a stream of aerosol is passed through the tortuous pathway of, say, a charcoal filter, as the stream goes through various twists and turns at a steady speed, centrifugal forces will be imposed on the particles, forcing them out along the "radius" of the curve and into the filter walls, where precipitation will occur.

A similar effect is seen when a stream of aerosol is forced to pass through a system of baffles. In such a system, as the aerosol is forced to change direction, due to the presence of the baffle plates, it must move through a curved path, setting up a net centrifugal force tangential to the stream and toward the baffle walls. As a result, the baffles will collect aerosol particles, and the gas stream leaving the system will carry a reduced particle load (Fig. 13.7a). Other methods of aerosol destruction have been developed over the years, and the problem remains one of great practical importance.

PROBLEMS

13.1. Estimate the critical drop size for the nucleation of rain droplets in a cloud formation at atmospheric pressure and air temperature (T) of 15°C. Assume that the surface tension of water is 73.0 mN m^{-1} at that temperature. Repeat the calculation for air temperatures of 45 and 90°C assuming that the surface tension of water over the temperature range 0–100°C is given by the formula $\sigma = -0.1664T + 75.98$.

13.2. If a drop of water in a cloud formation attains a diameter of 1 mm, what will its steady-state velocity be as it falls?

13.3. If water drops are formed over a boiling kettle over which there exists an upward convection current of 0.02 m s^{-1}, what will be the critical droplet diameter for which the drops will begin to fall?

13.4. Given a water drop falling through air at 20°C and 1 atm with a velocity of 0.25 m s^{-1}, calculate the Reynolds number of the air.

13.5. A dust particle of diameter 4×10^{-7} m settles in the air at 20°C and 1 atm. What is its steady-state velocity? How long will it take to fall one meter?

13.6. How much time is required for a spherical particle of density 2.0 g cm^{-3} and 1000 nm in diameter to fall 1 m under gravity at 25°C in (a) water; (b) air at 1 atm pressure; (c) in a high vacuum?

13.7. Aerosol-laden air at 25°C is purified by passing it between two parallel electrodes 6 cm apart at a velocity of 0.05 cm s^{-1}. The potential across the electrodes is 10^6 V cm^{-1}, the particle charge is $2e$, and the density is 2.5 g cm^{-3}. What is the minimum height of the electrodes that will ensure complete removal of the aerosol from the air?

13.8. Trials of a new process for preparing a powder of a temperature-ensitive material by using a spinning disk spray cooler result in an average particle size too large to meet product specifications. Suggest three process modifications that might result in a reduction of the average particle size that would not affect the production rate for the product.

13.9. With reference to Problem 13.8, if temperature sensitivity is not a problem, what additional step might be useful to solve the particle size problem?

13.10. In a nozzle atomization process, given a fixed nozzle diameter and liquid pressure, an increase in liquid temperature will (a) increase the average droplet diameter; (b) decrease the droplet diameter; (c) have no effect on droplet diameter; (d) the effect cannot be predicted based on available information. Explain your answer.

14 Polymers at Interfaces

As pointed out in Chapter 1, a colloid is characterized by a particle size range of about 10^{-7} to 10^{-3} cm. While that range is not absolute (most commonly encountered emulsions are larger on average), it serves as a good reference point with which to classify various heterogeneous systems. Using that size range as the only criterion, we can define two types of colloids: the lyophobic ("solvent-hating") colloids discussed in Chapters 10 and 11, and the lyophilic ("solvent-loving") colloids, to be discussed in this chapter.

The lyophobic colloids, as previously pointed out, are normally formed by the reduction or comminution of coarse particles to achieve the desired particle size or by controlled growth (by crystallization, condensation, etc.) from solutions of small molecules or ions. The lyophilic colloids, on the other hand, are composed of either solutions of large molecules (large relative to the size of the solvent molecules) or reversible associated or aggregated structures (association colloids) formed spontaneously in solutions of certain types of molecules. The large molecules (usually with molecular weights ranging from about 5000 to several million) are, of course, macromolecules or polymers and include a broad range of materials such as naturally occurring proteins, carbohydrates, gums, and other biocolloids, modified biopolymers such as gelatin and rayon, and the completely synthetic materials such as polyethylene, nylon, and polycarbonates. The best known association colloids are aqueous solutions of surfactants, although certain dyes and drugs may also form association structures. That very important class of lyophilic colloids composed of reversible association structures will be discussed in Chapter 15. This chapter will introduce in a very brief way the basic concepts of polymers as lyophilic colloids.

14.1. THE SOLUBILITY OF MACROMOLECULES

Small molecules of different chemical structures will mix to form homogeneous solutions if the mixing process results in a decrease in the free energy of the system. The total free energy change will, of course, have contributions from both the enthalpy of mixing, ΔH_{mix}, and the entropy of mixing, ΔS_{mix}. For systems in which the solute and solvent molecules are of comparable molecular size, ΔS_{mix} will always contribute to the decrease in ΔG_{mix}, so that a positive heat of mixing will not necessarily prevent the formation of a homogeneous solution. For a solution of macromolecules in which the solute molecule is much larger than that of the solvent, the concentration of the solute, in terms of moles per unit volume, will usually be relatively small so that the entropy of mixing will be small. Solubility or miscibility in such systems, therefore, will be determined almost exclusively by ΔH_{mix}. A negative ΔH_{mix} will result in a high degree of miscibility, while an even slightly positive value will result in almost complete insolubility. If ΔH_{mix} is near zero, the line between solubility and insolubility for a macromolecular system can be crossed from both sides, for example, by a small change in temperature or in the composition of a mixed solvent.

Because of the size of macromolecules, rates of diffusion and conformation changes accompanying dissolution (or precipitation) may be very slow, and significant amounts of time may be required for the system to reach equilibrium. The fine line between miscibility and immiscibility and the relatively long equilibrium times involved are very important to the comportment of macromolecules in solution and at various surfaces and interfaces. Understanding the effects produced by the special behavior of polymers is of great theoretical and practical importance.

14.1.1. Statistics of Polymer Chain Conformations in Solution

The conformation of polymer chains (in effect, their size) in solution is an important characteristic of a system in solution and at an interface. The detailed analysis of polymer conformations is a very complex process requiring powerful computer facilities. A simplified treatment based on random-flight (or random-walk) statistics allows for the estimation of chain dimensions to a degree adequate for most practical situations.

The bonds connecting atoms in a typical polymer chain have a specific length and are separated by an angle of about $110°$. Assuming free rotation about each bond, the polymer in solution can assume a large number of conformations. The mechanical and thermodynamic properties of a polymer or polymer solution will be determined, however, by the average or most common conformation, usually stated in terms of the average end-to-end distance of the polymer chain $(<r^2>^{1/2})$ or the root-mean-square distance of a chain element from the center of gravity of the coil $(<s^2>^{1/2})$.

If one draws a plane through the first three atoms of a polymer chain as in Figure 14.1, the next bond (bond 3–4) may align itself in any direction on a cone having a specific angle related to the bond angle, as shown. Bond 4–5 may then do the same on the basis of the possibilities for 3–4, and so on down the chain. Obviously, such a process rapidly produces a very large number of possible orientations so that an atom just a few steps down the chain may assume a position almost completely independent of that of the first two atoms (assuming the absence of intramolecular interactions). By assuming the statistical independence of the direction of interconnected chain elements, it

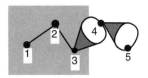

FIGURE 14.1. By using known bond lengths and bond angles of monomer units, one can calculate (or at least make a good estimate of) the most probable configuration of a polymer chain in solution.

is possible to derive an expression for the end-to-end distance for a chain with N elements (e.g., monomer units) of length A. Without going through the complete derivation process, the result is

$$<r^2>^{1/2} = (NA^2)^{1/2} \tag{14.1}$$

This equation indicates that the average end-to-end length is proportional to the square root of the stretched chain length, which means that it will also be proportional to the molecular weight since each segment of the chain will have a segment molecular weight.

The root-mean-square distance of a chain segment from the center of gravity of the coil will then be given by

$$<s^2>^{1/2} = \left(\frac{NA^2}{6}\right)^{1/2} \tag{14.2}$$

Other statistical data about a polymer chain can be determined using the same model but will not be discussed further here.

14.1.2. Problems with Random Walks

Some criticisms of the simple random-walk model are that (1) it is valid only for polymers of very high molecular weight, (2) it does not give sufficient weight to conformations in which the chain is stretched almost to its full length, and (3) it does not treat the problem of the interpenetration of various chain elements. The latter conflict can be avoided by a modification of the process to give a "self-avoiding walk," which predicts a slightly expanded coil where $<r^2>^{1/2}$ is no longer directly proportional to $N^{1/2}$. However, in many important instances, the consequences of such statistics are not greatly affected by such corrections.

The dimensions of a polymer chain in solution are important to the rheological properties of the system. More specific to the question of colloidal stability, however, such dimensions play a vital role in the ability of an adsorbed polymer to stabilize (or destabilize) a lyophobic colloid as discussed below and in Chapter 10.

14.2. ADSORPTION OF POLYMERS AT INTERFACES

Macromolecular species have played an indispensable role in the stabilization of colloidal systems since the first prelife protein complexes came into existence. We (humans) have consciously (although usually without knowing why) been making use of their properties in that context for several thousand years. Today macromolecules play a vital role in many important industrial processes and products, including as dispersants, stabilizers, and flocculants; as surface coatings for protection, lubrication, and adhesion; for the modification of rheological properties; and, of course, for their obvious importance to biological processes.

In order to understand the role of polymers in their various surface and colloidal applications, it is necessary to understand when, where, why, and how they adsorb at interfaces. While the interactions that control polymer adsorption at the monomer level are the same as those for any monomolecular species, the size of the polymer molecule introduces many complications of analysis that must be treated in a statistical manner, which means that we seldom really know what the situation is but must make educated guesses on the basis of the best available evidence.

In contrast to monomolecular species, it is highly unlikely that all or even most of the monomeric segments of a polymer chain will be simultaneously in contact with an interface. For an isolated polymer chain, the statistics based on allowed bond lengths, angles, and similar parameters dictate that there will be some equilibrium configuration that will describe the average situation. At an interface, that configuration will result from a balance of solution characteristics plus the net energy change on adsorption (whether positive or negative), the decrease in the entropy of the chain that must accompany adsorption, the gain in entropy due to freeing solvent molecules, and other changes. The latter effect is especially important because it explains why some polymers will adsorb at surfaces even when the adsorption process is endothermic overall. As a result of all combined factors, the adsorbed chain configuration will likely include loops, tails, and trains of monomer units (Fig. 14.2).

For a high molecular weight polymer, the equilibrium configuration will likely produce an adsorbed "layer" of typically 3–30 nm thickness. In general, one can assume that adsorption will be monomolecular, since the thickness of the first polymer layer will make attraction for a second negligible. The exceptions would be for polymers of low molecular weight, or for systems in which the polymer is close to the point of becoming immiscible (its so-called (θ point).

Because of their large chain size, it takes a relatively long time for polymer adsorption to reach equilibrium. One may assume, on the other hand, that adsorption of a high molecular weight polymer will be effectively irreversible. Although each polymer segment may be adsorbed reversibly, one must assume that many segments of a given chain will be adsorbed at any given moment,

(a)   (b)

FIGURE 14.2. When polymers adsorb at a surface, only a fraction of the monomer units need be involved for strong binding to occur. As a result, the configurations of the adsorbed chains can vary from the least likely end-group attachment (*a*) to the more random (and more probable) attachments as loops, trains, and tails (*b*).

and the probability of all adsorbed segments being desorbed at the same time become very small.

The irreversibility of adsorption is not the case for low-molecular-weight fractions in which there are only a few points of attachment. The concept of differential rates of adsorption based on chain length (or molecular weight) explains the commonly observed phenomenon that when a polymer of broad molecular-weight distribution is added to a colloidal system, the low-molecular-weight fraction adsorbs rapidly (i.e., it is more mobile and relatively quick to reach equilibrium), but slowly becomes displaced by high-molecular-weight chains that, once attached, will not be desorbed to any significant extent. It also helps explain the observation that high-molecular-weight polymers generally provide better stabilization for a given system than a low-molecular-weight polymer of the same composition. When particles with adsorbed polymer of low molecular weight approach, hydrodynamic forces may force the partial or complete desorption of the protecting molecules, decreasing their "protection," while high-molecular-weight molecules will not be so easily displaced (see Chapter 10 for a discussion of this phenomenon).

Just like their monomeric counterparts, adsorbed polymers exhibit characteristic adsorption isotherms. For macromolecular systems, those isotherms tend to be of the high-affinity type illustrated in Figure 14.3. In general, one

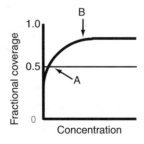

FIGURE 14.3. Because of the nature of polymer adsorption, monolayer coverage is almost assured and high-affinity isotherms are the norm.

finds that up to point *A* in the figure, all the polymer in the system is adsorbed more or less rapidly, while beyond that point more molecules can be accommodated only by changing the configuration of the adsorbed chains so as to reduce the surface area occupied by each one. That process is normally aided by the presence of favorable lateral interactions among neighboring chains.

At point *B*, the accommodation of additional chains has reached its maximum. The relative crowding of the chains at point *B* means that the molecules will not be able to attain their true statistical equilibrium conformations; instead, a pseudoequilibrium will be attained. In the region between *A* and *B*, it is usually found that the thickness of the adsorbed polymer layer increases, since more units of each chain are forced further away from the adsorbent surface (longer loops and tails are formed; see Fig. 14.4).

14.3. POLYMER–SURFACTANT INTERACTIONS

Surfactants constitute some of the most important (in terms of function, not quantity) ingredients in cosmetic and toiletry products, foods, coatings, pharmaceuticals, and many other systems of wide economic and technological importance. In many, if not most, of those applications, polymeric materials, either natural or synthetic, are present in the final product formulations or are present in the targets for their use. Other surfactant applications, especially in the medical and biological fields, also potentially involve the interaction of polymers (including proteins, nucleosides, etc.) with surfactant system.

Interactions between surfactants and natural and synthetic polymers have been studied for many years with varying degrees of understanding and experimental control. Although the basic mechanisms of surfactant–polymer interaction are reasonably well known, there still exists substantial disagreement as to the details of some of the interactions at the molecular level. Observations

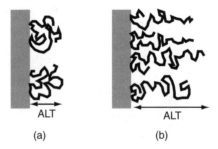

FIGURE 14.4. For strongly adsorbed polymers at low surface coverage (e.g., <0.5), each chain will usually have many points of contact (*a*). As the fraction of coverage increases, chains may begin to reaccommodate to allow more chains to adsorb with longer tails and loops. The result may be a thicker (and therefore more effective) adsorbed layer (*b*).

on changes in the interfacial, rheological, spectroscopic, and other physico-chemical properties of surfactant–polymer systems indicate that such interactions, regardless of the exact molecular explanation, can significantly alter the macroscopic characteristics of the system, and ultimately its application.

It is generally recognized that surfactant–polymer interactions may occur between individual surfactant molecules and the polymer chain (i.e., simple adsorption) or in the form of polymer–surfactant aggregate complexes. In the latter case, there may be complex formation between the polymer chain and micelles or premicellar aggregates. Other associations may result in the formation of so-called "hemimicelles" along the polymer chain. The term "hemimicelle" is relatively new to the field of surfactant science but is now encountered in several contexts, although the exact definition of the term is somewhat elusive. For now, it can be defined simply as a surfactant aggregate formed in the presence of a polymer chain or solid surface having many of the characteristics of a micelle, but is intimately associated with the locus of formation; hemimicelles do not exist as such in solution, although there is ample evidence for the formation of premicellar (or submicellar) aggregates in some systems. The formation of such structures in surfactant–polymer systems is often illustrated as resembling a string of pearls or water droplets on a spider's web (Fig. 14.5).

The forces controlling surfactant interactions with polymers are identical to those involved in other solution or interfacial properties, namely, van der Waals or dispersion forces, the hydrophobic effect, dipolar and acid–base interactions, and electrostatic interactions. The relative importance of each type of interaction will vary with the natures of the polymer and surfactant so that the exact characters of the complexes formed may be almost as varied as the types of material available for study.

Experimental methods for investigating polymer–surfactant interactions vary widely, but they generally fall into two categories: those that measure the macroscopic properties of a system (viscosity, conductivity, dye solubilization, etc.) and those that detect changes in the molecular environment of the inter-

✳ = Hemimicelle

FIGURE 14.5. In some polymer–surfactant interactions there is evidence for the formation of micelle-like or hemimicelle aggregates of surfactant molecules along the polymer chain—something like a "string of pearls."

acting species (nuclear magnetic resonance, optical rotary dispersion, circular dichroism, light scattering, etc.). A comparison of the experimental results of various studies can be complicated by variations in the sensitivity of experimental techniques and the physical manifestations of the interactions occurring, as well as differences in the purity and characterization of the experimental components. The results of each experimental approach, although useful in understanding the "symptoms" of surfactant–polymer interactions, do not always provide an unequivocal distinction among the possible mechanisms at the molecular level. Newer techniques such as small-angle neutron scattering, that can come close to "photographing" the relative relationships among polymer and surfactant units, promise to clarify many questions now in dispute.

14.3.1. Mechanisms of Polymer–Surfactant Complex Formation

The most generally accepted model for surfactant–polymer interaction is based on a stepwise sequence of binding between surfactant monomers (S) and the polymer chain (P), with each step governed by the law of mass action, and with unique rate constants, k, controlling each step:

$$P + S \leftrightarrow PS \qquad (k_1)$$
$$PS + S \leftrightarrow PS_2 \qquad (k_2)$$
$$PS_2 + S \leftrightarrow PS_3 \qquad (k_3)$$
$$PS_{n-1} + S \leftrightarrow PS_n \qquad (k_n)$$

The values of the various interaction constants and their dependence on experimental conditions (e.g., temperature, solvent, ionic strength, pH) serve as a basis for formulating feasible descriptions of the molecular processes involved in the interactions. The combination of macroscopic and molecular information can provide valuable insight into the overall process. In the model above, it is assumed that the stepwise binding process occurs initially through surfactant monomeric units, that is, that there is no significant direct association of micelles or other aggregates with the polymer chain. The formation of such aggregate–polymer complexes is not excluded, however, since they may form on the chain as the total concentration of bound surfactant increases. Alternatively, if polymer is added to a solution already containing micelles, a form of adsorption of polymer onto or into the micellar structure may possibly occur.

Surfactant–polymer interactions, as do all surfactant-related phenomena, involve a complex balance of factors encouraging and retarding association and can be understood only if those factors can be reasonably estimated. The dominating forces can be broken down into the categories of either coulombic attractions and repulsions, dipolar interactions (including hydrogen bonding or acid–base interactions), dispersion forces, and the hydrophobic effect. Combinations are, of course, possible and even likely, adding to the fun of interpreting the experimental results. While the electrostatic processes are fairly straightforward, involving the interaction of charged species on the polymer with those in the surfactant molecule, the remaining interactions are less easily quantified and can be quite complex. Polymers in particular add their own new twists since in solution they may possess secondary and tertiary structures that may be altered during the surfactant binding process in order to accommodate the bound surfactant molecules, thereby adding new energy terms to the total energy balance. The nature of the surfactant-polymer complex may significantly alter the overall energetics of the system so that major changes in polymer chain conformation will result. Any and all of those changes may result in alterations in the macroscopic and microscopic properties of the system such as changes in viscosity, polymer precipitation or solubilization, or changes in optical or electrical properties.

Forces opposing the association of molecules include thermal energy, entropic considerations, and repulsive interactions among electrical charges of the same sign. It is clear that the strength and character of surfactant–polymer interactions will depend on the properties of each component and the medium in which the interactions occur. However, even in systems where identical mechanisms are active for different surfactant and/or polymer types, the macroscopic symptoms of those interactions may be manifested in such a way that entirely different conclusions could easily be drawn.

Just as in the case of surfactants, four general types of polymer can be defined related to the electronic nature of the species: anionic, cationic, nonionic, and amphoteric. Not surprisingly, each polymer type will exhibit characteristic interactions with each surfactant class, with variations occurring within each group. It is little wonder, then, that surfactant–polymer interactions can become the subject of some very interesting discussions. With the understanding that a great deal remains to be learned about the subject as a whole, the following brief comments will introduce a few of the observed facts about this field of study.

Nonionic Polymers. The largest volume of published work in the field of surfactant–polymer interactions has involved surfactants and nonionic polymers such as polyvinylpyrrolidone (PVP), polyvinyl alcohol (PVA), polyethylene glycol (PEG), methyl cellulose (MC), polyvinyl acetate (PVAc), and polypropylene glycol (PPG) in water. The basic chemical structures of commonly used materials are shown in Figure 14.6. In general, the results indicate that the more hydrophobic the polymer, the greater is the interaction of anionic surfactants with it. For a given anionic surfactant interacting with typical polymers, it has been found that adsorption progresses in the order (PVA) < (PEG) < (MC) < (PVAc) < (PPG) < (PVP). In such systems, the primary driving force for surfactant–polymer interaction will be van der Waals forces and the hydrophobic effect. Dipolar and acid–base interactions may be present, depending on the exact nature of the system. Ionic interactions will be minimal or nonexistent, depending on the purity of the nonionic polymer.

If the primary mechanism of ionic surfactant–nonionic polymer interaction is hydrophobic or dispersion-related, the adsorption of surfactant will almost

−(CH₂ CH)ₙ−
|
N ⟶ O

Polyvinyl pyrrolidone (PVP)

−(CH₂ CH)ₙ−
|
OH

Polyvinyl alcohol (PVA)

−(CH₂ CH)ₙ−
|
OCCH₃
‖
O

Polyvinylacetate(PVAc)

HO-(CH₂ CH₂ O)ₙ-OH

Polyoxyethylene (POE) or
polyethylene glycol (PEG)

−(CH CH₂ O)ₙ−
|
CH₃

Polypropylene glycol (PPG)

Methyl cellulose (MC)

FIGURE 14.6. While a wide variety of polymers are used in systems with surfactants, those most studied and reported in the literature include common water soluble materials such as those illustrated.

certainly produce changes in the polymer chain conformation, expanding the coil as a result of repulsions between the ionic surfactant head groups (Fig. 14.7). The properties of the solution (e.g., viscosity) will be altered as a result of such changes. If a neutral salt is then added to such a system, repulsion between neighboring groups will be screened and the expanded coil will contract or collapse, again affecting various macroscopic properties of the solution. Such expansion and collapse of surfactant–polymer complexes as a

Native chain conformation   Expanded coil

FIGURE 14.7. If an aqueous polymer–surfactant interaction occurs via the surfactant tail, the resulting complex will generally exhibit modified solution characteristics, usually reflecting a greater interaction with water and an expanded chain conformation.

function of the extent of surfactant adsorption may be seen as being analogous to the solution behavior of polyelectrolytes as a function of the degree of dissociation and electrolyte content.

The bulk of the work on cationic surfactant–nonionic polymer interactions has involved the use of long-chain alkyl ammonium surfactants in aqueous solution. It has been found that the interactions between such species strengthen as the chain length of the surfactant increase. The drive to substitute surfactant-polymer for surfactant–water and polymer–water interactions, with the resulting increase in system entropy due to the released water molecules, becomes a dominating factor. The nature of the cationic head group seems to have some effect on polymer–surfactant interactions. For example, the viscosity of aqueous solutions of dodecylpyridinium thiocyanate–PVAc changes very little with variations in the surfactant concentration, whereas solutions of dodecyl ammonium thiocyanate–PVAc show considerable viscosity increases with increasing surfactant concentration. Such a result might be interpreted as reflecting a reduced extent of surfactant interaction with the polymer chain due to the greater hydrophilicity of the pyridinium ring relative to that of the simple ammonium group. The relative binding strengths between nonionic polymers and cationic or anionic surfactants are difficult to compare. The general trend is that anionics will exhibit stronger interactions with a given nonionic polymer than analogous cationic surfactants, all other things (e.g., chain length of the tail) being equal.

The interactions between nonionic surfactants and nonionic polymers have been much less intensively studied that those for ionic surfactants. The limited number of reports available indicate that there exists little evidence to indicate extensive surfactant–polymer association in such systems. Considering the size of the hydrophilic groups of most nonionic surfactants, their low cmc, and the absence of significant possibilities for head group–polymer interactions, the apparent absence of substantial interactions is not conceptually hard to accept. An assertion that binding does not occur under any circumstance, however, would be foolish, given the complexities of polymer and surfactant science in general. In food colloids especially, it has been shown qualitatively that many nonionic surfactants (monoglycerides, sorbitan esters, etc.) form rather strong complexes with starches and proteins, although the inherent complexity of such systems makes quantification of those effects difficult.

Ionic Polymers and Proteins. In practice, it is commonly found that surfactants will interact more strongly with charged polymeric species than with the nonionic examples discussed above. Many natural polymers, including proteins, some cellulosics, gums, and resins, carry some degree of electrical charge. Many of the most widely used synthetic polymers do as well. When one compares the possibilities for interactions between ionic polymers (polyelectrolytes) and surfactants with those for nonionic polymers, it is readily obvious that the presence of discrete electrical charges along the polymer backbone introduces the possibility (and probability) of significant electrostatic interaction, in addition to the nonionic factors mentioned previously. The polymers may be positively or negatively charged, or they may be amphoteric. In any case, they are commonly referred to as polyelectrolytes because of the multiple charges carried by each polymer molecule. The presence of charges on a polymer complicates the understanding of the solution properties of the polyelectrolytes. The potential for surfactant–polyelectrolyte interactions does so even more.

Polyelectrolytes, whether natural or synthetic, are of particular interest to surfactant users because of their potential for applications such as viscosity enhancers (thickening agents), dispersing aids, stabilizers, gelling agents, membrane components, and binders. They are also encountered, of course, as fibers and textiles. Common synthetic polyelectrolytes include polyacrylic and polymethacrylic acids, their copolymers, and their salts; cellulose derivatives such as carboxymethylcellulose (CMC—not to be confused with the critical micelle concentration, cmc), polypeptides such as poly-L-lysine, sulfonated polystyrenes and related strong-acid containing polymers, and polymeric polyalkylammonium and polyamine salts. Natural polyelectrolytes would include cellulose, various proteins, gum arabic, and lignins. In most cases, the charge on the polymer is fixed as either positive or negative, so that possible interactions with surfactants of a given charge type can be reasonably well defined. While such factors as pH, electrolyte content, and the nature of the polymer counterion will affect the extent of interaction in given systems, the sense of the interaction (e.g., anion–anion, anion–cation) will not change except where protonation or deprotonation of weak acids and bases occurs. Other polymers, proteins in particular, may be amphoteric; the net character of the charge is determined by pH.

Not surprisingly, interactions between surfactants and polymers of similar charge are usually found to be minimal, with electrostatic repulsion serving to inhibit the effectiveness of any noncoulombic attractions. This is especially true for polymers having relatively high charge densities along the chain. When opposite charges are present, however, the expected high degree of interaction is usually found to occur. In aqueous solution, the result of surfactant binding by electrostatic attraction is normally a reduction in the viscosity of the system, a loss of polymer solubility, at least to the point of charge reversal (see Fig. 14.8), and a reduction in the effective concentration of surfactant, as reflected by surface tension increases over what would be measured for that surfactant concentration in the absence of polymer.

Many naturally occurring random-coil polyelectrolytes of a single charge type, including some carbohydrates, pectins, and keratins, are anionic and exhibit the same general surfactant interactions as their synthetic cousins. Proteins, on the other hand, are amphoteric polyelectrolytes, which possess a net charge character (anionic or cationic) that depends on the pH of the aqueous solution. Unlike most synthetic polyelectrolytes, natural polyelectrolytes such as proteins and starches often have well-defined secondary and tertiary structures in solution that can affect, and be affected by, surfactant

Expanded polyelectrolyte chain

Collapsed coil

FIGURE 14.8. If a polymer-surfactant interaction occurs via the head group, the polymer will often exhibit solution characteristics reflecting a smaller, more compact configuration, even to the point of precipitation.

binding. When secondary and tertiary structures are present, complications arise as a result of alterations in those structures during surfactant adsorption. The denaturation of proteins by surfactants is, of course, just such a process of the disruption of higher orders of structure in the dissolved polymer molecule.

The question of exactly how a surfactant interacts with a protein molecule has been the subject of a great deal of discussion. In the case of interactions between bovine serum albumin (BSA) and sodium dodecylsulfate (SDS) the initial binding involves the electrostatic association of species of opposite charge, especially at bound surfactant levels (surfactant molecules per polymer chain) of less than 10. As such binding occurs, the electronic character of the protein changes, possibly resulting in changes in its secondary and tertiary structure. Such changes may then lead to the exposure of previously inaccessible charge sites for further electrostatic binding or of hydrophobic portions of the molecule previously protected from water contact by the higher-level protein structure. As charge neutralization occurs, precipitation of the protein may result (Fig. 14.9).

As the charges on a polymer are neutralized by surfactant adsorption, association between the hydrophobic tail of the surfactant and similar areas on the polymer becomes more favorable, again changing the net electrical character of the polymer complex. Reversal of the native charge of the protein may be the result at sufficiently high surfactant:polymer ratios (Fig. 14.9). Macroscopically, these events may lead to dramatic changes in the viscosity of the system, due to, first, collapse of the polymer coil, followed by a rapid expansion after charge reversal has taken place. In addition, a minimum in the solubility of the polymer may be encountered as evidenced by precipitation followed by repeptization.

When the bound surfactant level is high, exceeding approximately 20 surfactant molecules per high-molecular-weight polymer chain, evidence supports the view that both the head group and the hydrophobic portion of the surfactant molecule become involved in the binding process. In fact, there is some evidence that the bound surfactant molecules may be associated into micelle-like structures, forming a "string of pearls" along the polymer chain (see Fig.

FIGURE 14.9. As natural polyelectrolytes, proteins present special problems in that their solution characteristics can change significantly with changes in pH, in the presence of electrolytes, or upon interaction with surfactants, especially charged materials.

14.5). Alternatively, micelles may act as sites for polymer adsorption, much as is found for more "permanent" colloidal systems (Fig. 14.10). If such structures are present, they have the potential for altering the rheological properties of the system to a much greater extent than single molecular binding by bridging several protein molecules to produce a large polymer aggregate. Behavior suggesting such complex formation has been found for deionized bone gelatin in the presence of several anionic surfactants, and the mechanism has been suggested to explain the effect of some surfactants on the plasticity of bread dough (via the formation of crosslinks between protein chains in the gluten fraction of wheat flour) as illustrated in Figure 14.11. It is generally

FIGURE 14.10. In a manner similar to the bridging flocculation described earlier for other colloids, polymer chains in the presence of surfactant micelles may, under some circumstances, become associated with more than one micelle or other aggregate, producing unexpected results.

FIGURE 14.11. One of the interesting applications of surfactant–polymer interactions is that between gluten (wheat flour protein) and certain baking emulsifiers such as monoglycerides, sodium stearoyl-2-lactylate, or diacetyltartaric acid esters of monoglycerides. Such surfactants are employed worldwide as baking "improvers" because they act to reinforce the gluten, which, in turn, controls the rheology of the dough. It has been proposed that the interaction is between the protein and certain liquid crystal structures of the surfactant.

found that the extent of interaction as reflected by increases in the viscosity (or plasticity) of a system is highly dependent on the length of the hydrocarbon tail of the surfactant. For a series of sodium alkyl sulfates, the effect increases rapidly in the order $C_8 < C_{10} < C_{12} < C_{14} < C_{16}$, and so forth.

The interactions between cationic and nonionic surfactants and proteins has received less attention than the anionic case. Some alkylbenzene-polyoxyethylene surfactants appear to undergo limited binding with proteins, although there is little evidence for sufficient interaction to induce the conformational changes found in the case of anionic materials. The limited number of results published on protein–cationic surfactant systems indicates that little cooperative association occurs in those systems, even though the native protein charge may be of the opposite sign. Although a great deal is known about the interactions between polymers and surfactants, there is a distinct lack of good experimental data in the form of adsorption isotherms. While it is clear that the surfactant binding processes are controlled by the same basic forces as the other solution and surface properties of surfactants, the location of binding sites on the polymer molecule, the relative importance of the surfactant tail and head group, and the exact role of the polymer structure remain to be more accurately defined. In any case, anyone proposing to use a surfactant in a formulation containing polymers, or in an application where surfactant–polymer interactions will occur, must always consider the effect of each on the performance of the other.

14.3.2. Polymers, Surfactants, and Solubilization

As will be discussed in more detail in Chapter 16, a useful characteristic of many micellar systems is their ability to solubilize otherwise water-insoluble materials such as hydrocarbons, dyes, flavors, or fragrances. Some surfactant–polymer complexes have been shown to solubilize materials at surfactant concentrations well below the cmc of the surfactant in the absence of polymer. The effectiveness of such complexes differs quantitatively from that of conventional micelles.

In many instances it is found that complexes of surfactant with polymer solubilize various materials at lower total surfactant concentrations and have a greater solubilizing capacity (e.g., solubilized molecules per molecule of surfactant) than a surfactant solution alone. Unfortunately, our present state of knowledge in this area is not sufficient to allow quantitative predictions about the potential solubilizing properties of surfactant–polymer complexes based solely on chemical composition, although it is known that the effectiveness of a given combination depends on the nature of the polymeric component and the polymer : surfactant ratio.

14.3.3. Emulsion Polymerization

Surfactant–polymer systems have additional technological significance because surfactants are normally used in emulsion polymerization processes, often involving the solubilization of monomer (as well as low-molecular-weight oligomers) in micelles prior to particle formation and growth (Fig. 14.12). Surfactants have also been shown to increase the solubility of some polymers in aqueous solution. The combined actions of the surfactant as a locus for latex particle formation (the micelle), monomer solubilizer, and particle stabilizer (by adsorption), might lead one to expect quite complex relationships between the nature of the surfactant and that of the resulting latex, which is sometimes the case. Within a class, it is usually found that surfactants with high cmc produce latexes with larger particle sizes and broader size distributions, although no conclusive trend has been found for nonionic POE surfactants as a function of polyoxyethylene content.

The ability of surfactants to associate with (or adsorb onto) polymer chains may also affect the ultimate properties and stability of the resulting polymer, especially when the macromolecule exhibits some affinity for or reactivity with water. The best documented case of such a relationship involves polyvinyl acetate latexes, which have been found to differ greatly in stability depending on the surfactant used in their preparation. It is known, for example, that polyvinyl acetate can be dissolved in concentrated aqueous solutions of SDS, while cationic and nonionic surfactants have little or no solubilizing effect. In that case, solubilization presumably does not occur in the micelle, but extensive adsorption of surfactant onto the polymer chain is required. The fact that surfactants such as SDS can promote the solubilization of polyvinyl acetate has been used to suggest reasons for the observed increase in the rate of hydrolysis of polymers prepared with that surfactant relative to materials prepared with other, less strongly interacting surfactants. The assumption is that the solubilizing surfactant (SDS) can adsorb onto and solubilize the surface polymer units, causing swelling and greater exposure to water and catalyst for hydrolysis. There may also be a parallel loss of surfactant available for particle stabilization in the conventional colloidal sense. The nonsolubiliz-

(a)   (b)   (c)

FIGURE 14.12. Polymer–surfactant interactions are important in many areas of polymer science and technology, especially emulsion polymerization. In such processes surfactants and micelles perform several duties such as emulsification of monomers (*a*), solubilization of growing oligomeric free radical chains (*b*), and stabilization of growing and final polymer particles (*c*).

ing surfactant, on the other hand, would remain essentially fixed at the surface and available to perform its function as a colloid stabilizer.

In cases where there is little affinity of the polymer for water, as for polystyrenes or polyalkylacrylates and methacrylates, little effect of surfactant on water solubility would be expected. The action of the surfactant on such latex systems is then limited to its action as a monomer solubilizer during preparation and an adsorbed stabilizer afterward.

The complex relationships that can exist between polymers and surfactants raises a great many questions concerning the interpretation of data obtained from such mixed systems. They also open the door to possible new and novel applications of such combinations, however, and will no doubt provide many interesting hours of experimentation and thought for graduate students and industrial researchers in the future.

PROBLEMS

14.1. Other things being equal, the effect of changes in pH (range 2–12) on the viscosity of dilute solution of a high-molecular-weight sodium carboxylate polymer will be (a) negligible; (b) increase at higher pH; (c) decrease at higher pH; (d) reach a maximum at intermediate pH; (e) reach a minimum at intermediate pH. Explain.

14.2. Adsorption isotherms of polymers on surfaces usually exhibit a "high affinity" character. That is, at low polymer concentration virtually all the polymer is adsorbed, with very little left in solution (often immeasurable quantities). It is also common to find that the adsorption process is very slow and that adsorbed polymer cannot be readily removed by washing with the same solvent used for adsorption. Explain these observations using logical physical reasoning at the molecular level and, where possible, thermodynamic arguments as support.

14.3. It has been observed that for some sol–polymer systems the direct, rapid addition of the sol to a polymer solution containing some electrolyte results in little or no flocculation, while addition of the same sol to the same solution in portions produces effective flocculation. For example, if a particular AgI sol is added in one step to a dilute solution of polyvinyl alcohol (PVA) with electrolyte, little effect is observed. If the same sol is added in two portions—50% followed by the remaining 50%—complete flocculation occurred upon the second addition. Explain.

14.4. Referring to the preceding problem, what would you expect to be the effect of the PVA concentration and time between additions on the results? Explain.

14.5. Would you expect the sequence of mixing a gold sol, a gelatin solution, and an electrolyte solution to be important in determining the final result? Why?

14.6. A gentleman who has overindulged in the fruit of the vine starts to walk from a lamppost in the middle of a large plaza. He manages to take steps of equal length in the horizontal x and y directions, but can't remember between steps what direction he was heading in the previous step. Calculate the mean-square distance he covers from the lamppost after 50 (N) steps 90 cm (l) in length.

14.7. Explain how and why the molecular weight of an adsorbed polymer may affect its role in the steric stabilization of colloidal particles?

14.8. The use of additional surfactant in an emulsion polymerization usually leads to the formation of more latex particles. If the rest of the reaction mixture is unchanged, what will be the effect of the amount of surfactant on the final particle size, on the rate of polymerization, and on the average degree of polymerization?

15 Association Colloids: Micelles, Vesicles, and Membranes

Previous chapters have discussed the formation of colloidal particles by various mechanisms, including commutation, nucleation and growth, and emulsification. There exists another very important class of colloids that differ significantly from those discussed previously. Their formation, for example, does not result from the input of energy such as in commutation or emulsification; it is a spontaneous association process resulting from the energetics of interaction between the individual units and the solvent medium, as is crystallization. However, the size, shape, and basic nature of the associated structure are controlled by a complex series of factors distinctly different from those involved in crystallization. The size, in particular, will be much more limited than that of a normal crystal. This class of colloids is generally referred to as "association" or "self-assembled" colloids.

This class of association colloids can be further divided into several subgroups, which include micelles, vesicles, microemulsions, and bilayer membranes. Each subgroup of association colloids plays an important role in many aspects of colloid and surface science, both as theoretical probes that help us to understand the basic principles of molecular interactions, and in many practical applications of those principles, including biological systems, medicine, detergency, crude-oil recovery, foods, pharmaceuticals, and cosmetics. Before undertaking a discussion of the various types of association colloids, it is important to understand the energetic and structural factors that lead to their formation.

Association colloids form as a result of the unique character of the class of materials already described in Chapter 3—the surface-active agents or surfactants. Because of their chemical composition, surfactants have a "love-hate" relationship with most solvents, which results in a constant tug of war between forces tending toward a comfortable accommodation with a given solvent environment (i.e., solution), and a driving desire to escape to a more energetically favorable situation (an association of like-minded species). Surfactants, in other words, seem to feel that the grass is always greener on the other side of the fence, and as a result, they spend much of their time sitting on the "fence" between phases.

The adsorption of surfactants at interfaces has been, and will be, discussed in specific contexts. However, surfactants also have a life of their own within a given liquid environment: the formation of various associated structures such as micelles and vesicles. The exact behavior of a given surfactant in solution will depend on a number of internal (molecular) and external factors which will be discussed in turn below. At this point, however, it will be useful to take a general look at the possibilities open to a surface-active molecule.

15.1. SURFACTANT SOLUBILITY, KRAFFT TEMPERATURE, AND CLOUD POINT

The nature of surfactant molecules, which have both lyophilic and lyophobic groups, is responsible for their tendency to reduce the free energy of a system by adsorption at various interfaces. However, when all available interfaces are saturated, the overall energy reduction may continue through other mechanisms as illustrated in Figure 15.1.

The physical manifestation of one such mechanism is the crystallization or precipitation of the surfactant from solution—that is, bulk-phase separation. An alternative is the formation of molecular aggregates or micelles that remain in "solution" as thermodynamically stable, dispersed species with properties

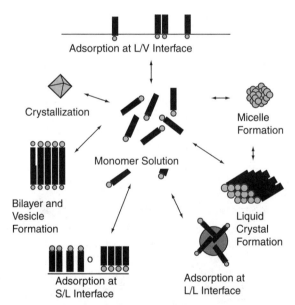

FIGURE 15.1. A surfactant in solution has various options in terms of its surface activity. Depending on the system composition, surfactants can (and usually do) "play the field," completing various functions at the same time. Usually, the multirole playing is advantageous, although there are situations in which such flexibility can be counterproductive. For that reason, surfactant selection can be an important decision in many applications.

distinct from those of the monomeric solution. Before turning our attention to the primary subject of micelles, it is useful to understand the relationship between the ability of a surfactant to form micelles and its solubility.

A primary driving force for the industrial development of synthetic surfactants in this century was the problem of the precipitation of classical fatty acid soaps in the presence of multivalent cations such as calcium and magnesium (hard-water soap films). While most common surfactants have a substantial solubility in water, that characteristic can change significantly with changes in the length of the hydrophobic tail, the nature of the head group, the valence of the counterion, and the solution environment. For many ionic materials, for instance, it is found that water solubility increases as the temperature increases. It is often observed that the solubility of the material will undergo a sharp, discontinuous increase at some characteristic temperature, referred to as the Krafft temperature, T_k. Below that temperature, the solubility of the surfactant is determined by the crystal lattice energy and heat of hydration of the system. The concentration of the monomeric species in solution will be limited to some equilibrium value determined by those properties. Above T_k, the solubility of the surfactant monomer increases to some point at which micelle formation begins and the associated species becomes the thermodynamically favored form. The concentration of monomeric surfactant will again be limited, as discussed below.

The micelle may be viewed simplistically as structurally resembling the solid crystal or a crystalline hydrate, so that the energy change in going from the crystal to the micelle will be less than the change in going to the monomeric species in solution. Thermodynamically, then, the formation of micelles favors an increase in solubility. The concentration of surfactant monomer may increase or decrease slightly at higher concentrations (at a fixed temperature), but micelles will be the predominant form of surfactant present above a critical surfactant concentration: the critical micelle concentration (cmc). The total solubility of the surfactant, then, will depend not only on the solubility of the monomeric material but also on the "solubility" of the micelles. A schematic representation of the temperature–solubility relationship for ionic surfactants is shown in Figure 15.2.

The Krafft temperature of ionic surfactants varies as a function of both the nature of the hydrophobic group and the ionic character of the head group. Nonionic surfactants, because of their different mechanism of solubilization, do not exhibit a Krafft temperature. They may, however, have a characteristic temperature–solubility relationship in water in that they may become less soluble as the temperature increases. In some cases, phase separation is found to occur, producing a cloudy suspension of surfactant aggregates. The temperature at which that occurs is referred to as the "cloud point."

Many nonionics, especially polyoxyethylene (POE) materials with a weight fraction of POE less than about 0.8, exhibit sharp, characteristic cloud points in water. As the solution temperature is increased, the clear, homogeneous micellar solution becomes turbid and a two-phase system results. The more

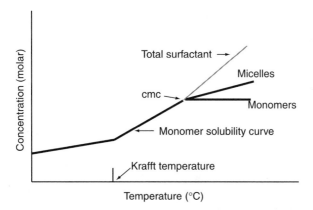

FIGURE 15.2. The temperature-solubility relationship for typical ionic surfactants illustrating the important characteristics such as the Krafft temperature, the monomer solubility curve, and the "limiting" monomer concentration at the critical micelle concentration.

dense lower phase consists of a surfactant-rich micellar phase, while the upper layer is a dilute solution of monomeric surfactant containing few, if any, micelles. The turbidity of the concentrated surfactant phase stems from the presence of very large micelles that scatter the visible light passing through the solution.

The cloud point of a surfactant depends on its chemical structure, with longer POE chains tending to increase the cloud point for a given hydrophobic group. For a given average POE chain length, the cloud point may be changed by (1) broadening the distribution of POE chain lengths (increases the cloud point), (2) branching in the hydrophobic chain (increases the cloud point), (3) nonterminal substitution of the POE chain along the hydrophobe (decreases the cloud point), (4) substitution of the terminal $-OH$ of the POE chain by a methoxyl group ($-OCH_3$, decreases the cloud point), and (5) other structural changes that may occur to the creative organic chemist.

The cloud point of a given surfactant can also be altered by the addition of various classes of materials. For example, the addition of neutral electrolyte usually lowers the cloud point, with the effect of a given salt depending on the hydrated radii of both ions. The addition of nonpolar organic materials that can be solubilized in the interior of the micelle (see Chapter 16) normally raises the cloud point, while polar materials have the opposite effect.

The existence of the cloud point phenomenon in nonionic surfactant systems carries with it a number of potential consequences—both aesthetic and functional—that must always be kept in mind. The appearance of cloudiness, while not necessarily altering the surface activity of a system, may detract from the subjective acceptability of a product. Functionally, the transition from small to large micellar aggregates may significantly alter the solubilizing capacity of a system, as well as altering the availability of free surfactant needed to complete a necessary function, etc.

15.2. SURFACTANT LIQUID CRYSTALS

Most discussions of surfactants in solution concern themselves with relatively low concentrations so that the system contains what may be called "simple" surfactant species such as monomers and their basic aggregates or micelles. Before entering into a discussion of micelles, however, it is important to know that although they have been the subject of exhaustive studies and theoretical considerations, they are only one of the several states in which surfactants can exist in solution. A complete understanding of surfactants requires a knowledge of the complete spectrum of possible states of the surfactant, including liquid crystalline phases, which can be important in the stabilization of emulsions and foams, as well in other areas.

As illustrated in Figure 15.1, the range of possible states for surfactants in the presence of solvents is quite wide. The possibilities range from the highly ordered crystalline phase to the dilute monomeric solution which, although not completely without structure, has order only at the level of molecular dimensions. Between the extremes lie a variety of phases whose natures depend intimately on the chemical structure of the surfactant, the total bulk-phase composition, and the environment of the system (temperature, pH, cosolutes, etc.). Knowledge of those structures, and the reasons for and consequences of their formation, influences both our academic understanding of surfactants and their technological application.

When surfactants are crystallized from water and other solvents that can become strongly associated with the polar head group, it is common for the crystalline form to retain a small amount of solvent in the crystal phase. In the case of water, the material would be a hydrate. The presence of solvent molecules associated with the head group allows for the existence of several unique compositions and morphological structures that, although truly crystalline, are different from the structure of the dry crystal.

As water or other solvent is added to a crystalline surfactant, the structure of the system will undergo a transition from the highly ordered crystalline state to one of greater disorder usually referred to as a liquid crystalline or mesophase. Such phases are characterized by having some physical properties of both crystalline and fluid structures. These phases will have at least one dimension that is highly ordered and, as a result, will exhibit relatively sharp X-ray diffraction patterns and optical birefringence. In other dimensions, the phases will behave in a manner more similar to nonstructured fluids.

Two general classes of liquid crystalline structures or mesophases are encountered depending on whether one is considering surfactants or other types of material. These are the *thermotropic* liquid crystals, in which the structure and properties are determined by the temperature of the system, and *lyotropic* liquid crystals, in which the structure is determined by specific interactions between the surfactant molecules and the solvent. With the exception of the natural fatty acid soaps, experimental data suggests that all surfactant liquid crystals are lyotropic.

Although liquid crystal (LC) theory predicts the existence of as many as 18 distinct structures for a given molecular composition and structure, Nature appears to have been kind in that only four of those possibilities have been identified in simple, two-component surfactant–water systems. The four LC phases usually associated with surfactants include the lamellar, hexagonal (normal and inverted), and cubic (Fig. 15.3). Of the four, the cubic phase is the most difficult to define and detect. It may have a wide variety of structural variations including a bicontinuous or interpenetrating structure that involve components of the other mesophases. The remaining types are more easily characterized and, as a result, better understood.

The lamellar liquid crystal can be viewed as a mobile or "plasticized" derivative of the typical surfactant crystalline phase. The hydrophobic chains in these structures possess a significant degree of randomness and mobility, unlike in the crystalline phase in which the chains are usually locked into the all-trans configuration (for terminally substituted n-alkyl hydrophobic groups). The level of disorder of the lamellar phase may vary smoothly or change abruptly, depending on the specific system. It is therefore possible for a surfactant to pass through several distinct lamellar phases. Because the basic unit is bilayer, lamellar phases are normally uniaxial. The lamellar phase resembles the bilayer and multilayer membranes to be discussed later, although they are formed as a result of changes in solvent concentration rather than the specific molecular structural features of the surfactant.

The hexagonal liquid crystal is a high-viscosity fluid phase composed of a close-packed array of cylindrical assemblies of theoretically unlimited size in the axial direction. The structures may be "normal" (in water) in that the hydrophilic head groups are located on the outer surface of the cylinder, or "inverted," with the hydrophilic group located internally.

Surfactant liquid crystals are normally lyotropic. The characteristics of the system, then, are highly dependent on the nature and amount of solvent present. In a phase diagram of a specific surfactant, the LC phases may span a broad region of compositions and may, in fact, constitute by far the major fraction of all possible compositions (Fig. 15.4). With the continued addition of water or other solvent, the system will eventually pass through the regions of the various mesophases into the more familiar isotropic solution phase. The solution is the most highly random state for mixtures of condensed matter and, as a result, tends to have fewer easily detected structural features. Surfactant solutions, however, are far from devoid of structure; it is only the scale of the structure that changes as dilution occurs.

15.3. MICELLES

The most intensely studied and debated type of association colloid is also perhaps the simplest in terms of the structure of the aggregate: the micelle.

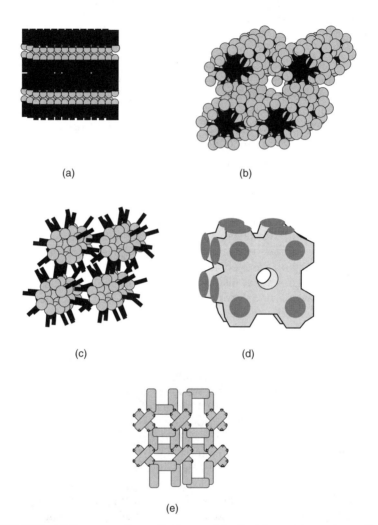

(a)

(b)

(c)

(d)

(e)

FIGURE 15.3. Of the many theoretically possible liquid crystal structures, five are most commonly encountered in surfactant systems. The lamellar phase (*a*) is "simply" alternating layers of surfactant molecules. The hexagonal phases (*b,c*) are "infinite" hexagonal close-packed structures of normal and inverted cylindrical micelles. The most complicated, and difficult to visualize and shown schematically here, are the cubic bicontinuous (or interpenetrating) network (*d*) and the cubic close packed ellipsoidal or finite cylindrical arrays (*e*).

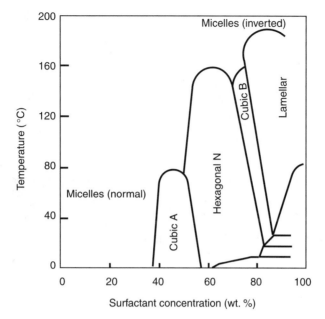

FIGURE 15.4. A typical, but simplified, phase diagram for an ionic surfactant will contain several liquid crystal phases in addition to the solution and micellar phases. Difficult-to-identify "intermediate" phases can, with careful experimental work, be found between the principal phases noted. The almost vertical boundaries reflect the low sensitivity of most ionic surfactant systems to temperature changes.

The number of publications related to micelles, micelle structures, and the thermodynamics of micelle formation is enormous. Extensive interest in the self-association phenomenon of surface active species is evident in such wide-ranging chemical and technological areas as organic and physical chemistry, biochemistry, polymer chemistry, pharmaceuticals, petroleum recovery, minerals processing, cosmetics, and food science. Even with the vast amounts of experimental and theoretical work devoted to the understanding of the aggregation of surface-active molecules, no unified theory or model has emerged that can unambiguously satisfy all the evidence and all interpretations of that evidence for the various association structures.

The solution behavior of surfactant molecules reflects the unique "split personality" of such species. The pushing and pulling that the molecules undergo in aqueous solution (or nonaqueous solution, for that matter) result from a complex mixture of effects, including (1) the interactions (both attractive and repulsive) of the hydrocarbon portion of the molecule with water; (2) the attractive interaction between hydrocarbon tails on separate molecules; (3) the solvation of the hydrophilic head group by water; (4) the interactions between solvated head groups (generally repulsive), and between the head groups and co-ions, in the case of ionic materials; and (5) geometric and packing constraints deriving from the particular molecular structure involved.

It is generally accepted that most surfactant molecules in aqueous solution can aggregate to form micellar structures with an average of 30–200 monomers in such a way that the hydrophobic portions of the molecules are associated and mutually protected from extensive contact with the bulk of the water phase. Not so universally accepted are some of the ideas concerning micellar shape, the nature of the micellar interior, surface "roughness," the sites of adsorption into (or onto) micelles, and the size distribution of micelles in a given system. Although increasingly sophisticated experimental techniques continue to provide new insights into the nature of micelles, we still have a lot to learn.

15.3.1. Manifestations of Micelle Formation

Early in the study of the solution properties of surface-active agents, it became obvious that the bulk solution properties of such materials were unusual and could change dramatically over very small concentration ranges. The measurement of bulk solution properties such as surface tension, electrical conductivity, or light scattering as a function of surfactant concentration will produce curves that normally exhibit relatively sharp discontinuities at comparatively low concentration (Fig. 15.5). The sudden change in a measured property is interpreted as indicating a significant change in the nature of the

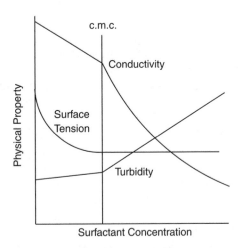

FIGURE 15.5. A number of important and accessible physical manifestations of micelle formation have been exploited in studies of surfactant aggregation. The most "classic" techniques include surface tension and conductivity titration, and turbidity (or light scattering), all of which show relatively sharp changes at the onset of micelle formation.

solute species affecting the measured quantity. In the case of the measurement of equivalent conductivity (top curve), the break may be associated with an increase in the mass per unit charge of the conducting species. That is, the primary conducting species changes from the small, mobile monomeric species to the relatively massive aggregate whose total charge will not usually be equivalent to the sum of the molecules in the aggregate due to forced ion binding (see discussion below).

For light scattering (bottom curve), the change in solution turbidity indicates the appearance of a scattering species of significantly greater size and optical properties than the monomeric solute. The effect on surface tension (middle curve) has already been introduced, but will be discussed in more detail below. These and many other types of measurement serve as evidence for the formation of aggregates or micelles in solutions of surfactants at relatively well-defined concentrations.

The results of studies of surfactant solution properties were classically interpreted in terms of a spherical association of surfactant molecules: the micelle. The structure was assumed to be an aggregate of from 50–100 molecules with a radius approximately equal to the length of the hydrocarbon chain of the surfactant Fig. 15.6). The interior of the micelle was described as being essentially hydrocarbon in nature, while the surface consisted of a layer or shell of the head groups and associated counterions, solvent molecules, and similar items.

Modern studies using techniques unavailable just a few years ago have produced more detailed information about the microscopic nature of the association structures. We now know, for example, that micelles are not static species. They are very dynamic in that there is a constant, rapid interchange

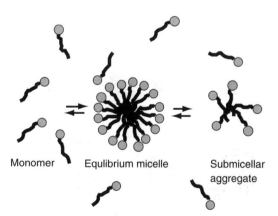

Monomer   Equilibrium micelle   Submicellar aggregate

FIGURE 15.6. Micelle formation is a rapid and dynamic process involving continuous movement of surfactant molecules into and out of the micelle and, perhaps, submicellar aggregates. The residence time of a given molecule in a micelle is estimated to be between 10^{-5} and 10^{-3} s.

FIGURE 15.7. The classic picture of smooth spherical, ellipsoidal, or cylindrical micelles is useful for visualizing and analyzing micellar systems. The reality, however, is that the micelle is a "living" aggregate much like a city whose constituent parts (monomers = citizens) come and go as the overall life of the organization continues. For that reason, any definition of the size or shape of a micelle is as accurate as a statement of the number of inhabitants in a large city at any given instant—an average guess.

of molecules between the aggregates and the solution phase. It is therefore unreasonable to assume that surfactant molecules pack into a micelle in such an orderly manner as to produce a smooth, perfectly uniform surface structure. If one could photograph a micelle with ultrahigh-speed film, freezing the motion of the molecules, the picture would certainly show an irregular molecular cluster more closely resembling a cocklebur than a golf ball (Fig. 15.7).

Although the classic picture of a micelle is that of a sphere, most evidence suggests that spherical micelles are not the rule and may in fact be the exception. Due to geometric packing requirements (to be discussed below) ellipsoidal, disk-shaped, and rodlike structures may be the more commonly encountered micellar shapes (Fig. 15.8). However, from the standpoint of providing

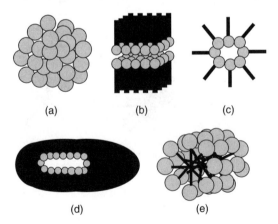

FIGURE 15.8. The most important micelle shapes include (*a*) normal spherical, (*b*) lamellar, (*c*) inverted spherical, (*d*) oblate ellipsoidal, and (*e*) prolate cylindrical or rod-shaped.

a concept of micelles and micelle formation for the nonspecialist, the spherical model remains a useful and meaningful tool.

15.3.2. Classic Thermodynamics of Micelle Formation

In the literature on micelle formation two primary models have gained general acceptance as useful (although not necessarily accurate) models for understanding the energetics of the process of self-association. The two approaches are the mass-action model, in which the micelles and monomeric species are considered to be in a kind of chemical equilibrium

$$nS \leftrightarrow S_n \tag{15.1}$$

where n is the number of molecules of surfactant, S, associating to form the micelle, and the phase separation model, in which the micelles are considered to constitute a new phase formed in the system at and above the critical micelle concentration, and

$$nS \leftrightarrow mS + S_n \downarrow \tag{15.2}$$

where m is the number of free surfactant molecules in the solution and the arrow $\downarrow$ indicates a new phase. In each case, classic thermodynamic approaches are used to describe the overall process of micellization. It is assumed that an equilibrium exists between the monomeric surfactant and the micelles with a corresponding equilibrium constant, K_m, given by

$$K_m = \frac{[S_n]}{[S]^n} \tag{15.3}$$

where brackets indicate molar concentrations and n is the number of monomers in the micelle, the aggregation number. Theoretically, one must use activities rather than concentrations in Equation (15.3); however, the substitution of concentrations for activities is generally justified by the fact that the critical micelle concentration occurs at such low concentrations that activity coefficients can be assumed to be unity.

It is usually observed that the critical micelle concentration for a surfactant is relatively sharp and characteristic. Although the detailed theory of micelle formation can become quite complex, the sharpness of the cmc can be explained conceptually in terms of the law of mass action. If C_t denotes the total concentration of surfactant in solution, C_s the fraction of surfactant present as free molecules, and C_m that in the aggregated state, Equation (15.3) may be written

$$K_m = \frac{C_m}{[C_s]^n} \tag{15.4}$$

In the process of micelle formation, there will be some concentration, C_{eq}, at which the number of surfactant molecules in the micellar form will be equal to that in the form of free surfactant molecules. At that concentration, $C_m = C_s = \frac{1}{2}C_{eq}$. Using Equation (15.4), one can then write that

$$K_m = (\tfrac{1}{2}C_{eq})^{-(n-1)} \tag{15.5}$$

At any value of C_t, the relationship between C_s and C_m can be found by substitution of Equation (15.5) into (15.4)

$$\frac{C_m}{(C_s)^n} = \left(\tfrac{1}{2}C_{eq}\right)^{-(n-1)} \tag{15.6}$$

where $C_t = C_s + C_m$. Rearrangement of this equation gives

$$\frac{C_m}{C_{eq}} = \tfrac{1}{2}\left(\frac{2C_s}{C_{eq}}\right)^n \tag{15.7}$$

Using this equation as a starting point, one can now estimate how the various concentrations vary in the neighborhood when $C_t = C_{eq}$ for a given aggregation number, n. Aggregation numbers for many surfactants lie in the range of 50–100; Table 15.1 gives the percentages of molecules in the associated state

TABLE 15.1. Percentage of Total Surfactant Molecules in Micellar Form Near $C_s = C_{eq}$[a]

	%C_t in Micellar Form		
C_s/C_{eq}	$n = 50$	$n = 75$	$n = 100$
0.45	0.57	0.04	0.003
0.47	4.6	1.01	0.22
0.49	27	18	12
0.495	38	32	27
0.50	50	50	50
0.505	62	68	73
0.51	73	81	88
0.52	87	95	98
0.53	95	99	99.7
0.54	98	99.7	99.95
0.55	99.1	99.9	99.99

[a] Calculated according to Equation (15.6) (n = aggregation number).

for $n = 50$, 75, and 100, calculated according to Equation (15.6). The results indicate that, while the cmc for a given system may not represent a truly sharp change in conditions, once the formation of micelles begins, any increase in surfactant concentration will be directed almost completely to the formation of more micelles. It is also obvious that the larger the aggregation number for a given system, the sharper will be the transition from monomolecular solution to predominantly micelles.

The alternative approach to modeling micelle formation is to think in terms of a phase separation model in which, at the cmc, the concentration of the free surfactant molecules becomes constant (like a solubility limit or K_{sp}), and all additional molecules go into the formation of micelles. Analysis of the two approaches produces the same general result in terms of the energetics of micelle formation (with some slight differences in detail), so that the choice of model is really a matter of preference and circumstances. There is evidence that the activity of free surfactant molecules does increase above the cmc, which tends to support the mass-action model; however, for most purposes, that detail is of little consequence.

15.3.3. Free Energy of Micellization

From Equation (15.3), the standard free energy for micelle formation per mole of micelles is given by

$$\Delta G_m^\circ = -RT \ln K_m = -RT \ln S_n + nRT \ln S \qquad (15.8)$$

while the standard free energy change per mole of free surfactant is

$$\frac{\Delta G_m^\circ}{n} = \left(\frac{-RT}{n}\right) \ln S_n + RT \ln S \qquad (15.9)$$

As shown above, at (or near) the cmc, $S \approx S_n$, so that the first term on the right side of Equation (15.9) can be neglected, and an approximate expression for the free energy of micellization per mole of surfactant will be

$$\Delta G_m^\circ = RT \ln \text{cmc} \qquad (15.10)$$

The situation is complicated somewhat in the case of ionized surfactants because the presence of the counterion and its degree of association with the monomer and micelle must be taken into consideration. For an ionic surfactant the mass-action equation is

$$nS^x + (n - m)C^y \leftrightarrow S_n^\alpha \qquad (15.11)$$

where C is the concentration of free counterions. The degree of dissociation of the surfactant molecules in the micelle, a, the micellar charge, is given by $\alpha = m/n$. The ionic equivalent to Equation (15.3) is then

$$K_m = \frac{[S_n]}{[S^x]^n} [C^y]^{n-m} \tag{15.12}$$

where m is the concentration of free counterions associated with, but not bound to the micelle. The standard free energy of micelle formation will be

$$\Delta G_m^\circ = \frac{RT}{n} \{ n\ln[S^x] + (n-m)\ln[C^y] - \ln[S_n] \} \tag{15.13}$$

At the cmc $[S^{-(+)}] = [C^{+(-)}] = $ cmc for a fully ionized surfactant, and this Equation can be approximated as

$$\Delta G_m^\circ = RT\left(1 + \frac{m}{n}\right) \ln \text{cmc} \tag{15.14}$$

When the ionic micelle is in a solution of high electrolyte content, the situation described by Equation (15.14) reverts to the simple nonionic case given by Equation (15.11).

In general, but not always, micelle formation is found to be an exothermic process, favored by a decrease in temperature. The enthalpy of micellization, ΔH_m, given by

$$-\Delta H_m = RT^2 \left[\frac{d \ln \text{cmc}}{dT} \right] \tag{15.15}$$

may therefore be either positive or negative, depending on the system and conditions. The process, however, always has a substantial positive entropic contribution to overcome any positive enthalpy term, so that micelle formation is primarily an entropy-driven process.

More recent approaches have employed more complicated treatments with more rigorous statements of the physical phenomena involved. However, they yield little information of value so far as understanding a given practical system is concerned. A different, and perhaps more useful, approach emphasizes the importance of molecular geometry in defining the characteristics of an aggregating system. Such a geometric approach would seem to be especially useful for applications in which the chemical structure of the surfactant is of central importance.

15.4. MOLECULAR GEOMETRY AND THE FORMATION OF ASSOCIATION COLLOIDS

A different and perhaps more conceptually useful approach to understanding surfactant aggregation phenomena emphasizes the importance of the molecular structure of the surfactant and geometry in defining the characteristics of an aggregating system. The classic thermodynamic models of micellization, while having served well over the years, have the important shortcoming that they cannot handle the larger aggregation structures such as vesicles, membranes, and liquid crystals, to be discussed later.

Surfactants can aggregate to form a range of structures as illustrated in Figures 15.3, 15.8, and 15.9. In addition, the characteristics of those basic structures such as size, shape, number of molecules involved, dispersity, and concentration at which aggregation begins (cmc), can change rapidly as solution conditions are altered. Simple thermodynamic analyses of the association process are seldom sufficient to explain the reasons for the observed transformations, nor do they serve to predict the formation of structures beyond simple micelles. In order to better understand the phenomena in question it is necessary examine the subtle interactions within the molecule, among the individual surfactant molecules making up the structure, and among neighboring elements in the system.

The classic picture of micelles formed by simple surfactant systems in aqueous solution is that of a sphere with a core of essentially liquidlike hydrocarbon surrounded by a shell containing the hydrophilic head groups along with associated counterions and water of hydration, for instance. Regardless of any questions surrounding the model, it is usually assumed that there are no water molecules included in the micellar core, since the driving force for micelle formation is a reduction of water–hydrocarbon contacts. Water will, however, be closely associated with the micellar surface; as a result, some water–core contact must occur at or near the supposed boundary between the two regions. The extent of that water–hydrocarbon contact will be determined by the surface area occupied by each head group and the radius of the core. It seems clear from a conceptual viewpoint that the relative ratio between the micellar core volume and surface area must play an important role in

(a)   (b)   (c)

FIGURE 15.9. In addition to the surfactant aggregate structures illustrated in Figures 15.3 and 15.8, such materials can also form extended bilayer structures (*a*), closed multiple bilayer vesicles (*b*), and closed single bilayer vesicles (*c*).

controlling the thermodynamics and architecture of the association process. Equally important is the need to understand the constraints that such molecular geometry places on the ability of surfactants to pack during the aggregation process to produce micelles, microemulsions, vesicles, and bilayers.

Tanford introduced the idea that there exist two opposing forces that control self-association or aggregation: hydrocarbon/water interactions that favor aggregation (i.e., pulling the surfactant molecule out of the aqueous environment) and head group interactions that work in the opposite sense. It was suggested that the two opposing actions could be viewed as an attractive interfacial tension term resulting from the basically fluid nature of the hydrocarbon tails and a complex repulsion term that depends on the nature of the head group. The overall phenomenon is termed the hydrophobic effect.

Israelachvili and others have quantified the basic ideas proposed by Tanford resulting in the concept of surfactant association controlled by the balanced geometry of the surfactant molecule. In brief, the geometric treatment of surfactant aggregation relates the overall free energy of association to three critical geometric characteristics of the molecule (Fig. 15.10): (1) the minimum interfacial area occupied by the surfactant hydrophilic or head group, a_0; (2) the volume of the hydrophobic tail or tails, v; and (3) the maximum extended chain length of the tail in a "fluid" environment such as the core of a micelle, l_c.

Using those three molecular parameters, all of which can be measured or calculated with some degree of accuracy, the geometric approach allows one to predict the shape and size of aggregates that will produce a minimum in free energy for a given surfactant structure.

Quantitatively, one defines a critical packing parameter, P_c, as

$$P_c = \frac{v}{a_0 l_c} \tag{15.16}$$

According to Israelachvili, a surfactant will be able to form spherical micelles only if the radius of the incipient micelle, R, is less than or equal to the l_c so that

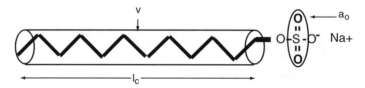

FIGURE 15.10. The geometric approach to the evaluation of surfactant aggregation processes is based on three molecular quantities: (*a*) the minimum interfacial area occupied by the head group, a_0; (*b*) the volume of the hydrophobic tail (or tails), v; and (*c*) the maximum extended length of the tail(s) in a fluid environment, l_c.

$$\frac{v}{a_0 l_c} \leq 0.33 \qquad (15.17)$$

Similar analyses for surfactants for which $0.33 \leq P_c \leq 0.5$ predict that cylindrical or disk-shaped micelles will result.

For sodium dodecylsulfate (SDS) with $v = 0.350$ nm^3, $a_0 = 0.62$ nm^2, and $l_c \approx 1.72$ nm, Equation (15.17) gives a value of $P_c = 0.33$, consistent with the formation of spherical micelles and in agreement with experimental observations. Solution conditions that alter one or more of the critical values (e.g., high salt concentrations that reduce the effective value of a_0 would, according to Equation (15.17), lead to cylindrical or disk-shaped micelles, again in agreement with observation. The predicted aggregation characteristics of surfactants covering the whole range of geometric possibilities are summarized in Table 15.2.

Although it is convenient to visualize the micellar core as a bulk hydrocarbon phase, the density may not be equal to that of the analogous true bulk material. However, X-ray evidence indicates that the molecular volumes of surfactants in micelles and other aggregate structures are essentially unchanged by the aggregation process. If a molecular volume for a hydrocarbon chain in the micellar core equal to that of a normal hydrocarbon is assumed, the core volume can be calculated from

$$V = m' \, (27.4 + 26.9 \, n_c') \times 10^{-3} \; (\text{nm}^3) \qquad (15.18)$$

where m' is an effective micellar aggregation number, and n_c' is the number of carbon atoms per chain in the core. In general, the value of n_c' will be one less than the total number of carbons in the hydrocarbon chain, n_c, since the

TABLE 15.2. Expected Aggregate Characteristics in Relation to Surfactant Critical Packing Parameter, $P_c = v/a_0 l_c$

P_c	General Surfactant Type	Expected Aggregate Structure
< 0.33	Simple surfactants with single chains and relatively large head groups	Spherical or ellipsoidal micelles
0.33–0.5	Simple surfactants with relatively small head groups, or ionics in the presence of large amounts of electrolyte	Relatively large cylindrical or rod-shaped micelles
0.5–1.0	Double-chain surfactants with large head groups and flexible chains	Vesicles and flexible bilayer structures
1.0	Double-chain surfactants with small head groups or rigid, immobile chains	Planar extended bilayers
> 1.0	Double-chain surfactants with small head groups, very large and bulky hydrophobic groups	Reversed or inverted micelles

first carbon after the head group is highly solvated and may be considered as a part of it. For normal surfactants with a single hydrocarbon tail, m' will be equal to the aggregation number, m, while for those that possess a double tail, $m' = 2m$.

If one assumes that the micellar core has no "hole" at its center, one dimension of the aggregate species will be limited by the length of the hydro-carbon chain when extended to its fullest. That maximum length can be calculated by assuming a distance of 0.253 nm between alternate carbon atoms of the extended chain and adding the value of the van der Waals radius of the terminal methyl group (= 0.21 nm) and half the bond distance between the first carbon in the core and that bonded to the head group (≈ 0.06 nm). The maximum extended length l_{max} for a normal hydrocarbon chain with n_c' core carbon atoms, therefore, is given by

$$l_{max} = 0.15 + 0.1265 n_c' \ (nm) \tag{15.19}$$

Since hydrocarbon chains in the liquid state are never fully extended, an effective chain length, l_{eff}, can be defined that gives the statistically most likely extension as calculated by the same procedure used for the calculation of polymer chain dimensions. For a chain with $n_c' = 11$, the ratio of l_{max} to l_{eff} will be approximately 0.75. In the micellar core, because of restrictions imposed by the attachment of the hydrocarbon tail to the head group bound at the surface, the mobility of the chains may be significantly limited relative to that of bulk hydrocarbon chains. The presence of "kinks" or gauche chain conformations, which may be imposed by packing considerations, will result in a calculated lmax amounting to only about 80% of the theoretical maximum.

Since hydrocarbon chains possess restricted bond angles as well as bond lengths, additional restrictions on the maximum extension of the chain arise beyond those mentioned previously. Chain segments located at the transition region from core to shell, for example, cannot assume arbitrary conformations in order to produce a perfectly "smooth," homogeneous surface. The micellar surface, therefore, must be assumed to be somewhat rough or irregular, al-though the dynamic nature of the aggregate may obscure any practical effect of such roughness.

Extension of the concepts of molecular geometry and aggregate structure has led to its use in predicting not only the structure to be expected (the shape to be expected (micelle, vesicle, extended bilayer, etc.), but also the size, size distribution, shape (spherical, ellipsoidal, disk, or rod-shaped), dispersity (or size distribution), critical micelle concentration, average aggregation number, and other such characteristics. The "rules" of association derived from the geometric analysis of molecular structure are summarized in Figure 15.11.

It has been found experimentally that the form of aggregate structure produced by a given surfactant also depends to a great extent on its solution environment. Geometric considerations explain fundamental processes op-erating in the aggregation process based on the various effects the solution

Critical packing parameter	Approximate molecular shape	Expected aggregate structure
<0.33		(Normal)
0.33 - 0.5		
0.5 - 1.0		
1.0		
>1.0		(Inverted)

FIGURE 15.11. The critical packing parameter, P_c, allows one to quickly determine the general type of aggregate structure to be expected for a given surfactant molecular composition.

environment has on a_0, v, and l_c. Some important effects to be expected can be summarized as follows:

1. Molecules with relatively small head groups, and therefore large values for P_c, will normally form extended bilayers, large (low curvature) vesicles, or inverted micellar structures (Fig. 15.9a). Such results can also be brought about in "normal" surfactant systems by changes in pH, high salt concentrations, the addition of multivalent cations, the addition of water miscible organic materials (e.g., alcohols), and other variables.

2. Molecules with long hydrocarbon chains ($\geq C_{16}$) tend to form rigid, almost crystal-like structures at ambient temperature. If held at temperatures above the melting temperature of the hydrocarbon chain, increased chain motion may occur allowing *trans–gauche* chain isomerization, thereby reducing the effective value of l_c and resulting in changes in aggregate structures. This effect may be of particular importance in understanding the effects of temperature on biological membranes, liposomes, and related structures.

3. Changes in the nature or solvation of cations through the addition of complexing or cryptating species (e.g., crown ethers) will increase the value of a_0 and perhaps alter the aggregate structure.

While the geometric approach to explaining surfactant aggregation phenomena shows great promise, it has not worked its way into the general thinking on micelles. As more experimental data become available and can be correlated with the predictions of geometric "rules", this approach may become the basis for the design of surfactant molecules with specific desirable aggregation characteristics. For commercial surfactants that are mixtures of a homologous series of various chain lengths, that contain potentially surface-active impurities (e.g., alcohols), or that have relatively high salt concentrations, care must be used in extrapolating from model to reality. Many surprises may await us in the twilight zone. Likewise, in multiple surfactant systems, a situation commonly encountered in cosmetics, for example, the geometric approach will require a great deal more investigation into possible synergistic interactions before its application becomes widespread.

15.5. SOME CORRELATIONS BETWEEN SURFACTANT STRUCTURE, ENVIRONMENT, AND MICELLIZATION

Although the model approaches to association phenomena discussed above are useful from a fundamental point of view, in practice, the association characteristics (the cmc, aggregation number, etc.) of a surfactant are very sensitive to external factors such as the isomeric purity of the sample, the presence of contaminants (especially potentially surface active ones such as alcohols), pH, electrolyte content, and temperature. A good working knowledge of micelle formation, therefore, must include some idea of how such factors will affect the behavior of the surfactant. The literature on those various topics is extensive and of varying quality; however, there have developed over the years a number of good generalizations that can be helpful in making "educated" extrapolations from ideal to reality. The following sections, then, will be devoted to the presentation of summaries and generalizations that illustrate many of the most significant effects of surfactant chemical structure and solution environment on the micellization process.

15.5.1. Aggregation Number

Aggregation numbers for many surfactants have been found to fall in the range of 50–100 molecules, although that can vary significantly according to structure and conditions. Some typical aggregation numbers for various surfactant types are given in Table 15.3.

Because the size and dispersity of micelles are sensitive to many internal (hydrophobic structure, head group type) and external (temperature, pressure, pH, electrolyte content) factors, it is sometimes difficult to place too much significance on reported values of m. However, some generalizations can be made that are usually found to be true are as follows:

TABLE 15.3. Aggregation Numbers for Representative Surfactants in Water

Surfactant	Temperature (°C)	Aggregation Number, n
$C_{10}H_{21}SO_3^-Na^+$	30	40
$C_{12}H_{25}SO_3^-Na^+$	40	54
$(C_{12}H_{25}SO_3^-)_2Mg^{2+}$	60	107
$C_{12}H_{25}SO_4^-Na^+$	23	71
$C_{14}H_{29}SO_3^-\ Na^+$	60	80
$C_{12}H_{25}N(CH_3)_3^+Br^-$	23	50
$C_8H_{17}O(CH_2CH_2O)_6H$	30	41
$C_{10}H_{21}O(CH_2CH_2O)_6H$	35	260
$C_{12}H_{25}O(CH_2CH_2O)_6H$	15	140
$C_{12}H_{25}O(CH_2CH_2O)_6H$	25	400
$C_{12}H_{25}O(CH_2CH_2O)_6H$	35	1400
$C_{14}H_{29}O(CH_2CH_2O)_6H$	35	7500

1. In aqueous solutions, it is generally observed that the longer the hydrophobic chain for an homologous series of surfactants, the larger will be the aggregation number.

2. A similar increase is seen when there is a decrease in the "hydrophilicity" of the head group—for example, a higher degree of ion binding in an ionic material or a shorter polyoxyethylene chain in nonionics.

3. Factors that result in a reduction in the hydrophilicity of the head group such as high electrolyte concentrations will also cause an apparent increase in aggregation number.

4. Temperature changes will affect nonionic and ionic surfactants differently. In general, higher temperatures will result in small decreases in aggregation numbers for ionic surfactants but significantly large increases for most nonionics.

5. The addition of small amounts of nonsurfactant organic materials of low water solubility will often produce an apparent increase in micelle size, although that may be more an effect of solubilization (see text below) than an actual increase in the number of surfactant molecules present in the micelle.

6. The addition of a water-miscible organic material such as an alcohol will generally reduce the apparent aggregation number. This may result from a change in the solvent quality of the aqueous phase and/or the formation of mixed micelles.

While the question of the size of micelles is of great theoretical interest, it is seldom very significant (as far as we know) in most surfactant applications, other than perhaps solubilization and microemulsion formation. Of more general importance is the concentration at which micelle formation occurs, the critical micelle concentration, since that is the time when many of the most useful surfactant properties come into play.

15.5.2. The Critical Micelle Concentration

Because there are many factors that have been shown to affect the observed critical micelle concentration strongly, the following discussion has been divided so as to isolate (as much as possible) the various important factors.

Any discussion of cmc data must be tempered with the knowledge that the reported values cannot be taken to be absolute but reflect certain variable factors inherent in the procedures employed for their determination. The variations in cmc found in the literature for nominally identical materials under supposedly identical conditions must be accepted as minor "noise" that should not significantly affect the overall picture (assuming, of course, that good experimental technique has been employed).

The Hydrophobic Group: The "Tail."   The length of the chain of a hydrocarbon surfactant is a major factor determining the cmc. The cmc for a homologous series of surfactants decreases logarithmically as the number of carbons in the chain increases. For straight-chain hydrocarbon surfactants of about 16 carbon atoms or less bound to a single terminal head group, the cmc is usually reduced to approximately one-half of its previous value with the addition of each $-CH_2-$ group. For nonionic surfactants, the effect can be much larger, with a decrease by a factor of 10 following the addition of two carbons to the chain. The insertion of a phenyl and other linking groups, branching of the alkyl group, and the presence of polar groups on the chain can produce different effects on the cmc.

The relationship between the hydrocarbon chain length and cmc for ionic surfactants generally fits the Klevens equation

$$\log_{10} \text{cmc} = A - Bn_c \tag{15.20}$$

where A and B are constants specific to the homologous series under constant conditions of temperature, pressure, and other parameters, and n_c is the number of carbon atoms in the chain. Values of A and B for a wide variety of surfactant types have been determined, and some are listed in Table 15.4. It has generally been found that the value of A is approximately constant for a particular ionic head group, while B is constant and approximately equal to $\log_{10} 2$ for all paraffin chain salts having a single ionic head group. The value of B will change, however, in systems having two head groups, or for nonionic systems.

For nonionic surfactants, in which the mechanism of solubilization of the surfactant molecule is basically hydrogen bonding, the relative importance of the tail and head groups to the overall process changes. An empirical relation-

TABLE 15.4. Klevens Constants [Eq. (15.20)] for Common Surfactant Classes

Surfactant Class	Temperature (°C)	A	B
Carboxylate soaps (Na$^+$)	20	1.85	0.30
Carboxylate soaps (K$^+$)	25	1.92	0.29
n -Alkyl-1-sulfates (Na$^+$)	45	1.42	0.30
n -Alkyl-2-sulfates (Na$^+$)	55	1.28	0.27
n -Alkyl-1-sulfonates	40	1.59	0.29
p -n -Alkylbenzene sulfonates	55	1.68	0.29
n -Alkylammonium Cl$^-$	25	1.25	0.27
n -Alkyltrimethylammonium Br$^-$	25	1.72	0.30
n -Alkylpyridinium Br$^-$	30	1.72	0.31

ship between the cmc and the number of oxyethylene (OE)$_y$ groups present in several nonionic surfactant series has the form

$$\ln \mathrm{cmc} = A' + B'y \tag{15.21}$$

where A' and B' are constants related to a given hydrophobic group. Examples of A' and B' for several commonly encountered hydrophobic groups are given in Table 15.5. In each case, the results are for one temperature and can be expected to vary significantly, given the sensitivity of such systems to changes in T.

For more complex surfactant structures, the following generalizations serve as a good guide:

1. Ionic surfactants having two or three ionic groups at one end of the hydrocarbon tail such as α-sulfonated fatty acids and their esters, alkyl malonates or alkyl tricarboxylates, exhibit a linear relationship between cmc and chain length similar to Equation (15.20), although they usually have a lower Krafft temperature and a higher cmc than the corresponding singly charged molecule of the same.

TABLE 15.5. Empirical Constants Relating CMC and Oxyethylene Content for Various Hydrophobic Groups in Nonionic Surfactants [Eq. (15.21)]

Hydrophobic Group	A'	B'
$C_{12}H_{25}OH$	3.60	0.048
$C_{13}H_{27}OH$	3.59	0.091
$C_{18}H_{35}OH$ (oleyl)	3.67	0.015
$C_{18}H_{37}OH$ (stearyl)	2.97	0.070
$C_9H_{19}C_6H_4OH$	3.49	0.065

2. For surfactants having branched structures, with the head group attached at some point other than the terminal carbon, such as, sodium tetradecane-2-sulfonate $CH_3(CH_2)_{11}CH(SO_3^-Na^+CH_3$, the additional carbon atoms off of the main chain contribute a factor equivalent to about one-half that for a main-chain carbons. Except for the lower members of a series, the relationship between carbon number and cmc follows a linear relationship similar to that expressed in Equation (15.20).

3. For surfactants that contain two separated hydrophobic chains, such as the sodium dialkylsulfosuccinates, it is generally found that the cmc values for the straight-chain esters follow the Klevens relationship, although the value of B is slightly smaller than that found for single-chain surfactants. The cmc for the branched esters of equal carbon number occur at higher concentrations.

4. In the alkylbenzene sulfonates, with various points of attachment of the alkyl group to the benzene ring, experimental data indicate that the aromatic ring has substantial hydrophilic character, with the benzene ring contribution being equivalent to about 3.5 carbon atoms.

5. For surfactants that contain ethylenic unsaturation in the chain, one generally finds that the presence of a single double bond increases the cmc by as much as a factor of 3–4 compared to the analogous saturated compound. In addition to the electronic presence of the double bond, the isomer configuration (*cis* or *trans*) will also have an effect, with the *cis* isomer usually having a higher cmc, presumably due to the more difficult packing requirements imposed by the isomer.

6. The presence of polar atoms such as oxygen or nitrogen in the hydrophobic chain (but not associated with a head group), usually results is an increase in the cmc. The substitution of an $-OH$ for hydrogen, for example, reduces the effect of the carbon atoms between the substitution and the head group to half that expected in the absence of substitution. If the polar group and the head group are attached at the same carbon, that carbon atom appears to make little or no contribution to the hydrophobic character of the chain.

7. A number of commercial surfactants are available in which all or most of the hydrophobic character is derived from the presence of polyoxypropylene (POP) groups. The observed effect of such substitution has been that each propylene oxide group is equivalent to approximately 0.4 methylene carbon.

Two classes of materials that cannot easily be fitted into the known schemes for conventional hydrocarbons are the silicone-based surfactants and those in which hydrogen has been replaced by fluorine atoms. The hydrophobic unit of the silicone-based surfactants consists of low-molecular-weight polyalkylsiloxane derivatives, usually polydimethylsiloxane. Possibly because of their "nonclassic" nature they have received little attention in the general scientific literature, although their unique surface characteristics have proved useful in many technological applications, especially in nonaqueous solvent systems.

The substitution of fluorine for hydrogen on the hydrophobic chain has produced several types of surfactants with extremely interesting and useful properties. The presence of the fluorine atoms results in large (i.e., orders of magnitude) decreases in cmc relative to the base hydrocarbon. Because of the electronic character of the carbon–fluorine bond, fluorinated materials have been found to have much lower surface energies and produce lower surface tensions than conventional materials. In general, a fully fluorinated surfactant with n_c carbon atoms will have a cmc roughly equal to that of a hydrocarbon material with $2n_c$ carbons.

The Hydrophilic Group. The effect of the hydrophilic head group on the cmc values of a series of surfactants with the same hydrocarbon chain will vary considerably, depending upon the nature of the change. In aqueous solution the difference in cmc for a C_{12} hydrocarbon with an ionic head group will lie in the range of 0.001 M, while a nonionic material with the same chain will have a cmc in the range of 0.0001 M. The exact nature of the ionic group has no dramatic effect, since the main driving force for micelle formation is the entropy gain on reduction of water–hydrocarbon interactions. The cmc values of several ionic surfactants are given in Table 15.6. Of the more common anionic head groups, the order of decreasing cmc values for a given hydrocarbon chain is found to be $-CO_2^-$ (containing one more carbon atom) $> -OSO_3^-$ $> -SO_3^-$. For cationic surfactants, one often finds that the cmc increases with methyl substitution on the nitrogen, probably due to increased steric

TABLE 15.6. Effect of Hydrophilic Group on the CMCs of Surfactants with Common Hydrophobes

Hydrophobe	Hydrophile	Temperature (°C)	cmc (mM)
$C_{12}H_{25}$	COOK	25	12.5
''	$-SO_3K$	25	9.0
''	$-SO_3Na$	25	8.1
''	$-NH_3Cl$	30	14
''	$-N(CH_3)_3Cl$	30	20
''	$-N(CH_3)_3Br$	25	16
$C_{16}H_{23}$	$-NH_3Cl$	55	0.85
''	$-N(CH_3)_3Cl$	30	1.3
''	$-N(CH_3)_3Br$	60	1.0
C_8H_{17}	$-OCH_2CH_2OH$	25	4.9
''	$-(OCH_2CH_2)_2OH$	25	5.8
C_9H_{19}	$-COO(CH_2CH_2O)_9CH_3$	27	1.0
''	$-COO(CH_2CH_2O)_{16}CH_3$	27	1.8
$C_{10}H_{21}$	$-O(CH_2CH_2O)_8CH_3$	30	0.6
''	$-O(CH_2CH_2O)_{11}CH_3$	30	0.95
''	$-O(CH_2CH_2O)_{12}CH_3$	29	1.1

requirements of the added methyl groups forcing an increase in ionization (i.e., less ion pairing).

Counterion Effects. In ionic surfactants micelle formation is related to the interactions of solvent with the ionic head group. The degree of ionization, in terms of tight ion binding, solvent-separated ion pairing, or complete ionization, will therefore influence the value of the cmc and the aggregation number. Since electrostatic repulsions among the ionic groups would be greatest for complete ionization, one finds that the cmc of surfactants in aqueous solution decreases as the degree of ion binding increases.

From regular solution theory it is found that the extent of ion pairing in a system will increase as the polarizability and valence of the counterion increase. Conversely, a larger radius of hydration will result in greater ion separation. It has been found that, for a given hydrophobic tail and anionic head group, the cmc decreases in the order $Li^+ > Na^+ > K^+ > Cs^+ > N(CH_3)_4^+ > N(CH_2CH_3)_4^+ > Ca^{2+} \approx Mg^{2+}$. In the case of cationic surfactants such as dodecyltrimethyl ammonium halides, the cmc are found to decrease in the order $F^- > Cl^- > Br^- > I^-$.

Although within a given valence the size of the hydrated counterion will have some effect on the micellization of an ionic surfactant, a more significant effect is produced by changes in valence. As the counterion is changed from monovalent to di- and trivalent, the cmc is found to decrease rapidly. The divalent and higher salts of carboxylic acid soaps generally have very low water solubility and are not useful as surfactants in aqueous solution. They do find use in nonaqueous solvents because of their increased solubility in those systems, especially in the preparation of water-in-oil emulsions.

15.5.3. The Effect of Additives

Many industrial applications of surfactants involve the presence in the solution of cosolutes and other additives that can potentially affect the micellization process through specific interactions with the surfactant molecules (thereby altering the effective activity of the surfactant in solution) or by altering the thermodynamics of the micellization process by changing the nature of the solvent or the various interactions leading to or opposing micelle formation.

Solution changes that might be expected to affect the association process include the presence of electrolytes, changes in pH, and the addition of organic materials that may be essentially water insoluble (e.g., hydrocarbons), water-miscible (short-chain alcohols, acetone, dioxane, etc.), or of low water solubility but containing polar groups that impart some surface activity although they are not classified formally as surfactants. The following generalizations are usually useful, although it must be remembered that each surfactant system can exhibit characteristics different from the general observations noted here.

Added Electrolyte. In aqueous solution the presence of electrolyte causes a decrease in the cmc of most surfactants; the greatest effect is found for ionic materials. Nonionic and zwitterionic surfactants exhibit a much smaller effect. For ionic materials, the effect of addition of electrolyte can be empirically quantified with the relationship

$$\log_{10} \text{cmc} = -a \log_{10} c_i + b \qquad (15.22)$$

where a and b are constants for a given ionic head group at a particular temperature and c_i is the total concentration of monovalent counterions in moles per liter. For nonionic and zwitterionic materials, the impact of added electrolyte is significantly less and the relationship in Equation (15.22) does not apply. Some observations have been reported in which specific cations appear to produce cmc changes in POE nonionics. One can speculate that such an effect may result from a complex formation related to crown ether effects (Fig. 15.12).

(a)

(b)

FIGURE 15.12. In the presence of large, polarizable cations such as calcium nonionic POE surfactants may, under some circumstances, form complexes with the ions producing a significant change in the configuration of the ether chain and surfactant solution properties.

pH. For most modern, industrially important surfactants consisting of long-alkyl-chain salts of strong acids, solution pH has a relatively small effect, if any, on the cmc of the materials. Unlike the salts of strong acids, however, the carboxylate soap surfactants exhibit a significant sensitivity to pH. Since the carboxyl group is not fully ionized near or below the pK_a, pH changes may result in significant changes in the cmc as well as the Krafft temperature. A similar result will be observed for the cationic alkylammonium salts near and above the pK_b. Changes in pH will have little or no effect on the cmc of nonionic surfactants except, perhaps, at very low pH where it is possible that protonation of the ether oxygen of OE surfactants can occur. Such an event would, no doubt, alter the characteristics of the system. Little can be found in the literature pertaining to such effects, however.

A number of amphoteric surfactants have pH sensitivity related to the pK's of their substituent groups. The possibilities can be grouped in the following way:

1. Quaternary ammonium–strong acid salts will show little or no significant pH sensitivity.
2. Quaternary ammonium–weak acid combinations will be zwitterionic at high pH and cationic below the pK_a of the acid.
3. Amine–weak acid combinations will be anionic at high pH, cationic at low pH, and zwitterionic at some pH between the respective pK values.
4. Amine–strong acid combinations will be anionic at high pH and zwitterionic below the pK_b of the amine.

Obviously, pH would be expected to produce significant changes in the cmc of surfactants in groups 2–4.

Organic Materials. Organic materials that have low water solubility can be solubilized in micelles to produce systems with substantial organic content where no solubility would occur in the absence of micelles. More details on the phenomenon of solubilization in surfactant micelles will be presented below. In any case it is usually found that immiscible hydrophobic materials will have relatively little effect on cmc, although evidence for slight decreases has been reported.

Small amounts of organic additives with substantial water miscibility such as the lower alcohols, dioxane, acetone, glycol, and tetrahydrofuran have relatively minor effects on cmc. As the alkyl group in the additive goes beyond C_3, the inherent surface activity of the alcohol can begin to become significant. Otherwise, it will be only at high concentrations, where the additive may be considered a cosolvent, that major effects on cmc will be evident. In general, large amounts of water-miscible organic material will increase the cmc by increasing the solubility of the tail, although the opposite effect may occur for highly ionized species, where the lower dielectric constant reduces head group repulsion.

The properties of a surfactant solution are found to change much more rapidly with the introduction of small amounts of long-chain alcohols, amines, and esters, especially for $C > 3$. Because so many classes of surfactants of importance academically and industrially are derived from raw materials containing alcohol or amine impurities, recognition of the effects of such materials can be very important. Most of the observed effects can be attributed to the inherent surface activity of the long chain alcohols and amines.

The interactions between surfactants and alcohols have become of great importance in recent years as a result of the intense interest in microemulsions and their potential application in various areas of technological importance. Some of the basic concepts in that area will be presented later. The presence of surface-active impurities in general can significantly alter (for better or worse) the solution characteristics of a system and create confusion if not properly taken into account.

Temperature. The effects of temperature changes on the cmc in aqueous solution have been found to be quite complex. It has been shown, for example, that the cmc of most ionic surfactants passes through a minimum as the temperature is varied from 0 to 70°C. Nonionic and zwitterionic materials are not quite so predictable, although it is has been found that some nonionics reach a cmc minimum around 50°C.

The temperature dependence of the cmc of polyoxyethylene nonionic surfactants is especially important since the head group interaction is essentially totally hydrogen bonding in nature. Materials relying solely on hydrogen bonding for solubilization in aqueous solution are commonly found to exhibit an inverse temperature–solubility relationship. As already mentioned, major manifestation of such a relationship is the presence of the cloud point for many nonionic surfactants.

15.5.4. Micelle Formation in Mixed-Surfactant Systems

When one discusses the solution behavior of many, if not most, industrially important surfactants, it is important to remember that experimental results must be interpreted in the context of a surfactant mixture rather than a pure homogeneous material. Studies of such systems are important both academically, assuming that the mixture can be properly analyzed, and practically, since most detergents and soaps contain homologues of higher or lower chain length than that of the primary component.

Determinations of the cmc of well-defined, binary mixtures of surfactants have shown that the greater the difference in the cmc between the components of the mixture, the greater is the effect of the chain length of the more hydrophobic member. The analysis of results for binary mixtures of an homologous series of surfactants must take into consideration the fact that at the cmc the mole fractions of the monomeric surfactants in solution are not necessarily equal to the mole fractions in the original mixture; each value must be decreased by the amounts of each incorporated into the micellar phase.

Interpretations may also be complicated by such effects as relatively small changes in the mole fraction of the smaller chain component due to preferential aggregation of the more hydrophobic material and the difficulty of inclusion of the longer chain into micelles of the shorter material. In some cases where the difference is very large, the component with the higher cmc may simply act as an added electrolyte, rather than becoming directly involved in the micellization process. When ternary surfactant mixtures are considered, it is usually found that the cmc of the mixture falls somewhere between the highest and lowest value determined for the individual components.

The presence of an ionic surfactant in mixture with a nonionic usually results in an increase in the cloud point of the nonionic component. In fact, the mixture may not show a cloud point, or the transition may occur over a broad temperature range, indicating the formation of mixed micelles. As a result of that effect, it is possible to formulate mixtures of ionic and nonionic surfactants for use at temperatures and under solvent conditions (electrolyte, etc.) in which neither component alone is effective.

Many mixtures of surfactants, especially ionic with nonionic, exhibit surface properties significantly better than do those obtained with either component alone. Such synergistic effects greatly improve many technological applications in areas such as emulsion formulations, emulsion polymerization, surface tension reduction, coating operations, personal care and cosmetics products, pharmaceuticals, and petroleum recovery, to name only a few. The use of mixed surfactant systems should always be considered as a method for obtaining optimal performance in any practical surfactant application.

15.5.5. Micelle Formation in Nonaqueous Media

The formation of micelle-like aggregates in nonaqueous solvents has received far less attention than the related phenomenon in water. In fact, there exists some controversy as to whether such a phenomenon, in fact, occurs in the same sense as in aqueous solutions. There can be no doubt, however, that some chemical species, many surfactants included, do associate in hydrocarbon and other nonpolar solvents.

The changes involved in surfactant aggregation in nonaqueous solvents must differ considerably from those already discussed for water-based systems. The orientation of the surfactant relative to the bulk solvent will be the opposite to that in water (hence the term "reversed" micelle). In addition, the micelle, regardless of the nature of the surfactant, will be un-ionized in solvents of low dielectric constant, and so will have no significant electrical properties relative to the bulk solvent, although electrostatic interactions will play an important role in the aggregation process, but in an opposite sense to that in aqueous solution.

The primary driving force for the formation of micelles in aqueous solution is the gain in entropy resulting from a reduction of the unfavorable interactions between water and the hydrophobic tail of the surfactant. In nonaqueous solvents, there will be little significant change in the interactions between surfactant tail and solvent, even if one is hydrocarbon and the other aromatic. A more significant energetic consequence of nonaqueous micelle formation is the reduction of unfavorable interactions between the ionic head group of the surfactant and the nonpolar solvent molecules. Or even more likely, the gain of favorable interactions among polar or charged groups.

Unlike the situation for aqueous micelles in which interactions between the hydrophobic tails contribute little to the overall free energy of micelle formation, ionic, dipolar, or hydrogen bonding interactions between head groups in reversed micelles are one of the primary driving forces favoring aggregation. In the face of factors favoring aggregation, there seem to be few obvious factors tending to oppose the formation of nonaqueous micelles (unlike head group steric and electrostatic interactions in the aqueous case). The possible exception is an unfavorable entropy decrease as a result of fewer degrees of freedom for monomers in the micelle relative to those free in solution.

In contrast to aqueous surfactant solutions in which micellar size and shape may vary considerably, small spherical micelles appear to be the most favored, especially when the reduction of solvent–polar group interactions is important. Similar to water-based systems, geometric considerations often play an important role in determining micelle size and shape. Many materials that commonly form nonaqueous micellar solutions possess large, bulky hydrocarbon tails with a cross-sectional area significantly greater than that of the polar head group. Typical examples of such materials are sodium di-2-ethylhexylsulfosuccinate and sodium dinonylnaphthalene sulfonate.

Since unambiguous experimental data are much less available on micelle formation in nonaqueous solvents than for aqueous systems, it is far more difficult to identify trends and draw conclusions concerning the relationships between chemical structures, cmc, and aggregation numbers. However, some generalizations can be made.

In hydrocarbon solvents, the nature of the polar head group is extremely important in the aggregation process. It has generally been found that ionic surfactants form larger aggregates than nonionic ones, with anionic sulfates surpassing the cationic ammonium salts. The aggregation number for an ionic surfactant in a given solvent will usually change little with changes in the counterion, indicating a lack of sensitivity to the nature of that species. The effect of the hydrocarbon tail length in a homologous series of surfactants is relatively small when compared to that in water. However, the aggregation number tends to decrease as the carbon number increases within a homologous series.

The presence of small amounts of water in a nonaqueous surfactant environment can have a significant effect on some systems. It can be presumed that the effects of water and other solubilized impurities on nonaqueous micelle formation stems from alterations in the dipolar interactions between head groups induced by the additive or impurity.

Although the vast majority of surfactants form micelles of some kind in aqueous solution, some materials, because of their special structure or composition, will not associate in the "normal" way described above. They will, however, take part in other association processes to form equally interesting and important association colloids, including especially vesicles and bilayer membranes.

15.6. VESICLES AND BILAYER MEMBRANES

As has been discussed previously, surfactants and related amphiphilic molecules, including the natural surfactants or lipids, tend to associate into a variety of structures in both aqueous and nonaqueous systems. In most cases, those assemblies can transform from one into the other as a result of sometimes subtle changes in the solution conditions (e.g., concentration, solvent composition, added electrolyte, temperature changes, pH). The basic concepts that govern self-association into micelles also apply to the formation of the larger, more extended aggregate systems consisting of vesicles, bilayers, and membranes. This section will present a limited discussion of some of the general aspects of molecular association into such structures as vesicles and bilayers, and the presentation of a few of the possible practical applications of such systems currently under investigation.

Those amphiphilic materials that cannot conveniently pack into compact structures, such as micelles in aqueous solution, generally associate to produce vesicles and extended bilayers. In general, such materials will have relatively small head groups or, more commonly, their hydrophobic groups will be too bulky to be packed in a manner necessary for normal micelle formation. Such a state of affairs is particularly common for molecules having more than one hydrocarbon chain, very highly branched chains, or structural units that produce molecular geometries incompatible with effective packing (e.g., large, flat ring structures such as steroids).

Although extended (essentially infinite) planar bilayers are a thermodynamically favorable option for the association of some bulky surfactants or amphiphiles in aqueous solution, there are some conditions under which it is more favorable to form closed bilayer systems, leading to the existence of various types of membranes and vesicles. Such a situation can be seen to arise from two basic causes. First, even large, highly extended planar bilayers possess edges along which the hydrocarbon core of the structure must be exposed to an aqueous environment, resulting in an unfavorable energetic situation. Second, the formation of an infinitely extended structure is unfavorable from an entropic standpoint. The formation of vesicles, then, addresses both of those factors—the edge effect is removed by the formation of a closed system, and the formation of structures of finite size reduces much of the entropy loss. As long as the curvature of the vesicle is gentle enough to allow the packed molecules to maintain close to their optimum area, vesicles will represent viable structures for the association of surfactants and related materials.

15.6.1. Vesicles

Many naturally occurring and synthetic surfactants and phospholipids that cannot undergo simple aggregation to form micelles will, when dispersed in water, spontaneously form closed bilayer structures referred to as liposomes or vesicles. They are constructed of alternating layers of lipid or surfactant bilayers spaced by aqueous layers or compartments arranged in approximately concentric circles (Fig. 15.13a). If the spontaneously formed multilayer vesicles are subjected to ultrasound or other vigorous agitation, the complex multilayer

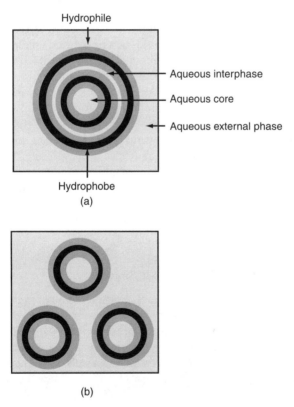

FIGURE 15.13. Surfactants that form closed bilayer aggregate structures such as vesicles usually produce multilayered systems such as (a). Smaller, single bilayer vesicles such as (b) can be formed by disruption of the multilayer systems.

structure may be disrupted to produce a single bilayer assembly consisting of a unilamellar vesicle in which a portion of the aqueous phase is encapsulated within the single bilayer membranes (Fig. 15.13b). Typically, a vesicle so produced will have a diameter of 30–100 nm, falling within the size range of classical colloidal systems.

Natural and synthetic amphiphiles that form vesicles are, by their nature, of limited solubility in aqueous systems, so that the exchange of individual molecules from the bilayer is often very slow. In addition, the bilayer structure has a significant degree of internal stability so that vesicles, once formed, can have a relatively long existence. Lifetimes of from a few days to several months have been reported. After extended periods, the unilamellar vesicles will begin to fuse to produce the more complex aggregate structures of the original systems.

One of the interesting and potentially useful characteristics of vesicles is their ability to entrap within the assembly a portion of the aqueous phase present at the time of their formation, along with associated solute. They therefore represent a unique microencapsulating technique, since residual solute located outside the vesicle can be removed by dialysis or some other related purification techniques. Oil-soluble materials can also be incorporated into vesicle systems, although they would then be located inside the hydrophobic portion of the membrane, much like materials solubilized in conventional surfactant micelles. The potential for the incorporation of both aqueous and nonaqueous additives into vesicles poses the interesting possibility of producing a system containing two active components, for example, a water- and an oil-soluble drug, for simultaneous delivery.

Other interesting and potentially useful physical characteristics of conventional vesicles include their activity as osmotic membranes, their ability to undergo phase transitions from liquid crystalline to a more fluid state, and their permeability to many small molecules and ions, especially protons and hydroxide. Because of their similarity to natural biological membranes, vesicles also have great potential as models for naturally occurring analogues that may be difficult to manipulate directly.

15.6.2. Polymerized Vesicles and Lipid Bilayers

Major barriers to the use of conventional vesicles in many applications include (1) the inherent long-term instability of the systems, (2) their potential for interaction with enzymes and blood lipoproteins, and (3) their susceptibility to the actions of other surface-active materials. For such critical applications as controlled-release drug delivery, even the most stable systems with a lifetime of several months do not begin to approach the shelf life requirements.

As a result of the potential utility and relatively low cost of vesicles, a great deal of effort has in been applied to the development of polymerized surfactant and phospholipid systems. The ability to crosslink the vesicle membrane after the encapsulation process should produce a system in which the basic nature of

FIGURE 15.14. Surfactant molecules containing polymerizable functionalities can be used to produce crosslinked vesicles for microencapsulation. The reactive functionality can be located at the extreme end of the hydrophobic tail (*a*), in the middle of the tail (*b*), or closely associated with the hydrophilic head group (*c*).

the vesicle as an encapsulating medium is retained while adding the structural integrity and increased stability of a crosslinked polymeric structure.

The general approach used to attain such structures has been the synthesis of conventional vesicle-forming amphiphilic materials containing polymerizable functionalities in the molecule, vesicle formation, and subsequent polymerization, preferably by some "nonintrusive" means such as irradiation. In principle, the polymerizable functionality can be located at the end of the hydrophobic tail, centrally within the tail, or in association with the ionic or polar head group (Fig. 15.14). The choice of a preferred structure will probably be determined by the final needs of the system and the synthetic availability of the desired materials.

15.7. BIOLOGICAL MEMBRANES

In the last few years there has been a dramatic increase in interest in the molecular structure of biological membranes. While model systems composed of artificially prepared (or isolated) amphiphilic materials and associated colloids serve a very useful purpose, a better understanding of the reality of biological systems would be invaluable in many areas of biochemistry, medicine, pharmaceuticals, and other fields. While it is reasonably easy to determine the constituents of a biological membrane, elucidating just how the various components are put together, how they interact, and their exact function within the membrane represents a decidedly more difficult task.

15.7.1. Membrane Surfactants or Lipids

The surface-active components of biological membranes are generally referred to as "lipids," with the majority consisting of double-chained phospholipids or glycolipids. The hydrophobic tails normally contain chains of 16–18 carbons, with one generally being branched or unsaturated. The combination of those factors guarantees that the lipids will be very surface active and easily form self-assembled bilayer membranes that can encapsulate or isolate different regions and functions in an organism. In addition, the long chain lengths ensure that the lipids will have a very low solubility in water (as the monomer), a low cmc, and therefore their assemblies will remain intact while contacting surrounding fluids. The presence of branching and unsaturation also guarantees that the membranes will remain fluid over a relatively wide temperature range, to insure the viability of the organism in the presence of varied environmental conditions.

The size, structure, and fluidity of membrane lipids are also important because those aspects of the molecules make it possible for them to pack efficiently into a variety of convoluted bilayer membrane structures with various degrees of curvature and flexibility. That flexibility also makes possible the inclusion of the various other important components of the cell wall, including proteins, carbohydrates, and cholesterol. In terms of the geometric concepts discussed previously (see Fig. 15.15), one can visualize where one class of lipid will have a critical packing factor, P_c ($= v/a_0l_c$) < 1, which will produce a truncated cone shape, while another will have $F > 1$ for an inverted truncated cone. Combinations of the two can then accommodate the inclusion of, for example, proteins and cholesterol, while maintaining an overall planar structure (or a given degree of curvature), or increase curvature to produce a smaller associated unit.

FIGURE 15.15. Using the molecular geometry of one or several surfactant molecules it becomes possible to visualize how a surfactant or mixture can produce the wide variety of aggregate structures encountered in nature.

15.7.2. Membrane Dynamics

Biological membranes, like micelles and vesicles, are dynamic structures in which the component lipids and proteins can move about relatively freely, even though the structure as a whole remains intact. In order to carry out its biological function, the cell membrane will also have heterogeneous regions of lipids, proteins, or other materials which may serve as specific binding sites or transport "channels," for example. The components of the entire structure, however, must all have one thing in common. They must be able to associate spontaneously to form the necessary assembly of molecules to do the job, even when all the components will not form such structures alone. It appears that an organism can "sense" the specific lipid structures needed in a given situation to produce the membrane structure, fluidity, or other criteria called for. When conditions such as temperature change, the organism synthesizes the new molecules (e.g., more or less saturated fatty acid chains) to fit the new conditions. Clearly, the creation and functioning of biological membranes cannot be a totally haphazard process of trial and error in selecting the proper lipids for a given cell structure and function. There must exist some feedback mechanism through which the organism can "know" what material is needed under given conditions so that it can be provided when and where called for.

Other aspects of the interactions of lipids and bilayer structures in biological systems can be understood in the context of molecular geometry, association phenomena, and general interfacial interactions. Unfortunately, those topics are too broad to be included here. It will be interesting to see how future research in molecular biology is able to incorporate the fundamentals of surface and colloid science into a better understanding of the function of membranes, cells, and entire organisms.

PROBLEMS

15.1. Estimate the size of spherical micelles that would be formed by a series of single-chain hydrocarbon surfactants with chains of 10, 12, 14, 16, and 18 carbons.

15.2. Assuming that the head groups of the examples in Problem 15.1 are sodium carboxylate, what would be the cmc of each material? Repeat the calculation for sodium sulfonate materials.

15.3. The following data for the cmc and aggregation number, N, were obtained for a typical straight-chain anionic hydrocarbon surfactant in solutions of various salt concentrations. Assuming a spherical geometry, calculate for each system the volume of the hydrocarbon core, the effective radius of the core, and the cross-sectional area per chain at the aggregate surface.

C (NaCl, M)	cmc (mM)	N
0.00	8.1	58
0.01	5.7	64
0.03	3.1	71
0.10	1.5	93
0.30	0.71	123

15.4. Given that the cmc of sodium dodecylsulfate is 8.9 mM at 45°C, calculate the expected cmc values for the C-14, C-16, and C-18 members of the homologous series.

15.5. A nonionic surfactant with the formula $C_{12}H_{25}(CH_3)_2NO$ was found to have the following micellar characteristics as a function of temperature:

T (°C)	cmc (mM)	N
1	0.124	77
27	0.092	76
40	0.080	78
50	0.076	73

(a) Assuming that the micellar radius is equal to the fully extended length of the hydrocarbon chain, calculate the area occupied by the head group.

(b) Using the equilibrium model for micellization, calculate the values of ΔG_{mic}, ΔH_{mic}, and ΔS_{mic} at 25°C.

15.6. It is usually found that the cmc of a homologous series of surfactants decreases by a factor of approximately 2 for every CH_2 added to the hydrophobic chain. Traubes rule states that adding a CH_2 to the chain changes the surface activity by a factor of 3, in the sense that a concentration one-third as large is required to obtain the same decrease in surface tension. What is the relation—if any—between these two phenomena?

15.7. Give (qualitatively) experimental evidence that the process of micelle formation in water is primarily an entropic rather than enthalpic phenomenon.

16 Solubilization, Micellar Catalysis, and Microemulsions

In addition to being a fundamental consequence of the nature of amphiphilic molecules, micelle formation also plays a significant part in the practical application of surfactants in various areas. Because they represent what might be considered a second liquid phase in solution, micelles are often found to facilitate the production of apparently stable, isotropic "solutions" of normally insoluble liquids and sometimes solids, quite distinct from the obviously two-phase emulsions and sols previously discussed. Depending on the system (and the observer), such "solutions" are said to result from either solubilization of a material in the continuous phase or from the formation of microemulsions. In addition, the unique character of the micelle makes it a potentially useful "transition zone" between phases in which the unique environment may facilitate (i.e., catalyze) chemical reactions difficult to achieve under normal two-phase conditions. The ability of a surfactant to carry out such functions is of great potential importance and warrants some closer attention.

16.1. SOLUBILIZATION

The increased solubility of organic materials in aqueous surfactant solutions is a phenomenon that has found application in many scientific and technological areas. It is only recently that a good understanding of the structural requirements for optimum solubilization has begun to develop as a result of extensive experimental and theoretical work.

The early work in this century addressing the mechanisms of micellar solubilization was, unfortunately, usually performed with surfactants of questionable purity. More recently, closer attention has been focused on using the purest or best characterized surfactant systems available, so that more confidence can be placed in the validity and interpretation of experimental results. That is not to say, however, that the pioneering work of the first half of this century is without merit; to the contrary, modern experimental techniques have done much to confirm the work of that era. Considering the relatively limited resources of the early investigators (compared to the modern chemical laboratory), one can only regard their results and interpretations with the highest respect.

There is some disagreement within the surfactant literature as to the exact definition of "solubilization," particularly as the ratio of surfactant to additive decreases, and one approaches the nebulous frontier between swollen micellar systems and the micro- and macroemulsion regions. For present purposes, solubilization will be defined as the preparation of a thermodynamically stable, isotropic solution of a substance (the "additive") normally insoluble or only slightly soluble in a given solvent by the addition of one or more amphiphilic compounds at or above their critical micelle concentration. By the use of such a definition, a broad area can be covered that includes both dilute and concentrated surfactant solutions, aqueous and nonaqueous solvents, all classes of surfactants and additives, and the effects of complex interactions such as mixed micelle formation and hydrotropes. It does not, however, limit the phenomenon to any single mechanism of action.

One problem with that definition is that it says nothing about the relative amounts of surfactant and additive in the system. That question will arise again in the context of microemulsions. For present purposes, we will say that in solubilization, the ratio of additive to surfactant will generally be less than two. The reasons for that limitation will be discussed a bit more later.

For a specified solvent system, water or aqueous solutions for example, there are two variables that must be considered in the solubilization process: (1) the molecular nature, purity, and homogeneity of the surfactant and (2) the chemical nature of the additive. From a technological viewpoint, it is important to understand exactly what surfactant structural features serve to maximize the desired solubilizing effect, and the best way to achieve that understanding is through a fundamental knowledge of the molecular and thermodynamic processes involved. In addition, since most technological applications of solubilization involve complex multicomponent systems, such factors as temperature, electrolyte content, and the presence of polymeric species and other solutes must be examined.

16.1.1. The "Geography" of Solubilization

In order to better understand the "why and wherefore" of solubilization, it is helpful to understand the "geography" of solubilization—that is, the possible positions or loci in (or on) the micelle that can serve as host sites for the additive molecules, and the factors that determine where solubilization will occur.

It is well established that the location of a solubilized molecule in a micelle relative to the different structural components of the surfactant molecule will be determined primarily by the chemical structure of the additive (Fig. 16.1). In aqueous solutions, nonpolar additives such as hydrocarbons are intimately associated with the core of the micelle (Fig. 16.1a), while slightly polar materials, such as fatty acids, alcohols, and esters, will usually be located in what is termed the palisades layer—the transition region between the hydrophobic core and surface head groups (Fig. 16.1b). The orientation of such molecules is probably more or less radial with the hydrocarbon tail remaining closely

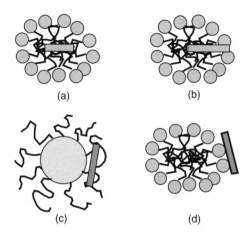

(a)   (b)

(c)   (d)

FIGURE 16.1. In micellar solubilization the solubilized additive structure will usually determine its location in the micelle. An essentially hydrocarbon additive will be found in the micellar core (*a*). A more polar material may orient itself so that it is primarily in the so-called palisades region where the nature of the environment changes from core to that of the hydrophilic region (*b*). In nonionic systems, polar additives may be solubilized in the region of the hydrated layer (*c*). In special cases of ionic surfactants, the additive may become directly associated with the electrical double layer (*d*).

associated with the micellar core. In some cases, that orientation can poten-tially have a significant effect on the energetics of the system, as will be discussed in the section on microemulsions.

Other structural factors, such as the charge on the surfactant head group, can significantly affect the locus of solubilization. Materials containing aro-matic rings, for example, may be solubilized in or near the core of anionic systems but in the palisades layer of cationic surfactants, due to electronic interactions between the ring and the cationic head group.

In addition to the solubilization of additives in the micellar core and the core–palisades boundary region, they may also be found entirely in the pali-sades region (Figure 16.1*c*) and on the micellar surface (Fig. 16.1*d*). The nature of the polar head group of nonionic surfactants, especially the polyoxyethylene materials, is such that a relatively large fraction of the micelle volume is in the palisades region. Because of the bulky nature of the POE chain and its attendant solvent molecules, it has been suggested that the hydrophilic chain is arranged in a spiral from the micellar core outward into the solution. As a result, areas of the palisades near the core will be sterically crowded with the POE chains, with relatively little room left for waters of hydration, or casual water molecules. As the distance from the core increases, the palisades layer becomes more hydrophilic, acquiring more characteristics of an aqueous solu-tion. The net effect is that, deep in the palisades layer, the chemical environ-ment may approximate that of a polyether, so that materials soluble in such solvents will be preferentially located in that region.

Even though chemical structures may dictate the preferred location for the additive, solubilized systems are dynamic, as are the parent micelles, and the location of specific molecules changes rapidly with time. It will always be important to remember that while a given region of the micelle may be preferred by an additive on chemical grounds, there is no guarantee that all phenomena related to the system (catalysis, for example) will be associated with that region.

In surfactant–nonpolar solvent systems where the "sense" of the micelle is reversed, the polar interactions of the head groups provide not only a driving force for the aggregation process but also an opportune location for the solubilization of polar additives. Water is, of course, one of the most important potential polar additives to nonaqueous systems, and it is located primarily in the core. The nature of such solubilized water is not fixed, however. The initial water added likely becomes closely associated with the polar head group of the surfactant (as waters of hydration), while subsequent additions appear to have the character of free bulk water. Other polar additives, such as carboxylic acids, which may have some solubility in the organic phase, are probably associated with the micelle in a manner analogous to that for similar materials in aqueous systems.

The effects of solubilized additives on the micellar properties of nonaqueous surfactant systems vary according to the structures of the components. Such changes, however, are often greater than those found in aqueous solutions, so that due care must be exercised in evaluating the effects of even small additions on the aggregation characteristics of surfactants in nonaqueous solvents.

16.1.2. The Solubilization Process

Just as molecular structure is important to such surfactant characteristics as the critical micelle concentration, aggregation number, and micellar shape, it also controls the ability of a surfactant to solubilize a third component. Conversely, the presence of a third component in a surfactant solution can often affect its aggregation characteristics. Whether micelles formed in the presence of a third component are the same as those formed in its absence is a subject of some controversy. It has been shown that micellar activity may be induced in surfactant solutions below the "normal" cmc in the presence of small amounts of solubilized additives. In some cases such effects have been attributed to additive-induced micellization. In others, effects have been seen at concentrations several orders of magnitude below the cmc, suggesting the presence in solution of submicellar species possessing some properties of the fully aggregated system. It has been suggested that many, if not most, surfactants in dilute solution undergo a low level of molecular aggregation at concentrations well below their cmc, during which dimers, tetramers, and other "premicellar" aggregates are formed. That may be especially true for surfactants having unusually large or bulky hydrophobic groups such as the bile acids and tetraalkylammonium halides.

Studies of micelle formation indicate that surfactant properties such as the cmc and aggregation number can be reasonably well correlated with the size and nature of the hydrophobic group. Unfortunately, comparably convenient relationships are not always so apparent in terms of surfactant structure and solubilizing power, probably because the structure of the additive can play such an important role in the overall aggregation process. Nevertheless, many of the factors discussed previously that cause an increase in micelle size might also be expected to increase the solubilizing power of the system.

16.1.3. Generalizations on Surfactant Structure and Solubilizing Power

As indicated in the previous chapter, an increase in the length of the hydrocarbon chain in a surfactant, for example, leads to a lower cmc and larger aggregation number. The usual result is that more of a nonpolar additive can be incorporated into the micellar core per mole of surfactant in the system. Branching of the hydrocarbon chain of the surfactant usually results in a decrease in the solubilizing power of the system relative to that of the analogous straight-chain material. That is presumably due to geometric and packing constraints, which limit the ability of the micellar core to accommodate the added bulk of the solubilized molecules. The addition of ethylenic unsaturation and aromatic groups also tends to decrease the maximum amount of additive that can be fitted into the core packing arrangement.

In the case of nonionic surfactants, the amount of aliphatic hydrocarbon that can be solubilized generally increases as the length of the hydrophobic tail increases and decreases as that of the POE chain increases. Those results parallel changes in the cmc and aggregation numbers of the respective materials. Divalent salts of alkyl sulfates quite often exhibit a greater solubilizing capacity than the corresponding monovalent salt for materials included in the micellar core, presumably reflecting the increased packing density attainable due to decreased head group repulsion.

If one considers the relative solubilizing powers of the different types of surfactant with a given hydrophobic tail, it is usually found that they can be ordered as nonionics > cationic > anionic. The rationale for such a result is usually related to the supposed looser packing of the surfactant molecules in the micelles of the nonionic materials, making available more space for the incorporation of additive molecules without greatly disrupting the basic structure.

The solubilizing power of amphoteric surfactants has not been as widely studied, or at least as widely reported, as that for the simpler ionic and nonionic materials. However, the available data indicate that they lie somewhere between the extremes in solubilizing capacity; the exact results are probably more sensitive to the nature of the additive than those for the other classes of surfactants.

16.1.4. Solubilization and the Nature of the Additive

The quantity of a substance that can be solubilized in surfactant micelles will depend on many factors, some of which have already been discussed. From the standpoint of the additive itself, such factors as molecular size and shape, polarity, branching, and the electronegativity of constituent atoms have all been found to be of some significance, depending on the exact system. One extensively explored factor relating the chemical structure of the additive to its solubilization is the relationship between the molar volume of the additive and the maximum amount of material that can be incorporated in a given surfactant solution. In general, one finds an inverse relationship between the molecular volume of the additive and the amount of material solubilized.

In general, increasing the chain length of an *n*-alkane or *n*-alkyl substituted benzene reduces its solubility in a given surfactant solution. While the presence of unsaturation or cyclic structures tends to increase solubility, branching appears to have little or no effect. More complicated additive structures fail to behave in such an orderly fashion. The addition of a benzene ring, for example, tends to increase solubility while a second, fused ring, such as in naphthalene derivatives, produces the opposite effect.

In summary, the relationship between the chemical structure of the additive and its ability to be incorporated into a surfactant solution is quite complex and has so far not lent itself to simple analysis and structural correlation. Perhaps, as our understanding of the geometric packing requirements of molecules in the micellar core and palisades layer improves, a more rational scheme for predicting solubilization results will emerge.

16.1.5. The Effect of Temperature on Solubilization

When one considers the effects of changes in temperature on the solubilization process, two areas of concern must be addressed. First, the ability of a given surfactant to solubilize an additive is intimately related to the characteristics of the micelle (size, shape, ionic nature, etc.). Since changes in temperature are known to affect some of those characteristics, it should not be surprising to find alterations in the solubilizing properties of surfactants as a result of modifications in micellar structure. Second, changes in temperature can affect the intermolecular interactions between solvent and solutes (e.g., hydrogen bonding), so that the overall solvent properties of the liquid for surfactant and additive may be significantly altered. In general, one can expect temperature changes that lower the cmc or increase the aggregation number of a surfactant to improve its solubilizing capacity.

16.1.6. Nonelectrolyte Solutes

Nonelectrolyte solutes that are not part of the primary solubilized system (solvent–surfactant–additive) can have a significant effect on the solubilizing power of micellar solutions as a result of their effects on cmc and aggregation number. It has become especially obvious that the addition of polar solutes such as phenols and long-chain alcohols and amines can greatly increase the solubility of nonpolar additives in ionic surfactant solutions. The mechanism for such enhancement likely involves the insertion of polar additive molecules between adjacent surfactant molecules in the micelle (Fig. 16.2). As a result of the "isolation" of the ionic groups, repulsive interactions and unfavorable contact between the aqueous phase and exposed hydrocarbon in the core can be reduced. Those two modifications of the micellar surface would allow a decrease in surface curvature of the micelle and a subsequent increase in the capacity of the core to accommodate solubilized nonpolar additives. Since such additives may also act as cosurfactants for microemulsion formation (see text below), it is possible that their function in each case is related.

Unlike polar cosolutes with relatively large hydrophobic tails, short-chain alcohols such as ethanol can significantly reduce the solubilizing power of a surfactant. In the earlier discussion of the effects of such materials on the micellization process, it was shown that the addition of significant quantities of short-chain alcohols, acetone, dioxane, and similar compounds could result in profound changes in the cmc and aggregation number of surfactants, even to the point of completely inhibiting micelle formation. It is understandable, then, that such solutes would also adversely affect the solubilization capacity of a surfactant solution.

From the above, it seems clear that the effects of an added nonelectrolyte on the solubilizing capacity of a given surfactant system may be quite complex and may not lend itself to easy analysis. It can be assumed, however, that the fundamental relationships that exist between the solutes and the micellization characteristics of the surfactant, in the absence of the solubilized additive, can be used to good advantage in predicting what may reasonably be expected in the four-component system.

16.1.7. The Effects of Added Electrolyte

For ionic micelles, the effect of addition of electrolyte is to decrease the cmc and increase the aggregation number. Such changes are predictable in micellar

Polar
additive

FIGURE 16.2. The inclusion of a polar additive such as a medium to long-chain alcohol often results in an enhancement of the solubilization capacity of a surfactant system. The mechanism of such action probably stems from an increase in the size of the micelles and/or a change in the micellar environment for the solubilized material.

systems and might be expected to produce parallel effects on solubilization. The results, however, are not always so clear cut. At surfactant concentrations near the cmc, it is usually found that the solubilizing power of a system will increase with the addition of electrolyte, as a result of the greater number and larger size of micelles available in the system. At surfactant concentrations well above the cmc, however, the simplicity of the relationship may disappear as a result of more fundamental changes in the nature of the aggregate structure.

Such inconsistencies might also be related to the nature of the additive and its potential location in the micelle. For nonpolar additives or those lying deep in the palisades layer of the micelle, it seems reasonable to expect the increased volume of the micellar core produced by electrolytes to lead to a greater capacity for solubilization, as is generally the case. For more polar materials that might be incorporated less deeply in the micelle, added electrolyte results in a closer packing of ionic head groups, which could reduce the available space for solubilized molecules. Changes in micelle shape, from spheres to rods, for example, would also result in less surface volume available in the palisades layer as a result of closer packing of the head groups.

In the case of nonionic surfactants, the effects of added electrolytes seem to parallel their effects on the micellization process. When such addition produces an increase in micellar aggregation number, an increase in solubilizing capacity for hydrocarbon additives is also found. The results for the solubilization of polar materials is, again, less clearcut.

16.1.8. Miscellaneous Factors Affecting Solubilization

Other factors that can affect the ability of a particular surfactant system to solubilize materials include pH and pressure. The effects of such factors, however, have not been as extensively reported in the literature as the factors discussed above, and they are often very specific to each surfactant system. Obviously, surfactants that show extreme sensitivity to pH such as the carboxylate salts can also be expected to exhibit significant changes in solubilization with changes in that factor. In addition, changes in pH can affect the nature of the additive itself, producing dramatic changes in its interactions with the micelle, including the locus of solubilization. Such effects can be especially important in many applications of solubilization, such as in the pharmaceutical field.

The effects of such a variable as pressure on micelle formation and solubilization is a relatively new field of investigation. It can be assumed that significant effects will be observed once sufficient pressure levels have been attained. However, such levels lie outside the normally available range of experimental conditions and are of little practical concern. Exceptions are highly pressurized products such as firefighting foams, shaving creams, and whipped toppings.

16.2. SOLUBILIZATION AND NUTRITION

An interesting and important phenomenon related to the micellar solubilization, and mixed micelles in particular, is that of the digestion and absorption of fatty nutrients by the body. Since fat is a major source of energy (~9 vs. 6 cal g^{-1} for proteins and carbohydrates), its transport through the digestive track and ultimate transfer through the intestinal walls is obviously of great importance.

While the body produces several natural surfactants, the most important in terms of digestion are the bile salts derived from cholesterol (Fig. 16.3a) and lecithin (Fig. 16.3b). The bile salts are synthesized in the liver and secreted

Lipids and natural surfactants from stomach

(a) Mechanical emulsification

(b) Introduction of lipases and bile salts

(c) Hydrolysis of fats to fatty acids ( ) and 2-monoglycerides ( )

(d) Micellization and solubilization

(e) Transport of solubilized fatty acids and other lipid components to the intestinal walls

(f) Adsorption of fatty acids and lipids

(g) Continued movement of bile acids, etc., down intestine and ultimate readsorption or excretion

FIGURE 16.3. The absorption of fatty nutrients in the intestines is an obviously important example of the solubilizing action of surfactant systems.

into the upper small intestine in response to stimuli indicating the presence of nutrients. When no nutrients are present, the acids are normally concentrated and stored in the gallbladder.

As fatty material, usually in the form of triglycerides, enters the intestine it is emulsified by muscular action of the duodenal wall which also causes the secretion of pancreatic enzymes that hydrolyze the triglycerides to produce fatty acids and 2-monoglycerides (Fig. 16.3c). At the same time the gallbladder releases bile acids into the intestine that, in combination with lecithin already present and the 2-monoglycerides (also a weak surfactant) produced by hydrolysis form mixed micelles that solubilize the essentially insoluble fatty acids.

Once incorporated into the micelles, the fatty acids and 2-monoglycerides are transported through the small intestine during which time random motion and diffusion bring the micelles into contact with the intestinal wall. Because the solubilized lipid/micelle system is a dynamic structure, while in contact with the wall, the fatty acids and 2-monoglycerides can be released and transferred through the membranes of the epithelial cells where the are reesterified to produce triglycerides again. The triglycerides then continue their journey to produce energy (good) or be processed and deposited for future use (usually not so good).

The bile salts themselves are too polar to be transferred along with the lipids and continue their journey through the intestines until they arrive at the last section of the small intestine. There they are actively transferred to the bloodstream and carried to the liver where they may be separated from other blood components and either returned to the intestine (if more fatty material is present to be processed) or sent to the gallbladder to be stored as mixed micelles with lecithin and free cholesterol.

When functioning normally, the human body uses the equivalent of approximately 30 g of bile acids per day. In fact, it has been determined that the process of fat digestion is repeated some 10 times per day, recycling a total of 3 g of bile in the process.

16.3. MICELLAR CATALYSIS

It is well recognized in all branches of chemistry that the rate of a chemical reaction can be very sensitive to the nature of the reaction environment. Reactions involving polar or ionic transition states can be especially sensitive to the polarity of the reaction medium. It should not be too surprising, then, that many chemical reactions, especially those in which one reactant may be soluble in water and the other in oil, can exhibit a significant enhancement in rate when carried out in the presence of surfactant micelles. The presence of the micellar species can provide a beneficial effect through two possible mechanisms:

1. The palisades region of the micelle represents a transition zone between a polar aqueous environment, which may be either the bulk phase or

the micellar core, and a nonpolar hydrophobic region. Such a gradient in polarity can serve as a convenient area of intermediate polarity suitable for increased reactant interaction or for optimizing the energetics of transition state formation.

2. The potential for the micelle to solubilize a reactant that would not normally have significant solubility in the reaction medium means that it can serve as a ready reservoir of reactant, in effect increasing the available concentrations of reactants. The rate enhancements that have been reported range as high as 10^5, which makes such systems very attractive for potential practical applications.

16.3.1. Catalysis in Aqueous Solvent

In aqueous media a micellar system can serve as a catalyst for organic reactions, but it is also possible for it to retard such reactions. Possible mechanisms for catalytic action are suggested in Figure 16.4; inhibitory actions may arise from unfavorable electrostatic interactions between reactants and changes in the distribution of reactants between the bulk and micellar phases. In the case of electrostatic inhibition, the presence of a charge on the micelle surface can have two effects on a reaction involving a charged species. In the base hydrolysis of water-insoluble esters, if the micelle charge is negative, the transport of hydroxide ion into or through the palisades layer will be retarded by charge repulsion. If the micelle is positively charged, the inclusion of the species of opposite charge will be facilitated. For nonionic and zwitterionic surfactants, there may be little or no effect observed. Although such a model of electrostatic effects is probably overly simplified, it has generally been supported by experi-

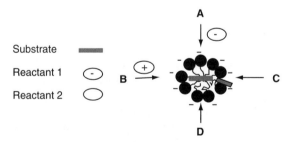

FIGURE 16.4. The presence of micelles in a two-phase reaction medium may produce several effects. A micelle with the same electrical charge as a dissolved reactant may slow its reaction with a solubilized component (path A), while one of opposite charge will usually enhance the reaction rate (path B). Alternatively, especially for nonionic surfactants, the micelle may provide an intermediate "solvent" environment that enhances the reaction rate (path C). Finally, the dynamics of micellar systems may provide a more readily accessible reservoir of insoluble reactant in the system thereby increasing the reaction rate (path D).

ment. The basic hydrolysis of esters, for example, is catalyzed by cationic and inhibited by anionic surfactants, while the opposite is true for the acid hydrolysis.

The ability of a micellar system to solubilize a reactant may affect its action as a catalyst or inhibitor in a reaction. When a surfactant system serves as a reservoir for increasing the availability of one reactant, any change that increases the solubilizing capacity of the micelle should also increase its effectiveness as a catalysis. If, on the other hand, the reaction must occur in the bulk phase, increased solubilizing power may remove reactant from the reaction medium and therefore decrease catalytic or increase inhibitor efficiency.

In aqueous solution, the effectiveness of micellar systems as catalysts is quite often found to increase with the length of the alkyl chain. For example, the rate of acid hydrolysis of methyl o-benzoate in the presence of sodium alkyl sulfates increases in the order $C_8 < C_{10} < C_{12} < C_{14} < C_{16}$. Such a result may be attributed to either electrostatic or solubilizing effects, or both. It might be expected that any effects due to electrostatic interactions would also increase. Alternatively (or additionally), the increasing aggregation number found in the series may result in a significant increase in the solubilizing power of the system. The importance of each mechanism will depend upon the specifics of the reaction.

In addition to the effects noted for increases in the charge density on the micelle, the charge on the individual surfactant molecules can also be important. One generally finds, for example, that cationic surfactants containing two charge groups are significantly better at increasing the rate of nucleophilic aromatic substitutions than analogous singly charged materials. Similar results have been reported for singly versus doubly charged anionic surfactants.

As might be expected, the structure of the reactive substrate can have as much influence on micelle-assisted rate enhancement as that of the surfactant. Since the catalytic effectiveness of the micelle can be related to the location and orientation of the substrate in the micellar structure, the more hydrophobic the substrate (and the surfactant), the more significant may be the catalytic effect.

When nonsurfactant solutes (electrolytes, etc.) are added to the micellar reaction mixture, the results can be quite unpredictable. It is often found that the presence of excess surfactant counterions (common ions), when added to a system in which an ionic reactant is involved, retards the catalytic activity of the micelle, with larger ions being more effective in that respect. The effect can probably be attributed to an increase in ion pairing at the micelle surface and a reduction of its "attractiveness" to charged reactants. In contrast, the addition of neutral electrolyte has been found to enhance micellar catalysis in some instances. It has been proposed that the retardation effect of excess common counterions is due to a competition between the excess ions and the reactive substrate most closely associated with the micelle for the available positions or "binding sites" on or in the micelle. The enhancing effect, however, has been attributed to the more general effects of added electrolyte on the properties of micelles, such as lowering of the cmc, and increasing the aggregation number, all of which often tend to increase catalytic activity.

16.3.2. Catalysis in Nonaqueous Solvents

Interactions between polar head groups in nonaqueous solvents provide the primary driving force for the formation of micellar aggregates in such media. The nature of such reversed micellar cores is such that they provide a unique location for the solubilization of polar substrates. While keeping in mind the potentially dramatic effects of additives on the properties of micellar solutions, it is obvious that such nonaqueous systems hold great potential from a catalytic standpoint. They are especially of interest as models of enzymatic reactions.

The fundamental principles controlling activity in nonaqueous systems are the same as those for aqueous solutions, except that the specificity of the micellar core for the solubilization of polar substrates is much greater than for the aqueous situation. The popularity of reversed micelles as models for enzyme catalysis stems from the fact that the micellar core is capable of binding substrates in concentrations and orientations that can be very specific to certain functionalities, much as an enzyme would do. As a result, reaction rate enhancements can be obtained comparable (with luck) to those of the natural systems, and far in excess of what can be explained on the basis of partitioning or availability of substrate.

Work in the area of micellar catalysis in both aqueous and nonaqueous solvent systems is certain to continue to grow in importance as a tool for better understanding the chemistry and mechanics of enzymatic catalysis, as a probe for studying the mechanistic aspects of many reactions, and as a route to improved yields in reactions of academic interest. Of more practical significance, however, may be the expanding use of micellar catalysis in industrial applications as a method for obtaining maximum production with minimum input of time, energy, and materials.

16.4. MICROEMULSIONS

Microemulsions are composed of two mutually immiscible liquid phases, one spontaneously dispersed in the other with the assistance of one or more surfactants and cosurfactants. While microemulsions of two nonaqueous liquids are theoretically possible (e.g., fluorocarbon–hydrocarbon systems), almost all of the reported work is concerned with at least one aqueous phase. The systems may be water continuous (o/w) or oil continuous (w/o), the result being determined by the variables such as the surfactant systems employed, temperature, electrolyte levels, the chemical nature of the oil phase, and the relative ratios of the components.

Most microemulsions, especially those employing an ionic surfactant, require the addition of a cosurfactant to attain the required interfacial properties leading to the spontaneous dispersion of one phase in the other. The structural character of the cosurfactant (chain length, head group, etc.) and its relationship to the primary surfactant have been a major focus of much of the research in the area. It is commonly found that microemulsions prepared using ionic surfactants require a relatively high ratio of surfactant to dispersed phase, while some nonionic surfactants have been found to produce microemulsions with much lower levels of surfactant and without the addition of a cosurfactant.

For a given oil phase, it is usually possible to optimize the oil:water ratio by varying the structure of the surfactant and cosurfactant, if needed. An aromatic oil phase, for example, may produce microemulsions in a given composition range using a short-chain alcohol as cosurfactant, while a hydrocarbon oil may require a longer-chain material to achieve comparable results. Although the process requires a great deal of tedious laboratory work, three-dimensional phase diagrams (oil–water–surfactant–cosurfactant) can be quite useful in determining the optimum composition for the production of a microemulsion. To simplify matters somewhat, it is possible to fix one pair of variables, say, the surfactant:cosurfactant ratio, and then prepare a simpler triangular phase diagram.

Microemulsions, like macroemulsions, can exist with either the oil or the water being the continuous phase. The characteristics of the system will, of course, be different in each case. The proper use of the phase diagram approach allows one to determine not only the component ratios necessary to produce a microemulsion but also the component forming the continuous phase.

16.4.1. Micelle, Microemulsion, or Macroemulsion?

The status of the systems commonly referred to as "microemulsions" among surface and colloid chemists is still somewhat uncertain. Various experimental approaches have been used in an attempt to ascertain the details of their structural and thermodynamic characteristics. As a result, new theories of the formation and stability of these interesting but quite complex systems are appearing. Although a great deal has been learned about microemulsions, there is much more to be learned about the requirements for their preparation and the relationships among the chemical structure of the oil phase, the composition of the aqueous phase, and the structures of the surfactant and the cosurfactant, where needed.

The distinction between microemulsions and conventional emulsions or macroemulsions is fairly clear. Although macroemulsions may be kinetically stable for long periods of time, in the end they will suffer the same fate: phase separation to attain a minimum in interfacial free energy. The actions of surfactants, polymers, and other stabilizing aids may shift the rate of droplet coalescence to extremely long times through decreased kinetic rate constants, but the thermodynamic driving force to minimize the interfacial area between immiscible phases remains unchanged. On the other hand, the microemulsions appear to be thermodynamically stable compositions with essentially infinite lifetimes, assuming no change in such factors as composition, temperature, and pressure.

In addition to the thermodynamic distinction usually drawn between macro-
and microemulsions, the two classes of colloids differ in several other more
tangible characteristics, including the size of droplets formed and the mechanical requirements for their preparation. As far as droplet size is concerned,
the conventional macroemulsions are generally found to have minimum diameters of 100–200 nm, meaning that such systems are usually quite turbid or
opaque. Microemulsions, however, normally have droplet diameters of 100
nm or less; many are only slightly larger than simple micellar systems. Because
those particles are much smaller than the wavelength of visible light, they are
normally transparent or slightly bluish.

The energy requirements for the formation of macroemulsions can be quite
substantial. The formation of small droplets requires that the system overcome
both the adverse positive interfacial free energy between the two immiscible
phases working toward drop coalescence and bulk properties of the dispersed
phase such as viscosity. Microemulsions, on the other hand, form spontaneously or with very gentle agitation when the proper composition is reached.

When one compares microemulsions and micelles, the demarcation line
can become quite blurred and, in some cases, does not exist. There is some
controversy as to the true definition of clear, isotropic solutions of oil, water,
and surfactant (and cosurfactant if needed) as microemulsions rather than
swollen micelles. Although the differences between the two systems may
appear to many to be more semantic than real, several arguments have been
presented that strongly support a differentiation of the two systems.

If one constructs a "spectrum" of the possible situations for the dispersion
of one liquid phase in another, the possible sizes of the dispersed phase units
range from the molecularly dispersed true solution where "droplet" sizes are
on the order of a few nanometers, to macroemulsions with droplet diameters
of hundreds or thousands of nanometers (Fig. 16.5). Lying between the ex-

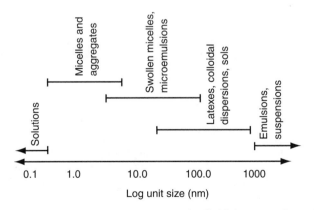

FIGURE 16.5. If one constructs a "spectrum" of colloidal systems, it can be seen that
the microemulsions and swollen micelles lie in the range between the "simple" aggregate structures such as micelles and the larger emulsion and dispersion systems.

tremes are simple micelles (a few tens of nanometers), macromolecular solutions (tens to hundreds of nanometers), and colloids of several hundred to several thousand nanometers. The systems typically referred to as "microemulsions" will normally have unit sizes between 5 and 100 nm, generally well beyond the range of spherical micelles in dilute solution, but overlapping significantly with larger assemblies, such as rod-shaped micelles and liquid crystals. The physical differences encountered among most of the different groups are sufficient to obviate any controversy as to their classification.

The question of differentiation between micelles and microemulsions is less simply answered. While it is undoubtedly true that, in the smaller size ranges especially, many systems classed as microemulsions are indistinguishable from swollen micelles, it is equally true that the larger microemulsion systems far exceed the solubilizing capacity of micelles as previously defined. Micelles will form under many circumstances, although the specifics of cmc, aggregation number, and other variables may change with the environmental conditions. The formation of microemulsions, on the other hand, has been shown to have very specific compositional requirements. This is due primarily to those specific demands on the composition of the system and the chemical structures of the various components that the nomenclature for this separate class of dispersed species has developed.

An additional argument for a distinction between micelles and microemulsions is that in all the literature on the solubilization of hydrocarbons, dyes, and other substances in micellar solutions, the ratio of solubilized molecules to surfactant molecules very rarely exceeds, or even approaches, 2. Many microemulsion systems, on the other hand, have been described in which the dispersed phase: surfactant (and cosurfactant) ratio exceeds 100! Because of the relatively low ratios of additive to surfactant obtainable in micellar systems, it is clear that there can exist no oil phase that can be considered separate from the body of the micelle. In many microemulsions, however, the size of the droplet and the high additive: surfactant ratio requires that there be a core of dispersed material that will be essentially equivalent to a separate phase of that material. The seemingly obvious conclusion is that microemulsion systems (in the latter case, at least) possess an interfacial region composed primarily of surfactant (and cosurfactant), analogous to that encountered in macroemulsions.

16.4.2. Negative Interfacial Tensions

Both theory and careful experimental work seem to indicate that the driving force for the spontaneous formation of microemulsion systems is the existence of a transient negative interfacial tension between the oil and aqueous phases, resulting in a rapid transfer of one of the two phases through the interface, producing the optimum droplet size for the given composition. It must be emphasized that the negative interfacial tension is a transient phenomenon, and at equilibrium must be zero or slightly positive.

Like many other topics covered here, microemulsions hold a great deal of promise in many practical applications. To date, most research in the area has been closely associated with the formation and destruction of microemulsions in the context of secondary and tertiary petroleum recovery, although more interest is being shown in cosmetic applications. While the concept of microemulsions is very attractive for potential use in many other areas, especially pharmaceuticals, their sensitivity to composition makes their application much more problematical. Since in many cases (i.e., drug delivery) one or more component (which may be somewhat surface-active) may be determined by the intended function of the system, the options for the formulator may be drastically reduced or at least greatly complicated.

PROBLEMS

16.1. Given the following chemical structures, predict the probable location of each of the following compounds if solubilized in aqueous micellar solutions of (*a*) sodium dodecyl sulfate, (*b*) *n*-hexadecyl-trimethyl ammonium bromide, (*c*) polyoxyethylene(7)dodecylphenol:

n-hexadecane	*n*-Octanoic acid *n*-butyl lactate	Chlorobenzene
Methyl oleate	2,5-Di-*tert*-butylhydroquinone	Di-*n*-butyl ether
Naphthalene	*N,N*-Dimethylhexadecyl amine	Tetrahydrofuran

16.2. Which of the following surfactants would be expected to be most efficient at solubilizing hexadecane: sodium *n*-nonylbenzene sulfonate, sodium hexadecylsulfate, benzyl trimethylammonium acetate, or SDS? For solubilizing cholesterol?

16.3. A system of aqueous micelles of a nonionic POE surfactant is found to solubilize an average of 2 molecules of a material per micelle at 25°C. If the temperature of the system is raised to 50°C, would you expect the solubilizing capacity per micelle to increase or decrease? From the information provided, for the same total surfactant concentration, what can you say about the total solubilizing capacity of the system at the two temperatures?

16.4. If an aqueous micellar solution of sodium tetradecylsulfate is employed to solubilize a polar dye, would you expect the addition of dodecyl alcohol to increase, decrease, or not affect the capacity of the system?

16.5. For the solubilization of high-molecular-weight hydrocarbons in anionic micelles, the addition of electrolyte would be expected to increase, decrease, or not affect the capacity of the system?

16.6. The base-catalyzed hydrolysis of esters is accelerated by the presence of 10^{-4} M POE(9)nonylphenylether nonionic surfactant, unaffected by

10^{-4} M SDS, and retarded by 10^{-3} M SDS. Explain the observed effects of surfactant concentration.

16.7. An amphoteric amine sulfate surfactant is tested as a catalytic system for the base-catalyzed hydrolysis of a triglyceride. What pH range (<5, 5–8, <8) would you expect to be most appropriate for maximum rate enhancement? Explain why.

16.8. Give three possible technological advantages one might expect a microemulsion to have over a conventional emulsion in an intravenous drug application system, assuming that the proper balance of constituents can be found to fit the pharmacological requirements of the system.

16.9. Microemulsions, while often being desirable, can become extremely bothersome when phase separation is the ultimate goal in an industrial process. That can be especially true in some advanced petroleum recovery systems that involve the use of aqueous surfactant systems. Suggest a chemical and a physical process that might be used to "break" a stable crude oil microemulsion resulting from the use of an anionic surfatant/ medium-chain-length alcohol extraction system.

17 Wetting and Spreading

The wetting of a surface by a liquid and the ultimate extent of spreading of that liquid are very important aspects of practical surface chemistry. Many of the phenomenological aspects of the wetting processes have been recognized and quantified since early in the history of observation of such processes. However, the microscopic details of what is occurring at the various interfaces and lines of contact among phases has been more a subject of conjecture and theory than of known facts until the latter part of this century when quantum electrodynamics and elegant analytical procedures began to provide a great deal of new insight into events at the molecular level. Even with all the new information of the last 20 years, however, there still remains a great deal to learn about the mechanisms of movement of a liquid across a surface.

There are (as usual) two aspects of the system that must be considered: equilibrium thermodynamics and kinetics. The former topic has been fairly well in hand for many years. The latter has received less general attention but has begun to enter the spotlight in recent years. The following discussion is intended to introduce the fundamental concepts underlying modern theories of wetting and spreading processes. Many of the topics discussed have already been introduced but will be reviewed where it seems useful. Other topics, especially kinetic aspects, which remain somewhat conjectural, will be presented as such. Before beginning the discussion of wetting, however, it will be useful to review somewhat a topic introduced in Chapter 6 in the context of capillary phenomena: the contact angle.

17.1. THE CONTACT ANGLE

One of the primary characteristics of any immiscible, two- or three-phase system containing two condensed phases, at least one of which is a liquid, is the contact angle of the liquid on the other condensed phases. The contact angle of one liquid on another, although of theoretical interest, is normally of little practical importance. An exception to that is in certain multilayer coating processes, in which several liquid layers are coated simultaneously on a solid surface (Fig. 17.1). In such processes, it is important that the wetting properties of each layer on the one below be such that smooth, uniform coverage is obtained. Such processes are especially important in the photographic and graphics arts industries, where small coating flaws will make a

FIGURE 17.1. In simultaneous, multilayer coating operations, the relative surface tensions (plus viscosity, contact angle, etc.) of the successive liquid layers must be carefully balanced in order to avoid coating defects due to poor wetting.

coated material useless. Of more practical and widespread importance is the contact angle of a liquid directly on a solid. For liquids on solids, the contact angle can be considered a material property of the system, assuming that certain precautions are taken in the collection and interpretation of data, and that liquid absorption is taken into solution.

When a drop of liquid is placed on a solid surface, the liquid will either spread across the surface to form a thin, approximately uniform film (Fig. 17.2a) or spread to a limited extent but remain as a discrete drop on the surface. The final condition of the applied liquid on the surface is taken as an indication of the wettability of the surface by the liquid or the wetting ability of the liquid on the surface, depending on your point of view. The quantitative measure of the wetting process is taken to be the contact angle, θ, which the drop makes with the solid as measured through the liquid in question (Fig. 17.2b).

In the case of a liquid that forms a uniform film (i.e., where $\theta = 0°$), the solid is said to be completely wetted by the liquid, or that the liquid wets the solid. Where a nonzero angle is formed, there exists some controversy as to how to describe the system. If a finite contact angle is formed ($\theta > 0°$), some investigators describe the system as being partially wetted. Others prefer to make a distinction based on the size of the contact angle. For example, a given worker may define as "wetting" any liquid that produces a contact angle of 30° or less on a solid. Between 30° and 89° the system would be "partially wetting," and 90° and above nonwetting. Alternatively, any system with $0° < \theta < 180°$ would be partially wetting, and only for $\theta = 180°$ would the nonwetting

(a)   (b)   (c)

FIGURE 17.2. A liquid placed on a solid surface can take one of three shapes. It may spread into a uniform duplex film (a, $\theta = 0$), it may form a convex lens with a section less than the radius of curvature (b, $0 > \theta > 90°$), or it may form a section greater than the radius of curvature (c, $\theta > 90°$).

term be applied. The terminology one prefers will often depend on individual circumstances and is of relatively minor importance. The important thing is to know what descriptive system is being employed in a given set of circumstances.

While the contact angle of a liquid on a solid may be considered a characteristic of the system, that will be true only if the angle is measured under specified conditions of equilibrium, time, temperature, component purity, and other parameters. Contact angles are very easy measurements to make (with a little practice) and can be very informative; but if the proper precautions are not taken, they can be very misleading.

The contact angle may be geometrically defined as the angle formed by the intersection of the two planes tangent to the liquid and solid surfaces at the perimeter of contact between the two phases and the third surrounding phase. Typically, the third phase will be air or vapor, although systems in which it is a second liquid essentially immiscible with the first are of great practical importance. The perimeter of contact among the three phases is commonly referred to as the "three-phase contact line" or the "wetting line." On a macroscopic scale, such terminology is meaningful and useful; however, when one begins to zero in on the situation in the region of three-phase contact, it must be remembered that one is talking about a transition zone of finite (even though very small) dimensions in which three phases merge and the characteristics of one phase change to those of another. It is conceptually and mathematically convenient to think in terms of "lines," but for a more complete understanding of the situation the real facts must be kept in mind.

The great utility of contact angle measurements stems from their interpretation based on equilibrium thermodynamic considerations. As a result, most studies are conducted on essentially static systems in which the liquid drop has (presumably) been allowed to come to its final equilibrium value under controlled conditions. In many practical situations, however, its is just as important, or perhaps more important, to know how fast wetting and spreading occurs as to know what the final equilibrium situation will be. That will especially be true in situations where the process in question requires that wetting bring about the displacement of one phase by the wetting liquid. Typical examples would be detergency, in which a liquid or solid soil is displaced by the wash liquid; petroleum recovery, in which the liquid petroleum is displaced by an aqueous fluid; and textile processing, in which air must be displaced by a treatment solution (e.g., dyeing or waterproofing treatments) in order to obtain a uniform treatment. Because of the economic importance of these and other processes, some emphasis will be placed on the dynamic aspects of the wetting processes.

As already mentioned, the interpretation of data on static contact angles must be done with the understanding that the system in question has been sufficiently well controlled so that the angle measured is the "true" angle and not a reflection of some contaminant on the solid surface or in the liquid phase of interest. Contact angles, for example, can be extremely useful as a spot test of the cleanliness of sensitive surfaces such as glass or silicon wafers for microelectronics fabrications. Both surfaces are "high energy" (see text below) and are completely wetted by pure water. If the surface is contaminated by something such as an oil that interferes with the processing of the material (e.g., the coating of a photoresist polymer), a drop of water will have a nonzero contact angle, and the contamination will be immediately apparent.

For systems that have "true" nonzero contact angles, the situation may be further complicated by the existence of contact angle hysteresis. Thus, the contact angle one observes may vary depending on whether the liquid is advancing across fresh surface (the advancing contact angle, θ_A) or receding from an already wetted surface (the receding contact angle, θ_R) (Fig. 17.3). As an operational convenience, many, if not most, static contact angles measured and reported are in fact advancing angles. For a given system, it will be found that $\theta_A \geq \theta \geq \theta_R$. In practice, very few systems exhibit a complete lack of hysteresis, so that the problem can be operational as well as philosophical. Some of the primary causes of hysteresis will be discussed further below. For now, it is sufficient to keep in mind that when discussing contact angle data, one must always be aware of how the angle has been measured in order to interpret its significance properly.

In dynamic contact angle studies, additional complications arise because the movement of the wetting line is not always a steady, continuous process. It is often observed that the movement is "jerky," with the drop or liquid front holding a position for a time and then jumping to a new configuration. This phenomenon is often referred to as a "stick–slip process" and is not fully understood as yet. It has also been observed that in dynamic systems, the values of θ_A and θ_R will vary as a function of the velocity of wetting line movement, with θ_A increasing with velocity and θ_R decreasing.

Obviously, contact angle measurements and their interpretation are not without their hidden pitfalls and blind alleys. However, because of the ease of making such measurements, the low cost of the necessary apparatus, and the potential utility of the concept, they should be seriously considered as a rapid diagnostic tool for any process in which wetting phenomena play a role.

Contact angle hysterisis = $\theta_A - \theta_R$

FIGURE 17.3. In dynamic liquid systems, a liquid front advancing across a new surface may exhibit a large contact angle (the advancing contact angle, θ_A), while the same liquid receding from an already wetted surface will have a much smaller contact angle (the receding contact angle, θ_R). The difference between θ_A and θ_R is termed the contact angle hysteresis.

17.1.1. Contact Angle Measurement Techniques

There are a variety of simple and inexpensive techniques for measuring contact angles, most of which are described in detail in various texts and publications and will be mentioned only briefly here. The most common direct methods (Fig. 17.4) include the sessile drop (a), the captive bubble (b), the sessile bubble (c), and the tilting plate (d). Indirect methods include tensiometry and geometric analysis of the shape of a meniscus. For solids for which the above methods are not applicable, such as powders and porous materials, methods based on capillary pressures, sedimentation rates, wetting times, imbibition rates, and other properties, have been developed.

17.1.2. Contact Angle Hysteresis

When used with Young's equation and other such relationships, the contact angle provides a relatively simple yet sensitive insight into the general chemical nature of a surface through such thermodynamic quantities as the work of adhesion. Unfortunately, as already mentioned, contact angles often exhibit hysteresis and cannot be defined unambiguously by experiment. It is always important to know as much as possible about the cleanliness, topography, homogeneity, and other characteristics of a solid surface, as well as the purity and composition of the liquid employed, when attempting to interpret contact angle data.

Although the existence of contact angle hysteresis has been recognized for at least 100 years, the root of the "evil" has not always been understood. In addition to the physicochemical adsorption process already mentioned, which leads to differences in advancing and receding contact angles, it is recognized

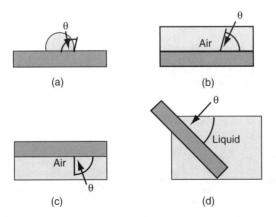

FIGURE 17.4. The contact angle is a relatively easy piece of information to obtain (with a little practice). The more common systems of measurement include the sessile drop (a), the captive bubble (b), the sessile bubble (c), and the tilting plate (d).

that several physical and kinetic factors also contribute to the overall problem. Two of the most important, from a practical point of view, are discussed below.

17.1.3. The Effects of Surface Roughness on Contact Angles and Wetting

The theoretical discussion of contact angle and wetting to this point has assumed implicitly that the solid surface in question is a smooth, ideal plane. In fact, of course, very few solid surfaces even begin to approach such a state. The finest polished glass surface, for example, will usually have asperities of 5 nm or more. Commonly encountered polished surfaces, will be much rougher by factors of 10–1000. The earliest, and still most useful, quantitative attempt to correlate the observed contact angle of a liquid on a solid with the surface roughness is the Wenzel relationship which proposes a thermodynamic relationship such that

$$\sigma_{12} \cos \theta = R_w (\sigma_{s2} - \sigma_{s1}) \qquad (17.1)$$

where R_w is defined as the surface roughness factor, the ratio of the true and apparent surface areas of the solid (Fig. 17.5b). Defining the apparent contact angle as θ' and substituting for from Equation (17.1) yields

$$\cos \theta = R_w \cos \theta' \qquad (17.2)$$

or

$$R_w = \frac{\cos \theta}{\cos \theta'} \qquad (17.3)$$

Equation (17.3) may be taken as a fundamental definition of the effect of surface roughness on wetting and spreading phenomena.

Since surfaces having $R_w = 1$ are essentially nonexistent, and highly polished surfaces usually have $R_w \geq 1.5$, the relationships shown above obviously have a great deal of relevance for exacting experimental work. Although the actual geometry of the surface roughness has no significance from a thermodynamic

FIGURE 17.5. The apparent contact angle of a liquid on a surface may differ from that expected, the "true" contact angle (a), due to irregularities—either physical or chemical—including surface roughness (b) or chemical heterogeneity (c).

standpoint, in practice, the type of topography present may carry with it certain practical consequences. For example, if the surface, rather than being randomly rough, has pores, crevices, capillaries, or other structures that have their own characteristic wetting and penetration properties, the apparent contact angle will be affected by the thermodynamics and kinetics associated with such structures.

There have been developed a number of more detailed and sophisticated treatments of the effect of surface roughness on wetting phenomena; however, the utility of such approaches is limited by the necessity of having good data concerning the roughness of the surface in question, data that are not always easy or convenient to obtain. If there exists some consistency in surface roughness (e.g., as for some manufacturing process), the ideas described above can still be employed with little concern for the details of surface topography, since R_w will presumably be a constant.

As a final note on the effects of surface roughness, examination of Equation (17.3) leads to a useful rule of thumb for some important applications of wetting and spreading phenomena; that is, if the "true" contact angle of a liquid (an adhesive, say) is less than 90 on the smooth surface, the angle will be even smaller on a rough surface. For a true contact angle $> 90°$, roughness will increase the apparent angle. Mathematically the situation can be described as

$$\text{For } \theta < 90°, \quad \theta' < \theta; \quad \text{for} \quad \theta > 90°, \quad \theta' > \theta$$

Practically, the preceding relationships indicate that if a liquid partially wets a surface, better wetting may be obtained if the surface is roughened in some way. Conversely, if wetting is not desired and a contact angle $> 90°$ can be attained, the situation can be further improved by roughening.

17.1.4. Heterogeneous Surfaces

Roughness represents just one aspect of the effects of the nature of the solid surface on contact angles and wetting phenomena. A second potentially important factor is that of the chemical heterogeneity of the surface (Fig. 17.5c). Working from the same basic approach as that used to arrive at Equation 17.2, it is possible to develope the following relationship relating apparent contact angle to the chemical composition of a surface

$$\cos \theta' = f_1 \cos \theta_1 + f_2 \cos \theta_2 \tag{17.4}$$

where f_1 and f_2 are the fractions of the surface having inherent contact angles θ_1 and θ_2. Since $f_2 = 1 - f_1$, Equation (17.4) can be written in terms of one component. Theoretically, if the inherent contact angles of a test liquid on the homogeneous surfaces are known, then the composition of a composite surface can be determined from a simple contact angle measurement. Obviously, such an approach must be accepted as being rather qualitative, considering the pitfalls inherent in contact angle data. However, experiments employing specially prepared composite surfaces have shown that contact angle data can give results that agree reasonably well ($\pm$ 15%) with more sophisticated surface composition data obtained using, for example, X-ray photoelectron spectroscopy (XPS).

17.1.5. Kinetic Aspects of Hysteresis

According to its definition, contact angle hysteresis should be concerned only with thermodynamic equilibrium situations and not with nonequilibrium kinetic events. However, because of experimental limitations, such as the time required for measurements, certain kinetic effects are almost unavoidable. Theoretically, such effects should fall under the heading of kinetic contact angles but are often reflected in "equilibrium" measurements.

Because the measurement of a contact angle must involve some movement of the wetting line, it is possible, or even probable, that the act of spreading of the liquid will displace certain surface equilibria that will not be reestablished over the time frame of the experiment. For example, the displacement of a second fluid may result in the establishment of a nonequilibrium situation in terms of the adsorption of the various components at the solid–liquid, solid–fluid 2, and liquid–fluid 2 interfaces. Time will be required for adsorption equilibrium to be attained, and it may not be attained during the time of the contact angle measurement if the transport and adsorption–desorption phenomena involved are slow. The kinetic effect may be especially significant for solutions containing surfactants, polymers, or other dissolved adsorbates.

A second potential kinetic effect may result from bulk interactions between the surface and the spreading liquid. For example, if the liquid can penetrate the surface (e.g., if the liquid can be absorbed as opposed to adsorbed), the rate of penetration may be so slow that the measured contact angle will not reflect the true equilibrium situation. Likewise, if the liquid swells the surface, the wetting line may lie on a ridge of swollen surface rather than on a flat surface, resulting in an error in θ' (Fig. 17.6).

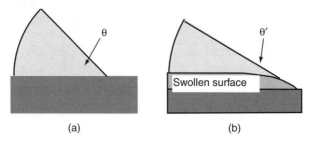

(a)   (b)

FIGURE 17.6. If the liquid used for measuring the contact angle is absorbed into the solid surface (i.e., it swells) the resulting contact angle will be smaller than the "true" value.

To summarize, while contact angle measurements represent a potentially powerful and practical tool for characterizing the nature and wettability of solid surfaces, variability leading to errors in interpretation can arise from various sources. That means that proper attention must be focused on experimental conditions, equilibria, solid and liquid purity, and so on, to ensure the best possible data. Even when all precautions have apparently been taken, interpretation must be done with the above-mentioned caveats in mind. Nevertheless, contact angle data should never be excluded from studies or processes in which wetting and spreading are involved.

17.2. THE THERMODYNAMICS OF WETTING

The basic framework for the application of contact angles and wetting phenomena lies in the field of thermodynamics. However, in practical applications it is often difficult to make a direct correlation between observed phenomena and basic thermodynamic principles. Nevertheless, the fundamental validity of the analysis of contact angle data and wetting phenomena helps to instill confidence in its application to nonideal situations.

17.2.1. Young's Equation (Again!)

If one considers the three-phase system depicted in Figure 17.7, in which the liquid drop is designated fluid 1, the surrounding medium fluid 2, and the

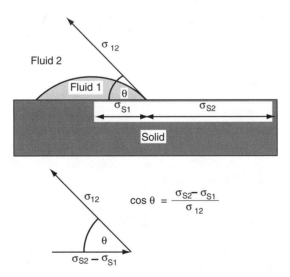

FIGURE 17.7. Young's equation for determining the contact angle was originally based on an analysis of the force balance among the three surface tensions involved.

solid surface S, then at equilibrium the contact angle θ will be given by Young's equation as

$$\sigma_{12} \cos \theta^\circ = \sigma_{S2} - \sigma_{S1} \tag{17.5}$$

where σ_{12}, σ_{S1}, and σ_{S2} are the interfacial tensions at the respective interfaces. Although this equation was originally proposed on the basis of a mechanical analysis of the resultant forces at the three-phase contact line, it has since been derived rigorously on the basis of fundamental thermodynamic principles.

While Young's equation provides a thermodynamic definition of the contact angle, its experimental verification is prevented by the fact that the values of s_{S1} and s_{S2} cannot be directly determined experimentally. In this sense, the contact angle of a liquid on a solid differs from that of a liquid on a second liquid since in the latter case all three interfacial tensions can be determined independently and the relationship can therefore be verified directly.

When the liquid of interest (1) is found to spread completely over the solid surface, $\theta = 0$, and Equation (17.5) reduces to

$$\sigma_{12} = \sigma_{S2} - \sigma_{S1} \tag{17.6}$$

In terms of the mechanical (hydrostatic) equilibrium derivation of Young's equation, this equation appears nonsensical, since the three-phase wetting line for which it describes the equilibrium does not exist. That is, there is no three-phase contact line at which the two fluid phases contact each other and the solid surface. Thermodynamically, however, one can show that the equation is exactly obeyed when spreading occurs.

If Equation (17.6) is rewritten to take into account the microscopic mechanism of spreading, the situation can be clarified. For example, spreading proceeds first by a molecular level spreading of liquid on the solid surface. If it is assumed that the second fluid (2) is air and the only component adsorbed at the S2 interface is the vapor of liquid 1, Equation (17.6) can be written

$$\sigma_{12} = \sigma_S - \pi_{S2,1} - \sigma_{S1} \tag{17.7}$$

where σ_S is the surface tension (energy) of the solid surface with no adsorbed molecules and $\pi_{S2,1}$ is the surface pressure of adsorbed liquid 1 at the S2 (see also Chapter 8) interface so that

$$\sigma_S - \pi_{S2,1} = \sigma_{S2} \tag{17.8}$$

If spreading occurs, the spreading pressure $\pi_{S2,1}$ will increase, reducing σ_{S2} until Equation (17.8) is exactly satisfied. At that point the thickness of the adsorbed film will be such that only the S1 and 1–2 interfaces exist. Further thickening of the spread film may then occur by spreading of liquid 1 on itself.

An interesting sidelight to the above-mentioned concept is the phenomenon generally referred to as autophobicity. In that situation, the spreading liquid is adsorbed onto the "bare" solid surface in the usual way (Fig. 17.8). However, the molecules in this case are adsorbed with a specific molecular orientation relative to the solid and the spreading fluid. As a result, when a complete monolayer of adsorbed molecules is formed, the spreading liquid no longer "sees" a surface tension σ_S or σ_{S2}, but rather a new surface tension $\sigma_{1,O}$ (liquid 1, oriented). If the spreading liquid is a polar organic material such as an alcohol or carboxylic acid and the solid surface one that can interact strongly with the polar group (glass, mica, metals and oxides, etc.) the orientation process will expose an essentially hydrocarbon surface (mostly $-CH_3$ groups) with $\sigma_{1,O} \ll \sigma_S$ (or σ_{S2}). In order for Equation (17.5) to be satisfied under those conditions, $\cos \theta$ must be < 1; that is, the spreading liquid will retract from the surface to produce a finite contact angle.

17.2.2. The Spreading Coefficient

Young's equation is usually found to be a very useful and adequate means of describing wetting equilibria in most circumstances. However, it is sometimes found useful to define another term that indicates from a thermodynamic point of view whether a given liquid–solid system will be wetting ($\theta = 0°$) or nonwetting ($\theta > 0°$). Such a term is the spreading coefficient, S.

For a spontaneous process (such as spreading) to occur, the free energy of the process must be negative. In terms of surface free energies, then, one can write the relationship

$$S_{S12} = \sigma_S - \sigma_{S1} - \sigma_{S2} \tag{17.9}$$

where the subscript S12 refers to the initial spreading coefficient for the spreading of liquid 1 over the solid S in the presence of (or displacing) fluid

(a)   (b)

FIGURE 17.8. An interesting phenomenon sometimes encountered in spreading is the "autophobic" effect. When the liquid is first applied to the surface, it has a small contact angle; however, as adsorption occurs at the solid-liquid interface, the orientation of the adsorbed molecules makes the interface more hydrophobic, leading to a larger contact angle.

2. At first sight, the autophobic phenomenon mentioned above would seem to be an exception to Equation (17.9). However, if a relationship like that in (17.8) is invoked, then

$$S_{S12} = \sigma_{S2} + \pi_{S2,1} - \sigma_{S1} - \sigma_{12} \tag{17.10}$$

and the autophobicity produced by the oriented monolayer adsorption of liquid 1 is easily accommodated.

An alternative to the relationship in Equation (17.10) is the so-called equilibrium spreading coefficient

$$S_{S12}{}^\circ = \sigma_{S2} - \sigma_{S1} - s_{12} \tag{17.11}$$

in which the solid surface tension σ_{S2} is now that of the solid with saturated adsorbed layer of fluid 2.

17.2.3. Classification of Wetting Processes

While the term "wetting" may conjure up a fairly simple image of a liquid covering a surface, from a surface chemical standpoint the situation is somewhat less clearcut. Classically, there are three types of wetting phenomena of importance: adhesion, spreading, and immersion wetting (Fig. 17.9). The distinctions may seem subtle, but they can be significant from a thermodynamic and phenomenological point of view.

"Adhesion wetting" refers to the situation in which a solid, previously in contact with a vapor, is brought into contact with a liquid phase. During the process, a specific area of solid–vapor interface, A, is replaced with an equal area of solid–liquid interface (Fig. 17.9a). The free energy change for the process is given by

$$-\Delta G = A\,(\sigma_{SV} + \sigma_{LV} - \sigma_{SL}) = W_a \tag{17.12}$$

where the σ terms refer to the solid–vapor (SV), liquid–vapor (LV), and solid–liquid (SL) interfacial tensions. The quantity in parentheses is the thermodynamic work of adhesion, W_a. From Equation (17.12), it is clear that any decrease in the solid–liquid interfacial energy σ_{SL} will produce an increase in the work of adhesion (and a greater energy decrease), while an increase in σ_{SV} or σ_{LV} will reduce the energy gain from the process.

"Spreading wetting" applies to the situation in which a liquid (L1) and the solid are already in contact and the liquid spreads to displace a second fluid (L2, usually air) as illustrated in Figure 17.9b. During the spreading process, the interfacial area between solid and L2 is decreased by an amount A, while that between the solid and L1 increases by an equal amount. The interfacial area between L1 and L2 also increases during the process. The change in

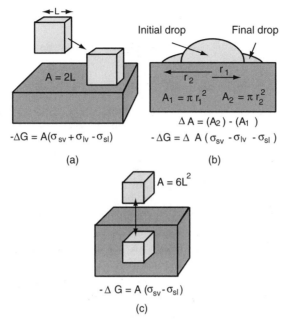

FIGURE 17.9. Three types of wetting phenomena may occur, each with its own thermodynamic relationship. They are (*a*) adhesional wetting in which two surfaces make contact displacing a third phase, but without flow; (*b*) spreading wetting, similar to (*a*), but with flow that changes the net interfacial area until equilibrium is attained; and (*c*) immersional wetting in which a body is completely submerged in a fluid, eliminating the original surface energy term.

interfacial area in each case will be the same, so that the total decrease in the energy of the system (for a spontaneous process), will be

$$-\Delta G = A(\sigma_{SL2} - \sigma_{SL1} - \sigma_{12}) \tag{17.13}$$

where σ_{12} is the interfacial tension between fluids 1 and 2. If the term in parentheses, the spreading coefficient S, is positive, then L1 will spontaneously displace L2 and spread completely over the surface (or to the greatest extent possible). If S is negative, the spreading process as written will not proceed spontaneously. Rewriting Young's equation in the form

$$\cos \theta = \frac{\sigma_{SL2} - \sigma_{SL1}}{\sigma_{12}} \tag{17.14}$$

and combining with that for the spreading coefficient S gives

$$S = \sigma_{12}(\cos \theta - 1) \tag{17.15}$$

It is clear from Equation (17.15) that for $\theta > 0$, S cannot be positive or zero, and spontaneous spreading will not occur.

The third type of wetting, "immersion wetting," results when a solid substrate not previously in contact with a liquid is completely immersed in liquid L1, displacing all of the solid–L2 interface (Fig. 17.9c). In this case, the free energy change at equilibrium is determined by two factors: a component related to the solid–air interface $A \, \sigma_{SL2}$ and that of the new solid–liquid interface $A \, \sigma_{SL1}$, where A is the total surface area of the solid. The free energy change is then given by

$$-\Delta G = A \, (\sigma_{SL2} - \sigma_{SL1}) \tag{17.16}$$

From the preceding relationships for wetting processes, it is clear that the interfacial energies between a solid and any contacting liquid, and the interfacial tension between the liquid and the second fluid (usually air), control the manner in which the system will ultimately perform. The ability to alter one or several of those surface energy components makes it possible to manipulate the system to attain the wetting properties desired for a given system. It is generally through the action of surfactants at any or all of those interfaces that such manipulation is achieved. A more specific discussion of the role of surfactants in the alteration and control of the wetting process is given below.

17.2.4. Additional Useful Thermodynamic Relationships for Wetting

Although the mathematical relationships encountered in wetting phenomena are usually quite simple, they are found to be very useful in many practical applications. Their combinations and variations have given rise to still more relationships, which further expand their utility without expanding the amount of information necessary for their application. Two thermodynamic relationships that can be useful in the analysis of wetting and spreading phenomena are the works of cohesion and adhesion.

The work of cohesion, W_c, is defined as the reversible work required to separate two surfaces of unit area of a material with surface tension s, given simply as

$$W_c = 2\sigma \tag{17.17}$$

The related quantity, the work of adhesion, W_a, is similarly defined as the work required to separate unit area of interface between two different materials or phases to leave a "bare" solid surface:

$$W_{aS12} = \sigma_S + \sigma_{12} - \sigma_{S1} \tag{17.18}$$

The subscript S12 is employed here to emphasize the fact that, as the two surfaces (S and liquid 2) are separated, two new interfaces are formed—S2 and 1–2. If the idea of the spreading pressure $\pi_{S2,1}$ is included, Equation (17.18) becomes

$$W_{aS12} = \sigma_{S2} + \pi_{S2,1} + \sigma_{12} - \sigma_{S1} \tag{17.19}$$

which takes into account the energetic effect of molecules of the spreading liquid being adsorbed on the solid surface ahead of the moving three-phase boundary.

Working from Equation (17.19), one can define a new quantity, the reversible work of adhesion , which includes $\pi_{S2,1}$:

$$W^{\circ}_{aS12} = \sigma_{S2} + \sigma_{12} - \sigma_{S1} \tag{17.20}$$

Comparing Equations (17.19) and (17.20), one can see that $W_{aS12} = W^{\circ}_{aS12}$ only if $\pi_{S2,1} = 0$. In strongly interacting systems in which $\pi_{S2,1}$ is significant, $W_{aS12} >> W^{\circ}_{aS12}$. Equations (17.18) and (17.20), can be useful in situations in which fluid 2 is a liquid that may also wet the solid (e.g., competitive wetting) or when it contains some component (a surfactant, for example) that can be adsorbed on the solid.

Some additional useful relationships based on the concepts already introduced include

$$S_{S12} = \sigma_{12} (\cos \theta^{\circ} - 1) + \pi_{S2,1} \tag{17.21}$$

$$S^{\circ}_{S12} = \sigma_{12} (\cos \theta^{\circ} - 1) \tag{17.22}$$

$$W_{aS12} = \sigma_{12} (1 + \cos \theta^{\circ}) + \pi_{S2,1} \tag{17.23}$$

$$W_{aS12} = \sigma_{12} (1 + \cos \theta^{\circ}) \tag{17.24}$$

$$\cos \theta^{\circ} = -1 + 2 \left(\frac{W^{\circ}_{aS12}}{W_{c12}} \right) \tag{17.25}$$

A close look at Equation (17.25) shows that if $W_{c12} = W^{\circ}_{aS12}$, $\theta = 0^{\circ}$; if $W_{c12} = 2W^{\circ}_{aS12}$, $\theta = 90^{\circ}$; and finally, if $W_{c12} > 2W^{\circ}_{aS12}$, $\theta > 90^{\circ}$. Phenomenologically, this says that the contact angle observed for a given liquid–solid system will reflect the competition between the drive of the liquid to "stick to its own kind," and the pull of the solid surface. It also says that it is impossible for a liquid–solid–vapor system to have a 180 contact angle since that would require that (W°_{aS12}/W_{c12}) = 0, which is physically impossible. Such a situation would require, according to the definition of W°_{aS12}, that there be no net attractive interaction between liquid 1 and the solid in the presence of a non-condensed third phase.

17.3. CONTACT ANGLES AND CALCULATION OF SOLID SURFACE ENERGIES

It is generally a rather simple task to measure the surface tension of a liquid. However, from the standpoint of acquiring a better theoretical handle on surface and interfacial process (including wetting), it would be useful if one could calculate from first principles the surface tension of a system based on an understanding of the intermolecular forces giving rise to it. If such a calculation could be made for a pure liquid (σ_{LV}), it could conceivably be carried out for more complex liquid–liquid (σ_{12}), and solid–liquid (σ_{SL}) systems, opening the way for analyses of wetting systems requiring terms that are not experimentally accessible.

Using a statistical treatment of the variation of local intermolecular forces as the interface is traversed from the liquid to the vapor phase, it is possible to calculate the surface tension of a simple liquid (e.g., argon) that agrees well with experiment. However, such exact methods become quite complex or (currently) impossible for calculating surface tensions in most practical systems.

As is usually the case when theoretical and experimental science meet, it is necessary to make some simplifying assumptions in order to apply theories to practical systems. Good and Girifalco proposed an empirical approach to the problem based on the Berthelot principle that the interaction constant between two different surfaces or particles will be the geometric mean of the interaction constant for the individual surface units, an approach already introduced in Chapter 4. Good and Girifalco suggested that the work of adhesion between two different liquids could be expressed as a similar function

$$W_{a12} = \Phi(W_{c1}W_{c2})^{1/2} \tag{17.26}$$

where the constant Φ takes into account differences in the molecular size and polar content of the two materials involved. Combination of Equations (17.17), (17.18), and (17.26) gives

$$\sigma_{12} = \sigma_1 + \sigma_2 - 2\Phi(\sigma_1\sigma_2)^{1/2} \tag{17.27}$$

For nonpolar liquids with $\Phi \approx 1$, results agree reasonably well with experiment. For dissimilar substances, such as water and alkanes or water and mercury, values of Φ between 0.32 and 1.15 must be used to obtain agreement.

The Good–Girifalco approach has been extended to the use of contact angles in the computation of surface tension values for solid–liquid interfaces. Considering a system where the fluid 2 is either vapor or air (in which case it can be ignored), and combining with Young's equation, one obtains the expression

$$\cos\,\theta = -1 + 2\Phi\left(\frac{\sigma_S}{\sigma_1}\right)^{1/2} - \frac{\pi_{S,1}}{\sigma_1} \qquad (17.28)$$

Equation (17.28) is found to give reasonable values of σ_S for nonpolar solids and represents a potentially useful tool for characterizing solid surfaces, empirically, at least.

The logical next step in the process of extending the utility of theory to practical systems is to include polar molecular interactions. For this step, Fowkes suggested that the intermolecular forces contributing to surface and interfacial tensions, and subsequent phenomena such as wetting, could be broken down into independent and additive terms. For example, a polar molecule such as an ester would have two terms making up its surface tension—dispersion forces (d) and dipolar interactions (p)—so that

$$\sigma = \sigma_d + \sigma_p \qquad (17.29)$$

where σ_d and σ_p are the dispersion and dipolar contributions to the total surface tension. Application of the principle of Equation (17.26) to Equation (17.29) produces a reasonable approximation for the work of adhesion for interactions involving only dispersion and dipole forces

$$W_{a12} = 2(\sigma_1\sigma_2)^{1/2}\,[(d_1d_2)^{1/2} + (p_1p_2)^{1/2}] \qquad (17.30)$$

where d and p represent the respective fractions of dispersion and polar components contributing to the energy densities of the adjacent phases in which $p = (1 - d)$.

For nonpolar materials involving only dispersion forces such as simple alkanes $d = 1$ and $\sigma = \sigma_d$. Fowkes developed an expression for the dispersion force contribution to the total work of adhesion of the form

$$W_{a12}^d = 2(\sigma_1^d\sigma_2^d)^{1/2} \qquad (17.31)$$

If one combines the suggestion of Good and Girifalco with the modifications proposed by Fowkes, one may rewrite Equation (17.30) in the form

$$W_{a12} = W_{a12}^d + W_{a12}^p \qquad (17.32)$$

in which the work of adhesion is divided into dispersion and polar components. Identification of the separate components of surface tension (σ^d and σ^p) with the corresponding works of adhesion has important theoretical and practical consequences.

Equations similar to (17.27) and (17.28) can be written for solid–liquid interfaces and, when combined with Young's equation, give an experimentally accessible relationship between the contact angle a liquid will make on a solid surface and the attractive components of the surface tensions of the two phases.

$$\cos \theta = -1 + 2\left[\frac{(\sigma_s^d \sigma_1^d)^{1/2}}{\sigma_1}\right] - \left(\frac{\pi_{s,1}}{\sigma_1}\right) \tag{17.33}$$

The great potential utility of this Equation lies in the fact that, since all the variables are accessible through reasonably simple experiments, a solid interacting by dispersion forces alone (e.g., a pure hydrocarbon wax or polymer) can be used to determine the dispersion component of the surface tension of a liquid. Alternatively, a standard nonpolar liquid can be used to characterize the surface tension of a solid, or if the "clean" solid has been previously characterized, the contact angle can be used to assist in determining the nature of a surface contaminant.

When using Equation (17.33), it must be kept in mind that it was derived primarily for dispersion and dipolar interactions. It has been used with some success for strongly hydrogen bonding systems (a special class of dipolar interaction), but with greater variability than found for simpler systems. Other types of intermolecular interactions are also possible, including ionic interactions and metallic forces. Whether such forces can be reasonably incorporated into the Good–Girifalco–Fowkes (GGF) theory via the geometric mean approach remains an active question. More complex equations have been proposed to include such interactions. Such extended equations have been used to investigate hydrophilic and biological surfaces with some success. Because of the relationship between intermolecular interactions, cohesive energy densities, and solubility parameters, solubility parameters have also been used in the context of wetting phenomena. For most practical systems, the GGF approach remains one of the best tools for estimating the interactions at solid–liquid interfaces. A workable alternative, however, is that of the critical surface tension discussed in the next section.

17.3.1. The Critical Surface Tension of Wetting

Before the introduction of the theory of Good and Girifalco, Zisman and coworkers developed a useful and practical systematic method for characterizing the "wettability" of solid surfaces. The system is based on the observation that for solid surfaces having a surface tension $(\sigma_s) < 100$ mN m^{-1} (generally classified as "low-energy surfaces"), the contact angle formed by a drop of liquid on the solid surface will be primarily a function of the surface tension of the liquid, σ_{12} (where phase 2 is air saturated with the vapor of liquid 1). They found, in particular, that for a given solid surface and a homologous series of related liquids (e.g., alkanes, dialkyl ethers, alkyl halides), $\cos \theta$ was an approximately linear function of σ_{12}. The relationship for several liquid

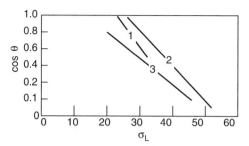

FIGURE 17.10. The critical surface tension of wetting, σ_c, is usually found to be consistent for a homologous series of hydrocarbon liquids (curve 1), a series of alkyl halides (curve 2), and miscellaneous polar liquids (curve 3) on a nonpolar surface such as Teflon. Results on polar surfaces are less unequivocal, but still can be useful.

types is illustrated in Figure 17.10. For nonpolar liquids, the relationship holds very well, while for high-surface-tension, polar liquids, the correlation begins to break down somewhat and the line begins to curve.

The general limitation of the technique to the so-called low-energy surfaces must be made because such materials as metals, metal oxides, or ionic solids, which have surface free energies in the hundreds and thousands, are almost always covered with an adsorbed layer of a low-energy substance such as water or oils from the atmosphere. Under rigorously controlled experimental conditions, the technique may be applied to such materials, but interpretation of the results can be difficult .

From a plot of $\cos \theta$ versus σ_{12}, one can obtain the value of the liquid surface tension at which $\cos \theta = 1$, a value that has been termed the "critical surface tension of wetting," σ_c. It is defined as the surface tension of a liquid that would just spread on the surface of the solid to give complete wetting. In other words, if $\sigma_{12} \leq \sigma_c$, the liquid will spread; if $\sigma_{12} \geq \sigma_c$, the liquid will form a drop with a nonzero contact angle. Typical values of σ_c for commonly encountered materials are given in Table 17.1.

While σ_c is an empirically determined value, depending to some extent on the nature of the liquids used for its determination, attempts have been made to identify it with such theoretical terms as σ_s or σ_s^d. According to the Good–Girifalco equation [Eq. (17.27)], for $\pi_{s,1} = 0$ and $\cos \theta = 1$

$$\sigma_{12} = \sigma_1 = \Phi^2 \sigma_s = \sigma_c \tag{17.34}$$

For nonpolar liquids and solids, $\Phi \approx 1$, therefore $\sigma_s = \sigma_c$. Starting with the Fowkes postulate

$$\sigma_{12} = \sigma_1 = (\sigma_s^d \sigma_1^d)^{1/2} = \sigma_c \tag{17.35}$$

TABLE 17.1. Values of Critical Surface Tension of Wetting (σ_c, mN cm^{-1}) for Various Materials

Solid	σ_c	Solid	σ_c
Teflon	18	Copper	60
Polytrifluoroethylene	22	Silver	74
Polyvinylidene fluoride	25	Silica (dehydrated)	78
Polyvinyl fluoride	28	Anatase (TiO$_2$)	92
Polyethylene	31	Graphite	96
Polystyrene	33	Lead	99
Polyvinyl alcohol	37	Tin	101
Polyvinyl chloride	39	Iron	106
Polyvinylidene chloride	40	Iron oxide(Fe$_2$O$_3$)	107
Polyethyleneterephthalate	43	Silica (hydrated)	123
Nylon 6,6	46	Rutile (TiO$_2$)	143

Therefore, when $\sigma_1^d = \sigma_{12}$, then $\sigma_c = \sigma_s^d$. For nonpolar solids, for which only dispersion interactions occur, the same might be expected to hold regardless of the nature of the liquid employed. However, experiment has shown that the condition of $\pi_{s,1} = 0$ is unlikely to hold for systems in which q is near or equal to zero.

Young's equation can also be employed to test the theoretical significance of the σ_c concept. If $\cos\theta = 1$, then

$$\sigma_c = \sigma_{12} = \sigma_s - \pi_{s,1} - \sigma_{s1} \qquad (17.36)$$

where it is clear that $\sigma_c = \sigma_s$ only if both σ_{s1} and $\pi_{s,1} = 0$. It must be concluded, then, that the value of σ_c is not a characteristic property of the solid alone but of the solid–liquid combination. That does not, however, greatly diminish the practical utility of the concept as a method for characterizing the wettability of a surface, which was, after all, its purpose all along.

17.3.2. Some Practical Drawbacks

While the critical surface tension concept is attractive as a practical tool for characterizing solid surfaces, it does have several drawbacks. Of particular significance is the fact that the procedure requires the use of several liquids of different surface tensions in which $\sigma_{12} > \sigma_c$. This means that a relatively large number of measurements are required for each solid surface. In addition, it is sometimes difficult to obtain a sufficient quantity of pure liquids with the required range of σ_{12}, leading many investigators to employ liquid mixtures (e.g., alcohols or glycols in water) or solutions of surfactants as their test liquids. While such a procedure is attractive, making available a wide range of surface tensions, from 72 mN m^{-1} for water to about 20 mN m^{-1} for solutions of some fluorinated surfactant solutions, it carries with it its own pitfalls.

While the solution method may give results in agreement with pure hydrocarbon liquids for surfaces such as paraffin or polytetrafluoroethylene (PTFE), the agreement is not general. For example, while solutions of hydrocarbon surfactants on PTFE and polyethylene give good agreement, solutions of fluorocarbon surfactants give good agreement for PTFE, but deviate significantly on hydrocarbon surfaces. Low values of s_c may also be obtained for solutions on slightly polar surfaces such as polyesters, polyacrylates, or polystyrenes. The generally accepted explanation for the problems with surfactant solutions is that surfactant molecules will adsorb at the σ_{s1} and σ_{s2} interfaces and alter the nature of the surface being measured. The general topic of the effects of surfactants on the wetting of solid surfaces is treated in a following section.

From a practical applications point of view, both the critical surface tension approach and the use of contact angles with the Good–Girifalco–Fowkes equation represent handy tools for the characterization of the wettability, and therefore something of the chemical nature, of solid surfaces. The choice of technique is basically one of preference and convenience.

17.4. THE KINETICS OF WETTING

So far the discussion of wetting and contact angles has been essentially limited to "equilibrium" situations. In many practical applications, the wetting phenomena of interest are "dynamic" in nature, involving a moving wetting line at which equilibrium is never fully attained. The contact angle of a moving wetting line is generally called a dynamic contact angle. It is generally found that the dynamic contact angle for a given system will differ from the equilibrium value, even for very slow rates of movement, with the difference usually being velocity-dependent.

Although exceptions have been reported, it is generally true that dynamic contact angles are dependent on both direction and speed of movement of the wetting line; that is, the velocity of the movement. Experimentally, it is observed that advancing contact angles increase and receding angles decrease as the velocity of the wetting line increases. When thought of in terms of the contact angle hysteresis discussed previously, this indicates that the dynamic contact angle is a function of thermodynamically irreversible processes. In the case of the static angle, hysteresis may at times be attributed to the movement of the wetting line between the so-called metastable states. In the dynamic case, it must be assumed that the wetting line never reaches even a metastable "equilibrium" and irreversibility results from the constant drive toward an equilibrium that can never be reached.

A "typical" curve shape for the relationship between dynamic contact angle and wetting velocity v_w is shown in Figure 17.11. It is generally found that at low wetting rates, the contact angle θ_d is a rapidly changing function of velocity, v_w. As the velocity increases, the slope of the curve decreases until, at still higher values of v, it again changes rapidly to approach 180°. The velocity at

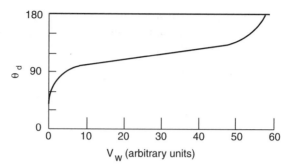

FIGURE 17.11. In dynamic wetting processes, the dynamic contact angle, θ_d, will increase with the wetting velocity, V_w, making effective coating more difficult, until at some critical velocity wetting will become irregular.

which $\theta_d = 180°$ is often denoted as v_{w180}. At $v_w > v_{w180}$, it is observed that the wetting line becomes irregular (Fig. 17.12a) and air becomes entrained at the trailing vertices leading to a loss of contact between wetting liquid and solid surface. In a practical coating process, such an event obviously leads to the production of a spotty, irregular coating which is unlikely to be of use. A similar effect results in dewetting operations in which the fluid being displaced from the solid surface stops being displaced in a smooth process but begins to be entrained by the displacing fluid as drops or rivulets (Fig. 17.12b).

The practical aspects of dynamic contact angle phenomena, then, center around determining the maximum wetting rates that can be attained before entrainment or wetting failure occurs, and how a system can be modified to increase that maximum velocity, since in many, if not most, cases, speed is money. Liquid coating operations are obviously impacted by the dynamic restraints of a system. If one can increase the maximum speed of coating a substrate from, say, 200 to 400 m min^{-1}, productivity gains will be impressive.

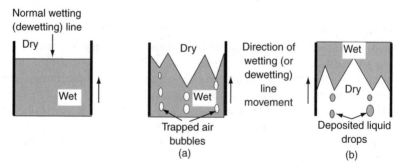

FIGURE 17.12. When the wetting velocity, V_w, is such that $\theta_d = 180°$, the normally uniform wetting line develops a sawtooth shape and air bubbles become entrained in the coating, the bubbles originating at the vertices of the sawtooth pattern.

Less obviously, petroleum recovery by various flooding techniques is affected because, if the flooding fluid is injected at a velocity that surpasses v_w180 for the system, petroleum may be trapped by the flooding fluid resulting in a loss in recovery efficiency and ultimately total reservoir yield.

17.4.1. Factors Affecting Dynamic Wetting Phenomena

It is generally observed that the main factors affecting v_{180} for a given solid substrate are the viscosity and surface tension of the coating liquid. The dynamic contact angle at a given value of v is found to increase with increasing viscosity, η, and decreasing surface tension, s. In quantitative terms, the relationship between q_d and those two values is correlated with the capillary number, C_a

$$C_a = \frac{\eta v}{\sigma} \tag{17.37}$$

so that θ_d always increases with C_a.

The effect of viscosity can be rationalized somewhat in that viscous forces would be expected to work against wetting by slowing the rate at which the wetting line approaches its equilibrium (or metastable) position. The effect can be seen, for example, in the fact that even for "static" contact angles, the rate at which the contact angle of viscous liquids such as molten polymers change is a function of viscosity. In general, then, viscous forces tend to oppose wetting, so that a higher viscosity leads to an increased velocity dependence of θ_d and a smaller v_w180.

The observed effect of surface tension is less obvious. For static or equilibrium wetting, it is usually found that a lower liquid surface tension will improve wetting (i.e., reduce θ^o). In the dynamic case, however, the velocity dependence of θ_d, and therefor v_{180}, for liquids of the same viscosity but different surface tensions was in the direction of greater dependence on v with lower s.

It has been suggested that the role of surface tension forces in the dynamic wetting process may be represented by

$$F_w = \sigma_{12} (\cos \theta^o - \cos \theta_d) \tag{17.38}$$

where F_w is the nonequilibrium surface tension force acting at the wetting line in the direction of wetting (Fig. 17.12), which must be a constant for a given set of circumstances. The work performed by F_w, then, is the irreversible work done by surface forces leading to wetting at some finite velocity, as opposed to infinitely slowly. According to Equation (17.38), if σ_{12} is decreased, θ_d must increase to maintain a constant value of F_w. On the other hand, reducing σ_{12} may also alter θ^o, so that the overall change may be more complicated than the simple relationship given by Equation (17.38).

If the reduction in surface tension for the liquid is brought about by the addition of a surfactant, the net affect on θ_d will depend on levels and rates of adsorption of the surfactant molecules at the various interfaces involved. Obviously, an analysis of such a complex situation becomes very involved (or impossible), and experiment and experience become the best tools of the trade.

17.5. COMPETITIVE WETTING

The preceding discussion of wetting phenomena was restricted to the situation in which fluid 2 was air. In many practical situations, such as detergency and petroleum recovery, the second fluid will also be a liquid. Not surprisingly, such systems may exhibit even more complex behavior than those having one vapor phase. Qualitatively, one can analyze the situation as follows: at the interface between water and a nonpolar solid in the presence of a nonpolar liquid (SW2), one expects that the work of adhesion between water and solid, W°_{aSw2} will be smaller than that between the nonpolar liquid and the solid, W°_{aS2w}. The reason for that, of course, is that the only attractive interactions possible between water and solid are the dispersion forces, which will be of similar magnitude for each side of the system (≈ 22 mN m^{-1} for water and 24–28 mN m^{-1} for a nonpolar hydrocarbon solid). In comparison, the internal binding in water, including hydrogen bonding, will be much greater ($W_{cw} = 2\sigma_w \approx 144$ mN m^{-1}). As a result of this balance (or more accurately, imbalance) of forces, θ°_w will be quite large, with values of well over 110° being common.

Beginning with the basic definition of the reversible work of adhesion [Eq. (17.20)] and the Fowkes relationship [eq. (17.31)], one can place the above situation on a more quantitative footing, yielding the relationship

$$W^{\circ}_{aSw2} = 2[\sigma_2 - (\sigma^d_w\sigma_2)^{1/2} - (\sigma^d_s\sigma_2)^{1/2} + (\sigma^d_s\sigma^d_w)^{1/2}] \qquad (17.39)$$

If this equation is evaluated for the system water–tetradecane–PTFE, a value of $W^{\circ}_{aSw2} = 0.4$ mJ m^{-2} is obtained. For the same system, $W^{\circ}_{aS2w} = 99.8$ mJ m^{-2}. Clearly, any wetting process that requires the displacement of an oil from a nonpolar solid surface by an aqueous solution must work against a considerable thermodynamic barrier.

Of particular practical importance is the primary mechanism of detergency for oily soils. In that case, the main role of the detergent solution is to displace or "roll up" the oily soil from the solid surface so that it can be more easily and completely removed from the surface by mechanical action. Obviously, for a nonpolar surface, such action must be limited by the above balance of adhesive and cohesive forces. Due to the complexity of the situation it is difficult to generalize as to how or even if, the addition of a surfactant will improve a given situation. (That point is discussed in the following section.) However, one can venture the following propositions:

1. If significant surfactant adsorption occurs at the oil–solid interface, displacement of the oil by aqueous solution will be hindered.
2. If little or no adsorption occurs at the solid–water interface, detergency will be adversely affected.
3. The most effective surfactants will be those which lower s_{12} the most, while not coming into conflict with statements 1 and 2.

In other words, the best surfactant from the point of view of detergency should be one that adsorbs well at the solid–water and water–oil interfaces, but not at all at the solid–oil interface.

17.6. EFFECTS OF SURFACTANTS ON WETTING PROCESSES

Having mentioned several times the use of surfactant solutions in wetting studies, we now consider specifically some of the effects their presence can have on contact angles and wetting. The action of surfactants derives from their adsorption at the various interfaces and the resultant modification of interfacial tensions. In terms of the Gibbs equation, the relationship between the specific adsorption of a solute, G, and surface tension is given by

$$\left(\frac{d\sigma}{d \ln c}\right)_{T,P} = -RT\,\Gamma \tag{17.40}$$

where T, P, R, and c have their usual significance. Looking again at Young's equation

$$\cos\,\theta° = \frac{\sigma_{S2} - \sigma_{S1}}{\sigma_{12}} \tag{17.41}$$

one can see that $\theta°$ will decrease if either σ_{S1} or σ_{12} or both are reduced and σ_{S2} remains essentially unchanged. The effect of such changes will be greater if s_{S2} is larger, that is, if the second fluid in the system is air. The contact angle will increase with surfactant addition only if the surfactant is adsorbed at the S2 interface. Such a situation requires some sort of transport mechanism for carrying surfactant from the solution to that interface. For most surfactants of low volatility, such a mechanism is not readily available when fluid 2 is a vapor. If the second fluid is a liquid, transport of surfactant from liquid 1 through liquid 2 can result in significant adsorption at the S2 interface. For more mobile surface-active materials such as alcohols, molecular diffusion may be sufficient. That is, the mechanism already mentioned leading to the phenomenon of autophobicity.

A general relationship between contact angle, surfactant concentration, and specific adsorption can be obtained by differentiating Young's equation with respect to $\ln c$

$$\frac{d(\sigma_{12} \cos \theta^\circ)}{d \ln c} = \frac{d\sigma_{S2}}{d \ln c} - \frac{d\sigma_{S1}}{d \ln c} \tag{17.42}$$

Combining with the Gibbs equation gives

$$\sigma_{12} \sin \theta^\circ \left(\frac{d\theta^\circ}{d \ln c}\right) = RT \left(\Gamma_{S2} - \Gamma_{S1} - \Gamma_{12} \cos \theta^\circ\right) \tag{17.43}$$

where $\sigma_{12} \sin \theta^\circ$ will always be positive, so that $(d\theta^\circ/d \ln c)$ must always have the same sign as the right-hand side of the equation. Using that relationship, one can define three situations for changes in contact angle and wetting:

1. The addition of surfactant lowers θ° and improves wetting. This situation corresponds to the inequality $\Gamma_{S2} < \Gamma_{S1} - \Gamma_{12} \cos \theta^\circ$.
2. The addition of surfactant increases θ° and dewetting occurs. In that case it must be that $\Gamma_{S2} > \Gamma_{S1} - \Gamma_{12} \cos \theta^\circ$.
3. If $\Gamma_{S2} = \Gamma_{S1} - \Gamma_{12} \cos \theta^\circ$, the addition of surfactant has no net effect on θ° and wetting is unaffected.

In some practical situations it is found that the effect of surfactant addition on wetting is variable, with behavior 1 observed in one concentration range and 2 in another. Such an example will be discussed below. In general, however, one finds that situations 1 and 3 are most commonly encountered in systems where the solid substrate is a low-energy, nonpolar material. Situation 2 is usually observed only with higher-energy, more polar substrates.

In some cases, one finds that solutes that adsorb strongly at solid–liquid interfaces, are not as strongly adsorbed at the liquid–fluid (1–2) interface. Such materials, including many polymers, will affect wetting depending on how the adsorbed layer interacts with liquid 1. If the adsorbing polymer presents a lower energy surface to the liquid, dewetting will be observed; if a higher energy surface is developed, improved wetting will result. In practice, it is often found that in order to coat a low-energy polymeric substrate (e.g., a polyester) effectively with an aqueous solution, it is necessary first to apply a very thin layer—a primer or mordant ("biting") coating—of a more polar polymer (e.g., gelatin or a lightly carboxylated vinyl polymer) to improve wetting and adhesion of the coating. Alternatively, one can modify the surface by chemical etching or corona discharge to improve wetting and adhesion.

17.6.1. Surfactant Effects on Nonpolar Surfaces

When one considers the effects of low-molecular-weight surfactants on wetting, it is helpful to divide the subject into two regimes: the effect on nonpolar surfaces and that on polar surfaces; the reason is that, for nonpolar surfaces, it is often possible to assume that adsorption at the S2 interface will be negligible. If surfactant adsorption at the S2 interface is small (e.g., $\Gamma_2 \approx 0$), as is usually observed for the situation where fluid 2 is air, Equations (17.41) and (17.42) indicate that

$$\frac{d\sigma_{12} \cos \theta^\circ}{d \ln c} = - \left(\frac{d\sigma_{S1}}{d \ln c} \right) = RT \, \Gamma_{S1} \qquad (17.44)$$

Theoretically, one can determine the adsorption at the S1 interface from the change in θ° with surfactant concentration. Using the Gibbs equation, it is possible to determine the adsorption of surfactant at the 1–2 interface (from $d\sigma_{12}/d \ln c$). It is found experimentally that for most hydrocarbon surfactants on nonpolar surfaces, the adsorption at S1 is the same as that at the 1–2 interface, or that $\Gamma_{S1} \approx \Gamma_{12}$.

A more quantitative picture of the situation can be obtained by looking at the effect of variations in σ_{12} on the reversible work of adhesion, W°_{aS12}. Differentiating Equation (17.20) by σ_{12} yields

$$\frac{dW_{aS12}^\circ}{d\sigma_{12}} = \left(\frac{d\sigma_{s2}}{d\sigma_{12}} \right) + 1 - \left(\frac{d\sigma_{s1}}{d\sigma_{12}} \right) \qquad (17.45)$$

Assuming that adsorption at the S2 interface is zero, and setting $(d\sigma_{S2}/d\sigma_{12}) = 0$ and $(d\sigma_{S1}/d\sigma_{12}) = 1$, the result is

$$\frac{dW_{aS12^\circ}}{d\sigma_{12}} = 0 \qquad (17.46)$$

Thus, the reversible work of adhesion will be independent of surfactant concentration. On the basis of these assumptions, a number of interesting relationships can be derived relating the effects of surfactant concentration on wetting phenomena.

It should be kept in mind that the ideas expressed above are based on the assumption that $\Gamma_{S2} = 0$, which is probably not true in some cases. However, it can be shown that if Γ_{S2} is significant, but still proportional to σ_{12}, then useful linear relationships between, say, W°_{aS12} and σ_{12} can be obtained.

A variation of Equation (17.44) useful for predicting whether the addition of a surfactant will improve wetting is to differentiate Young's equation and combine with the Gibbs equation to yield

$$\frac{d\sigma_{12} \cos \theta^\circ}{d\sigma_{12}} = \frac{\Gamma_{S2} - \Gamma_{S1}}{\Gamma_{12}} \tag{17.47}$$

If $\Gamma_{S2} = 0$, then a plot of $\sigma_{12} \cos \theta^\circ$ versus σ_{12} will have a slope of $-(\Gamma_{S1}/\Gamma_{12})$. From Equation (17.47), it can be seen that if $\Gamma_{S1}/\Gamma_{12} = 1$, complete wetting of a surface should occur if the surfactant is capable of lowering σ_{12} to σ_c or below. If $\Gamma_{S1}/\Gamma_{12} = 0$ (e.g., no surfactant adsorption at the S1 interface), then complete wetting cannot be obtained for any value of σ_{12}. Finally, if $\Gamma_{12}/\Gamma_{S1} = 0$ (no adsorption at the 1–2 interface), complete wetting can occur with no change in σ_{12}. Obviously, in order for wetting improvement to be obtained with surfactant addition, it is necessary for adsorption to occur at the S1 interface.

Experimentally, it is found that few systems adhere completely to the "pure" behavior described by Equation (17. 47). As pointed out in Chapter 9, many surfactant–solid combinations can show a complex pattern of adsorption, resulting in equally complex patterns of wetting. In any case, with a little thought, the qualitative concepts related to adsorption and wetting can serve as useful guides to understanding the possible role of surfactants in controlling wetting processes.

17.6.2. Surfactants and Wetting on Polar Surfaces

When one extends the preceding discussion to include more polar solid surfaces, several things can be expected to change as a result of the possibility of additional interactions between the aqueous solution and the polar surface. For present purposes, one can include as polar surfaces not only ceramics and minerals, but also polar organic solids. The importance of being able to modify the wetting properties of such solids is seen in many important industrial processes including detergency, petroleum recovery, mineral ore flotation, the wetting of powders and pigments prior to dispersion, and the wetting of stone by road tars.

Inorganic polar solids (e.g., minerals, ceramics, metals, and metal oxides) generally have relatively high surface energies, ranging from a few hundred to several thousand mJ m^{-2} (millijoules per square meter). Organic polar solids, on the other hand, normally have surface energies of 30–50 mJ m^{-2}. As a result, the inorganic materials generally are completely wetted by high-surface-tension liquids such as water, while the organic surfaces are only partially wetted. The situation is complicated by the fact that strongly polar surfaces can undergo specific interactions with polar liquids that can alter their wetting characteristics once liquid–solid contact has taken place. A good example of such a situation is that of the autophobic effect mentioned previously.

When surfactants are added into the equation, wetting can become even more complicated because of the many specific interactions that can occur between surfactant and solid, surfactant and water, and surfactant and oil.

The exact effect of a given surfactant on a system will be determined by the degree and mode (i.e., orientation) of its adsorption at the various interfaces, and the reversibility of that adsorption. Some of the factors that affect those variables include

1. The degree and nature of the polarity of the surface (dipole moment, polarizability, etc.)
2. The presence and nature of surface charges (charge density, degree of ionization, nature of charge-determining ions, etc.)
3. The pH and ionic strength of the aqueous solution, both of which may affect charge density, ionization of the surface, ionization of the surfactant, and other variables.
4. The presence of ions that may specifically interact with the solid surface or surfactant molecules, resulting in the formation of chelates or insoluble salts, such as Al^{3+}, Ca^{2+}, and Cu^{2+}.
5. The degree of surface hydration.

While surfactant adsorption on weakly polar surfaces such as polyesters and polymethylmethacrylate is often sufficiently nonspecific to allow the use of models based on nonpolar solids, interactions with more polar ionic surfaces tend to be more complicated. Even those cases, however, can be successively analyzed in terms of the concepts described above, so that the modification of wetting characteristics by surfactant adsorption can be predicted with reasonable confidence, possibly saving a great deal of time (= money) in various processes.

Of particular interest is the case of surfactant adsorption onto surfaces of opposite charge, in which a complex relationship between surfactant concentration and wetting is often encountered (Fig. 17.13). At low surfactant concen-

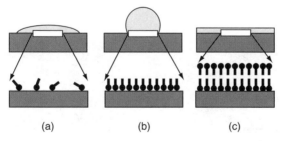

(a)   (b)   (c)

FIGURE 17.13. When an ionic surfactant in water adsorbs on the surface of an oppositely charged solid, the orientation of the molecules will be such that one may observe a contact angle cycle as the adsorption increases. The initial surface will be hydrophilic and have a small contact angle (*a*). As adsorption proceeds, the surface becomes more hydrophobic and the contact angle increases to a maximum (*b*). Further adsorption (if it occurs) will reverse the process leading to a more polar surface and a smaller contact angle.

trations (stage 1), for example, the strong electrostatic attraction between the surface and surfactant will produce a significant level of adsorption at S1 (i.e., a significant increase in σ_{s1} due to the orientation of the adsorbed molecules) while there will be only limited adsorption at the 1–2 interface (i.e., little change in σ_{12}). The reduced attraction between the aqueous solution and solid surface (as reflected in a "new" W°_{aS1}) results in an increase in q and a retraction of the wetting line (e.g., "dewetting").

As the surfactant concentration increases, adsorption at the S1 interface approaches completion while that at the 1–2 interface continues to increase, reducing σ_{12}. Eventually, θ will reach a maximum value, usually corresponding closely to the zero point of charge for the surface (stage 2). As the surfactant concentration continues to increase, θ will often begin to decline as a result of the formation of a second adsorbed monolayer in which the heads of the surfactant molecules are now oriented outward to the aqueous phase (stage 3). This results in a reduction of σ_{S1} and another "new" value for W°_{aS1}. In some surfactant–solid systems, especially for surfactants with long, straight-chain hydrocarbon tails, a cycle of complete wetting to nonwetting ($\theta > 90°$) to complete wetting can be obtained.

From a practical standpoint, the phenomenon just described can have significant impact on various processes. For mineral ore flotation, for example,

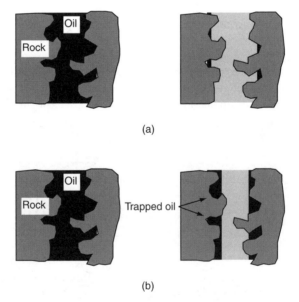

FIGURE 17.14. In petroleum recovery processes an aqueous displacing fluid may be used to force crude oil from a rock formation. Optimum results will usually be expected if the displacing fluid effectively wets the petroleum bearing rock (*a*). If good wetting is not obtained, petroleum may remain trapped in the rock structure never to be recovered (*b*).

it is generally desirable to obtain a large contact angle so that air bubbles will adhere to the ore particles and "float" them to the surface. If the concentration of the surfactant used as the flotation agent is too low or too high, optimum results will not be obtained. Likewise, if the charge and adsorption characteristics of the various minerals present in an ore are properly evaluated, it becomes easier to formulate the surfactant system that will give the maximum separation of ores.

For processes such as petroleum recovery, on the other hand, the maximum degree of wetting is desired so that oil attached to the rock formations can be displaced more efficiently. If poor wetting by the aqueous fluid occurs, significant amounts of oil will be left "stranded" as the plug of displacing fluid passes (Fig. 17.14).

A number of similarly important processes can be found that rely on the proper balance of adsorption and wetting phenomena for optimum operation. While wetting is obviously a complex process, careful evaluation of the examples and guidelines presented above, and the application of a bit of intuition and common sense, can help one arrive at a pretty good analysis of most situations.

PROBLEMS

17.1. An air bubble 2×10^{-6} m in diameter is attached to a hydrophobic surface. What is the expected contact angle, θ, given the following data: $\sigma_{LV} = 72.5$ mN m^{-1}, $\sigma_{LS} = 45$ mN m^{-1}, and $\sigma_{SV} = 22$ mN m^{-1}?

17.2. The surface and interfacial tensions for a series of liquids are given in the table below. From on that information, predict whether octyl alcohol will spread at the water mercury interface. Will hexane? If the alcohol spreads at the water–mercury interface, what molecular orientation do you predict for the alcohol?

Interface	σ (mN m^{-1})	Interface	σ (mN m^{-1})
Water–air	72	Mercury–water	375
Octyl alcohol–air	28	Mercury–octyl alcohol	348
Hexane–air	18	Mercury–hexane	378
Mercury–air	476	Water–octyl alcohol	9
Water–hexane	50		

17.3. An experimenter found the contact angle of water on paraffin wax to be 110° at 25°C. A solution of 0.1 M butylamine, with a surface tension of 56.3 mN cm^{-1}, was found to have $\theta = 92°$. Calculate the spreading pressure of butylamine at the water–paraffin interface. Clarify any assumptions required for the calculation.

17.4. The surface tensions of sodium and mercury at 100°C were measured as 220 and 460 mN cm^{-1}, respectively, and their contact angle on quartz were measures as 66° and 143°, respectively. Calculate a value for the surface tension (energy) of the quartz sample, clarifying all assumptions.

17.5. Given the following information, estimate the contact angle, θ, of gallium on glass and the corresponding adhesion tension. The contact angle, θ_{SLL}, for the system glass–mercury–gallium is approximately (but not exactly) 0° measured through the mercury phase. The interfacial tension of gallium–mercury is 37 mN m^{-1}. The surface tension of gallium is 700 mN m^{-1}. The surface tension of mercury is 484 mN m$_{-1}$. The contact angle of mercury on glass is 140°. All data are 20°C.

17.6. The contact angle is proportional to $(\sigma_{SG} - \sigma_{SL})$, therefore addition of a surfactant that adsorbs at the SL interface should decrease σ_{SL}, increase the quantity in parentheses and reduce θ. However, in flotation systems such addition increases θ. Explain what is incorrect or misleading about the opening statement.,

17.7. Where contact angle hysteresis is present, which do you think is more critical to bubble adhesion in flotation: the advancing or receding angle? Explain.

17.8. Explain qualitatively why a minimum in the entropy of adsorption may be expected at or near monolayer coverage.

17.9. The contact angle for n-decane on Teflon is 35° at 25°C. The film pressure of adsorbed n-decane may be as high as 9 mN cm^{-1}. Calculate the contribution of this film pressure to θ, specifically what would be the value of θ if the film pressure were zero but all other interfacial properties were the same?

17.10. Given the following data and assumptions: (1) the surface tensions of n-decane and n-hexadecane are 24 and 28 mN cm^{-1}, respectively; (2) the surface tension of mixtures of the hydrocarbons is linear in mole fraction, and the mixture behaves as if it were a single hydrocarbon of that average chain length; (3) the critical surface tension for Teflon and for polyethylene are 20 and 30 mN cm^{-1}, respectively, and are linear for all compositions of mixtures of the two; (4) the contact angle for n-hexadecane on Teflon is 46°, and all the Zisman cos °/σ plots are parallel, regardless of the polymer surface composition employed. Calculate (a) the contact angle for 50% n-hexadecane-n-decane on a 50% mixed Teflon–polyethylene polymer; (b) the chain-length hydrocarbon that will give the same angle as in (a), but on a 60% Teflon polymer mixture; (c) the percent of Teflon containing polymer on which *n*-decane will just spread; and (d) if the polymer surface is roughened to an *r* value of 1,1, what should now be the answers to (a) and (b)?

17.11. Explain how contact angle considerations are involved in each of the following observations or phenomena: (a) the rule of thumb that water drains evenly from adequately clean laboratory glassware; (b) the use of paraffin on the edges of a Langmuir trough; (c) the shape of raindrops sliding down a glass window pane (vertical and inclined).

18 Friction, Lubrication, and Wear

One of the most important natural phenomena we must constantly battle in our modern mechanized world Is that of friction and wear between the multitude of moving systems that function to maintain our lifestyles. While most aspects of the three related areas—friction, lubrication, and wear—lie outside the true domain of surface and colloid chemistry, a few of the most important practical aspects of how to attack the problems they produce do fall into that area. For that reason, the subjects have been included here in the form of a brief descriptive discussion. Each of the three fields has developed its own basis of theory and practice along with an extensive literature. The following discussion is intended only as a general introduction, with the goal of trying to integrate those aspects most directly concerned with surface chemistry so that the novice can begin to see how some knowledge of interfacial phenomena may help clarify problems, and hopefully solutions, in practical applications.

18.1. FRICTION

When two solid objects are in contact under a normal load W, a certain finite amount of force will be required to initiate and maintain tangential movement with respect to one another. When at rest, no recoverable energy is stored at the interface between the two, so that when force is applied and work is done, most of that work is dissipated as heat. The force which must be overcome in order to make the two objects move is known generally as friction. In general, one finds that two frictional forces will be involved in such a process: the force necessary to initiate movement or that to overcome static friction, and that necessary to maintain movement or kinetic friction.

The general nature of frictional forces was recognized as early as the time of Leonardo da Vinci (in fact, earlier, but not recorded). Since then they have been rediscovered several times and formulated into "laws" of friction that have served well, even though they are generally found to be limited in their application. The three "laws" of friction, generally known as Amonton's law, can be stated as follows:

1. The frictional force is proportional to the load, W.
2. The force is independent of the geometric area of contact between the two objects.

3. The kinetic frictional force for a system is approximately one-third of the value of the normal load.

The first two statements are generally found to be true over a relatively wide (and useful) range of conditions. The third is much more limited in application but, in the absence of other data, may be found useful.

The most general modern model used to describe frictional phenomena assumes that the friction between two unlubricated surfaces arises from two sources. The first and generally most important is that of adhesion between points of actual contact between the surfaces. We have seen on various occasions that real solid surfaces are almost never smooth. A very smooth surface will normally have asperities of between 5 and 10 nm so that the "true" area of contact between surfaces will be less that the apparent area (Fig. 18.1). At those areas of contact, the two surfaces will be bound by a certain adhesion force arising from the interaction between the materials at the molecular level—the same basic forces we have encountered before plus, in some cases, more physical interactions due to mixing, interpenetration, or "locking." For the two surfaces to move tangentially, the points or areas of adhesion, welds, or junctions must be sheared or broken. If the real area of contact is A and the shear strength of the weld or bond is s, then the frictional force due to adhesion will be

$$F_{ad} = As \qquad (18.1)$$

The adhesional friction, then, can be considered to be truly a "surface" as well as a bulk phenomenon.

A second contribution to the total friction is that due to deformation (F_{def}) and is more of a bulk physical contribution. It arises as a result of deformation,

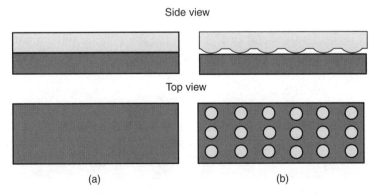

FIGURE 18.1. (*a*) "Ideal" interfacial contact area, A_i; (*b*) real contact area, A_r. In considering friction problems, it is almost always true that the actual area of contact between surfaces will be considerably less than the "ideal" or projected area taken from simple geometric considerations.

cracking, or plowing caused by penetration of the asperities of one surface into that of the other. Without going into detail about the exact processes involved, the deformation process may be summed up as a nonspecific term, P. If it is assumed that the two terms act independently, then the total frictional force for a system, F, may be written as

$$F = F_{ad} + F_{def} = As + P \tag{18.2}$$

It can be seen that, in order to understand friction, the important unknowns include the real area of contact between surfaces, A, the shear strength of the points of contact, s, and the deformation component, P. If the various unknowns do not operate independently, their mutual interrelationships obviously becomes important.

If a normal force W is applied to the system, it may be expected that the added pressure on the asperities will cause some deformation leading to an increase in the real area of contact as illustrated in Figure 18.2. If the materials respond to the added pressure by plastic deformation (i.e., a permanent change in the shape of the asperities brought about by the application of the mechanical force), the real area of contact can be written as

$$A = \frac{W}{\tau_o} \tag{18.3}$$

where τ_o is the yield strength of the weld or bond. For two objects of the same material, the coefficient of friction, μ_f, can be defined as

$$\mu_f = \frac{F}{W} = \frac{s}{\tau_o} \tag{18.4}$$

For metals, it is usually found that the yield strength, p_o, is about five times the shear strength, s, which helps explain the fact that the coefficient of friction between unlubricated metal surfaces is quite often found to be about 0.2.

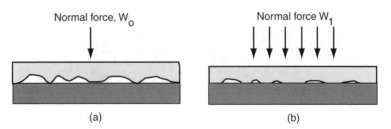

(a)   (b)

FIGURE 18.2. Where the physical characteristics of the materials permit, the application of a normal force to the interface will result in an increase in the real area of contact between surfaces.

FIGURE 18.3. When two solid surfaces of differing hardness make contact, asperities in the harder material may cut into or "plow" the softer surface. Such a process will increase the apparent friction between the two surfaces. The plowing contribution can be estimated using a cone model in which the angle of the plow cut is taken as 2θ and the coefficient of friction calculated according to Equation (18.5).

If a material responds to the applied load by elastic deformation (i.e., temporary, reversible deformation under the load), the contact area will vary as $W^{2/3}$; if response is viscoelastic (i.e., the deformation is reversible, but time-dependent), the exponent of W will be between $\frac{2}{3}$ and 1 and will have a time component so that μ_f will vary with the speed of the tangential movement.

If the two surfaces are of different materials, with different hardness, asperities on the hard surface will tend to plow into the soft surface, forming grooves. The force required for such a process will depend on the cross section of the groove as well as the yield strength of the soft material. The cross section will depend on the geometry of the asperity and the depth it penetrates into the soft surface. If the asperity is assumed to be a cone with an apical angle of 2θ (Fig. 18.3), the plowing contribution to the coefficient of friction, μ_p, will be

$$\mu_p = \left(\frac{2}{\pi}\right) \cot \theta \tag{18.5}$$

The total coefficient of friction will, of course, be a combination of adhesional and plowing contributions. However, where both are found to be important, it is usually difficult to separate the relative roles of each; that is, the two are not necessarily directly additive in their contribution. If the asperities on a surface are known to be relatively blunt, it can usually be assumed that they will not make a significant contribution to μ_f in terms of the plowing mechanism.

18.2. FRICTION AND THE NATURE OF THE SURFACE

When trying to understand friction between two surfaces, it is necessary to know specific details about those surfaces. As stated above, the frictional force between surfaces will depend on the forces of interaction between them (i.e., their chemical nature) and the hardness and strength of each material (their physical nature, so to say). In a general way, one can divide commonly encountered surfaces into four types: metals and their oxides, relatively isotropic crystalline materials, layered crystals, and polymeric (e.g., amorphous) materials. Each type of surface may have its own unique frictional characteristics related to the chemical and physical nature of the relevant interactions.

18.2.1. Metals and Metal Oxides

From a practical standpoint, metal–metal friction is probably the most intensely studied system. In fact, the study of metal friction is in reality the study of two topics—metal friction and metal oxide friction—because except in very special (and seldom very practical) systems, a truly clean metal–metal contact is never encountered. In the "best" of situations, a normal metal surface will be covered with at least a monolayer of adsorbed gas molecules or other contaminants. As a result, frictional forces will be less than would be expected in their absence.

For truly clean metal surfaces that have, for example, been treated by electron bombardment at elevated temperatures and ultrahigh vacuum, coefficients of friction in the range of 3–6 have been recorded. Sometimes, when good contact is achieved between clean surfaces of the same metal, the two surfaces will in fact weld or seize, to the point that the union is as strong as the bulk metal. For dissimilar metals, a similar result may be encountered if there exists a degree of mutual miscibility. If the two metals are mutually immiscible, there may or may not be such seizure.

The behavior of metal surfaces in the presence of air will be quite different from the clean surfaces. When oxygen is present, all but the most noble of the metals will rapidly develop a layer of metal oxide on the surface. The oxide, then, will have frictional properties significantly different from those of the metal. It is found, for example, that a clean copper surface has a coefficient of friction of about 6.8. After exposure to air, the coefficient falls to the range of 0.8.

In the case of friction between oxide layers, several patterns of behavior may be observed, depending primarily on the strength of the oxide and its bonding to the bulk metal surface. At very light loads and/or low sliding velocity, an oxide layer may completely separate the metal surfaces, resulting in a μ_f in the range of 0.6–1. As the load is increased, the oxide layer, if it is relatively weak, will begin to break or detach, allowing for more direct metal–metal contact and a significant increase in μ_f. The exact effect of load, in that case, will depend on various factors. In aluminum, for example, the oxide layer is much harder than the metal. The frictional interaction therefore easily deforms (plastic deformation) the metal supporting the oxide, the latter, which is a thin film with little inherent strength, then breaks under even small loads. Friction, then, is relatively independent of load. Copper, on the other hand, generally shows a dependence of friction on load, indicating that the oxide layer is relatively strong and better matched to the characteristics of the underlying metal.

In practice, then, the friction between unlubricated metal surfaces should probably be considered that between oxide layer or composite surfaces in which there are oxide–oxide, oxide–metal, and metal–metal components. If one considers only two of the three—metal–metal and oxide–oxide friction— one may estimate surface composition from the frictional force using the relationship

$$F = A \left[(\alpha s_m + (1 - \alpha)s_o \right] \tag{18.6}$$

where α is the fraction of metal–metal contact, and s_m and s_o are the respective shear strengths of the metal and the oxide. Clearly, the effect of an oxide layer, assuming that it does not chemically "bridge" the two metal surfaces, may be considered a special instance of boundary lubrication by adsorbed films, which will be discussed further below.

18.2.2. Crystals with Relatively Isotropic Structures

Examples of materials falling into this class include salts such as sodium chloride, diamond, sapphire and other similar minerals, and solid nonmetallic elements such as krypton. The softer members of the class are generally found to obey Amonton's laws with frictional coefficients falling in the range of 0.5–1.0. The harder, more brittle substances such as sodium chloride tend to suffer extensive surface damage due to cracking but still hold more or less to "normal" behavior.

Diamond and sapphire differ in that they have lower than normal μ_f values in the area of 0.1, and its value depends on the load, as might be expected for materials that deform elastically rather than plastically. Such materials also begin to show surface damage beyond a certain load. Under very clean conditions, μ_f for diamond has been found to rise to 0.6, suggesting that some mechanism such as the adsorption of a monomolecular water layer or slight surface oxide formation may act to lubricate the diamond surface naturally.

18.2.3. Anisotropic or Layered Crystalline Materials

Many commonly encountered solid lubricants are in fact highly anisotropic crystalline materials that can form layered structures. These include graphite, molybdenum disulfide, and talc, all of which are known as good lubricants under certain conditions. What might one expect to be the primary mechanism of lubricating action is such systems?

A schematic representation of layered crystalline materials is shown in Figure 18.4. Because the anisotropic structure is the "natural" scheme of things for graphite, MoS_2, and similar compounds, one must assume, based on the discussion of surface energies and crystal faces given in Chapter 7, that

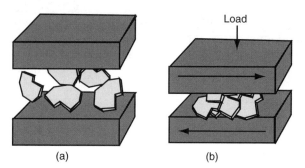

FIGURE 18.4. Certain crystalline structures such as graphite and molybdinum sulfide have flat, disklike structures. Such materials make good lubricants because they can orient themselves and facilitate the sliding of one surface past the other.

the larger faces represent surfaces of lower surface energy. It would seem safe to assume, therefore, that the intermolecular forces involved in the adhesional component of friction along the large, flat surfaces would be significantly less than those at the narrow edges. The result should be a lower shear strength along those faces and a lower coefficient of friction. If it could be accurately measured, one would then expect to find a higher μ_f for edge-to-edge encounters and something intermediate for edge-to-face ones.

For molybdenum disulfide, which has a structure like that of graphite, the intermolecular forces in the crystal are relatively weak, so that the picture described above, along with some contribution by a more complete breakdown of the crystals, seems a reasonable explanation for its low m_f, independent of such factors as the "cleanliness" of the system. Graphite, on the other hand, has a $\mu_f = 0.1$ in air, but in vacuum it rises to about 0.6. That indicates that for graphite, the adsorption of gases at the interface probably plays some role in weakening the adhesional forces between the parallel faces. Other mechanisms may also be involved that have yet to be fully explored, such as the "rollup" of graphite layers in front of the sliding face producing a "ball-bearing effect" on the motion (Fig. 18.5). New forms of carbon, the buck-

FIGURE 18.5. In frictional interactions between a hard and a soft surface, heat, pressure, and/or plowing may cause the softer material to roll up or dislodge, reducing the real area of contact and, in some cases, producing "bearings" that effectively lower the apparent coefficient of friction—at the cost of damage to the softer surface.

minsterfullerenes or bucky ball dodecahedrons (and similar structures), may represent a new class of molecular ball bearings for lubrication in critical situations as illustrated in Figure 18.6.

18.2.4. Polymeric (Amorphous) Materials

In the last few decades, the importance of friction between polymeric materials has rapidly increased. While the coefficients of friction for such materials are usually found to fall in the "normal" range, their behavior as a function of load indicates that the deformations occurring at the points of contact are elastic rather than plastic in nature.

For many polymers it is found that the static friction is significantly greater than the kinetic component, and that value will be very load-dependent. For an interface between two polymers, because of the relative mobility of polymer units at an interface, some small degree of chain interpenetration between surfaces may be expected to occur with time, depending on temperature, load, and the mutual miscibility of the two materials, leading to a relatively strong adhesive joint—strong relative to a "fresh" joint, that is. The result would be the observed high static μ_f. Once sliding is initiated, it has been suggested that the relatively low shear strength of most polymers allows for the detachment of a thin layer of polymer from the sliding surface (Fig. 18.7), which then acts as a boundary layer lubricating film (see that below).

The effect of such a transfer mechanism between two polymer surfaces might be expected to be small. For metal–polymer interfaces, on the other hand, the effect may be significant, since one is now going from a situation of metal–polymer sliding contact to one of polymer–polymer contact, which would be expected to have a much smaller inherent coefficient of friction. Because polymers are relatively soft materials, the questions of plowing contributions and elastic and viscoelastic work loss must be considered.

FIGURE 18.6. The discovery of new forms of carbon, the spherical "bucky balls," opens the possibility of new and more effective lubrication in critical areas. One can imagine the presence of a layer of "bucky ball bearings" lubricating moving surfaces by rolling along with the flow.

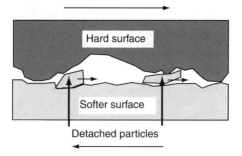

FIGURE 18.7. For some polymers in contact with a harder surface, small sections of the surface material may detach and slide along the surface with the harder material. The apparent coefficient of friction will then be that of polymer against polymer.

Obviously, the question of friction and its various contributing factors can become quite complicated, which explains the large volume of scientific and technical literature on the subject. So far, the discussion has been limited to unlubricated (formally, at least) surfaces. From a surface chemical point of view, the more interesting subject, perhaps, is not friction but how to combat it, or perhaps in a few cases increase it. With that in mind, we now turn to the subject of lubrication.

18.3. LUBRICATION

Functionally, one may define lubrication as the reduction of the friction between two surfaces by providing some mechanism(s) for the reduction of the various chemical (e.g., adhesional) and physical (e.g., plowing) interactions between them. A lubricant may be a solid, plastic, liquid, or gaseous substance entrained between two sliding or rolling surfaces. The materials may be further classified as a wetting lubricants, which interact strongly with one or both surfaces (i.e., adsorb to or "wet" them), leaving any intervening film relatively fluid; or strictly hydrodynamic, in which case there is no specific interaction between lubricant and the sliding surfaces, but the physical presence of the lubricant acts to separate the two surfaces and reduce friction.

Of the solid lubricants, the most common include graphite, molybdenum disulfide, and talc, although as we shall see, spontaneously formed oxide layers also serve that purpose in some cases. "Plastic" lubricants include soaps and fatty acids, and petroleum residues (greases). Liquids include animal, vegetable, and mineral oils, although almost any liquid may serve the purpose in a given situation. Gaseous lubricants would include air or, in principle, any inert gas.

18.3.1. Mechanisms of Lubrication

Mechanisms for the reduction of friction between moving surfaces can be conveniently broken down into four regimes: (1) hydrodynamic, in which an intervening, relatively thick layer of material physically prevents contact between sliding surfaces, thereby reducing friction; (2) elastohydrodynamic, in which the lubricating film has thinned to the point where the properties of the lubricant begin to exhibit the special characteristics of systems only a few molecules thick (i.e., bulk fluid dynamics relationships; e.g., viscosity) no longer hold; (3) boundary layer, in which the lubricating action is a result of the existence of an adsorbed monomolecular film at one or both solid surfaces that reduce the adhesional forces acting at points of contact; and (4) what may be called "chemical lubrication," in which the lubricant effectively weakens or destroys the welds at points of contact by chemical attack. Although from the point of view of surface chemistry boundary lubrication is the most directly applicable, it is of interest to discuss each category briefly in order to better understand their overall significance.

Hydrodynamic Lubrication. Under many important operating conditions, it is possible for moving parts to operate with a relatively thick, continuous film of material separating them, such films being of sufficient thickness that the bulk properties, primarily viscosity, of the lubricating material are maintained. In such conditions, the friction is a result of the work needed to overcome the viscosity of the lubricant, and no contact or wear between surfaces results (in principle, at least). In a flow stream of a lubricating liquid in laminar flow, the velocity of the fluid is a maximum at the point equidistant from the two confining surfaces (Fig. 18.8a). As one moves closer to either surface the velocity of the lubricant molecules decreases until, in theory, those contacting the surfaces have zero velocity. The friction (i.e., resistance to movement),

FIGURE 18.8. A lubricant fluid in lamellar flow will offer minimum resistance to the movement of the opposing solid surfaces (*a*). In turbulent flow (*b*), eddies and voids in the fluid will impede the smooth flow of liquid and may translate that resistance into an increase in the apparent coefficient of friction for the system.

therefore, is between molecules of the lubricant passing one another and does not involve the solid surfaces. In a turbulent flow regime, frictional forces would increase because of the presence of eddies and voids that would interfere with the smooth passage of adjacent molecular layers.

While we normally think of such lubricating systems as containing some liquid material, in fact solids and gases can also act in that capacity. A simple example of hydrodynamic lubrication would be that pictured in Figure 18.9, in which an asymmetric cam rotates in a bearing. The lubricant forms a film between the cam and the bearing, but as the cam rotates, the thickness of the film varies from point to point due to the eccentricity of the cam movement. In the regions of closest approach between cam and bearing surfaces, the thickness of the lubricating film is reduced and the pressure between the surfaces (the frictional load) increases. The friction is found to be a linear function of the viscosity of the lubricant, h; the number of revolutions sc^{-1} (reciprocal seconds), ω; and the nominal pressure between the surfaces, P:

$$\mu_f = f\left(\frac{\eta\omega}{P}\right) \tag{18.7}$$

For hydrodynamic lubrication, it is obviously best to maintain a relatively thick film of lubricant between the surfaces. In order to maintain such a situation, it is necessary to keep the lubricant viscosity and relative speed of movement as high as possible, and the load as small as possible. The problems of load and speed are usually variables set by the operation of the device in question, so that their control is somewhat limited by use. That leaves viscosity as the primary handle on the friction problem.

The optimum viscosity for a lubricant will be determined by the configuration and running conditions of the device. However, as movement occurs and heat is generated, there is a tendency for the viscosity of most liquids to decrease, which could be dangerous in most systems. The temperature dependence of the viscosity of a liquid is given by a relationship such as

FIGURE 18.9. In hydrodynamic lubrication it is usually convenient to maintain the thickest lubricating film possible (and practical) between the moving parts in order to take maximum advantage of the relationship between pressure and viscosity given in Equation (18.7).

$$\eta = \frac{\eta_0 e^Q}{RT} \tag{18.8}$$

where Q is the activation energy for the viscous process. In order to counteract the natural tendency of viscosity to decrease with increases in temperature, critical lubricants usually contain materials termed viscosity "modifiers" or "improvers." Such materials are normally polymeric species that, under normal conditions are tightly coiled, thereby adding little to the viscosity of the system. As heating due to friction or other causes occurs the tight coils begin to expand (i.e., the polymer becomes more "soluble"), thereby increasing, or in this case, maintaining, the viscosity of the system (Fig. 18.10). Such is the case as long as the temperature does not reach a level at which chemical degradation of the system or some components begins.

In some situations it is found convenient to use a gas rather than a liquid as a lubricant. Gaseous lubricants are convenient because they are cheap (air in an "air bearing," e.g.,), their viscosity naturally increases with temperature, so that additives are not required, and there is little chance of chemical degradation. Such systems are limited, however, by the fact that gas lubricants cannot support large load factors. In addition, for a gas lubricating system to function properly, the volume between the moving surfaces must be relatively

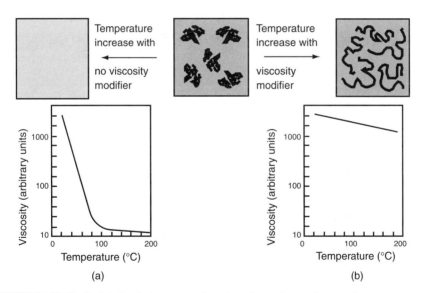

FIGURE 18.10. In lubricating systems that experience large changes in temperature, viscosity control is important. A lubricant that has no viscosity additive (*a*) will usually experience a rapid viscosity loss (illustrated by lighter shading) with increased temperature. The addition of a viscosity modifier (usually a polymer) will help maintain the viscosity by the expansion of the polymer chains as temperature increases (*b*).

small, calling for a very fine surface finish (i.e., small asperities) and very good alignment of the parts.

Elastohydrodynamic Lubrication. With liquid lubricants, as the thickness of the lubricant film decreases, one may reach the point at which the properties of the lubricant are no longer those of the bulk material, but rather those of a special film only a few molecules thick, which is penetrated by large surface asperities leading to excessive wear. Such a situation does not correspond to normal hydrodynamic lubrication, but neither does it indicate the onset of classic boundary lubrication (see discussion below). This intermediate regime is referred to as "elastohydrodynamic lubrication."

In a system in which at least one of the surfaces in question is relatively soft, such as a polymer, when the lubricating film thickness becomes very small, the pressure being applied to the lubricant film can become quite large, leading to elastic deformation of the softer surface. For a system of a smooth rubber roller on a glass surface, one can describe the situation as shown schematically in Figure 18.11. Under light loading conditions, the lubricating

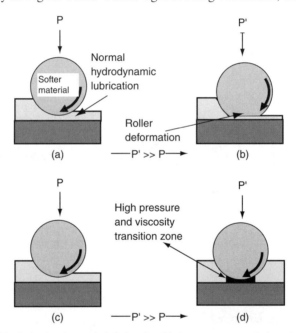

FIGURE 18.11. In hydrodynamic lubrication high pressures can bring about significant changes in the viscosity of the lubricant. In the case of a relatively soft material moving against a harder one (*a*), the pressure may lead to a deformation of the softer element (*b*), spreading the pressure over a wider area and reducing its impact on the characteristics of the lubricating fluid. For two hard materials (*c*), deformation is not possible so that the total load is applied to the fluid causing a significant change in viscosity, often to the point of producing a waxy or solid consistency, with the accompanying change in friction coefficient.

system operates according to normal hydrodynamic rules (Fig. 18.11a). As the load increases, the film thickness decreases to the point where the rubber deforms or flattens, giving a configuration with greater load carrying capacity (Fig. 18.11b). If the load is removed the pressure drops suddenly and the rubber elastically rebounds to its original condition.

For two hard surfaces, the situation is somewhat different. In that case, elastic deformation as in the rubber example is no longer a significant factor. However, the pressure increase on the lubricant is still present. In a finely machined and aligned system, the pressures on the lubricant film may be on the order of 700–7000 kg cm^{-2}. The approximate relationship between the viscosity of a liquid and pressure is

$$\eta = \eta_0 e^{\beta P} \tag{18.9}$$

where the constant β for most mineral oils is on the order of 2.8×10^{-3} cm^2 kg^{-1}. For the pressure range indicated above, the viscosity changes to be expected for the lubricant are something like the following:

$$
\begin{aligned}
P = 700 \quad &\text{kg cm}^{-1} \Rightarrow \eta \approx 7 \; \eta_0 \\
= 1000 \; &\text{kg cm}^{-1} \Rightarrow \eta \approx 16 \; \eta_0 \\
= 3000 \; &\text{kg cm}^{-1} \Rightarrow \eta \approx 4450 \; \eta_0 \\
= 5000 \; &\text{kg cm}^{-1} \Rightarrow \eta \approx 1.2 \times 10^6 \; \eta_0
\end{aligned}
$$

Obviously, at very high contact pressures, the lubricating liquid between the two surfaces rapidly increases in viscosity until it must attain the consistency of a solid or wax rather than a liquid. In such a case, it is easy to see why some lubricating oils that exhibit such thickening behavior show better performance than would be predicted for classic hydrodynamic theories. It also helps explain why other materials (e.g., silicone oils), which have less dramatic viscosity increases with pressure, do not perform as well under extreme conditions. In the viscosity range where elastohydrodynamic lubrication occurs, fluids may begin to exhibit non-Newtonian behavior leading to a more complicated relationship in terms of lubricant effectiveness.

When one considers this mechanism of elastohydrodynamic lubrication, the question may arise, as to why one should ever see the failure of a lubricant. While the answer has yet to be determined definitively, two possible explanations are (1) even at the extremely high pressures involved, the extreme local temperatures that also exist surpass the "critical" point so that the lubricant is effectively evaporated away; or (2) shear forces near points of contact are sufficient to effectively break up the "solidified" lubricant film, leaving bare spots that can have direct contact, leading to excessive wear and possible seizure.

If a lubricant contains a small amount of a surface-active material, that is, one that has a specific adsorptive interaction with the surfaces, it is often found that the conditions under which a system may operate without excessive wear or failure may be extended significantly. While it can be shown that the added material has no significant effect on the bulk or viscous properties of the lubricant, its presence provides an important degree of added protection not anticipated by the models of hydrodynamic or elastohydrodynamic lubrication. From the available evidence, it can be concluded that the added surface-active material must form a thin, probably monomolecular, film at each surface that does not affect the viscosity of the lubricant, yet affords lubrication protection when the elastohydrodynamic film breaks down. Such action leads to another of the important lubrication mechanisms, that most directly related to surface chemistry: boundary lubrication.

Boundary Lubrication. The addition of small amounts of materials that adsorb specifically at the moving surfaces affords an added protection against excessive friction due to the presence of the adsorbed monomolecular film. In other words, the lubricating action derives not from the bulk viscous properties of the lubricant, but from the specific interactions between adsorbed films. In boundary lubrication, materials that are most effective are those that have relatively long hydrocarbon tails and strong specific interactions with the surfaces, especially those containing such groups as hydroxyl ($-OH$), amino ($-NH_2$), or carboxyl ($-COOH$). Groups containing phosphorous (e.g., $-PO_4$) and sulfur ($-SH$) are also found useful.

While boundary lubrication is usually encountered as the third in a series of lubricating mechanisms under increasingly harsh conditions, it can also be employed as a "standalone" mechanism by the direct application of a monomolecular film to a surface in situations in which hydrodynamic or elasto-hydrodynamic processes are impractical or impossible. An example would be in the lubrication of magnetic tapes that must continuously pass over metal surfaces (recording and playback heads, guide posts, etc.) but that cannot suffer much wear or abrasion without rapid loss of signal quality. Effective lubricating action can be attained in such a case by including in the oxide coating a surfactant that produces the required low coefficient of friction while being strongly adsorbed to the coating surface or directly to the oxide. Some polymeric materials, especially those containing silicones and fluorinated hydrocarbons, also serve well in that situation.

Because an adsorbed monomolecular film will have a thickness on the order of 2.5 nm, while the surface asperities present on all but the finest surfaces will seldom be less than 5–10 nm, it is important to have a clear picture of the mechanism of boundary lubrication at the molecular level. A "typical" situation is shown schematically in Figure 18.12, where it can be seen that there are two types of contact between the two surfaces in the total contact area A: contact between the adsorbed lubricant films (area αA the figure) and that between the actual surfaces where the adsorbed film has broken down (area βA). The total frictional force between the two will be the sum of each contribution

FIGURE 18.12. In "perfect" boundary lubrication (*a*) contact is exclusively between adsorbed layers so that the coefficient of friction corresponds to those materials. In reality, the natural irregularities in even the most polished surfaces will produce areas of direct contact between surfaces (*b*). The apparent coefficient of friction, therefore, will be some average value based on the relative areas of contact αA and βA.

$$F = \alpha A s_l + \beta A s_s = \alpha A s_l + (1 - \alpha) A s_s \qquad (18.10)$$

where subscripts l and s refer to the lubricant and bare surface, respectively. Obviously, the greater the fraction of contact between adsorbed films (αA), the lower will be the friction.

The situation may be somewhat complicated when one surface is lubricated and the other is not. In that case, one often finds that with each pass of the unlubricated surface, some amount of lubricant film is transferred to the bare surface, leaving bare spots on the lubricated surface. With excessive transfer, the effectiveness of the lubrication will clearly decline. The rate of such transfer will depend on several factors, the most important of which are the applied load and the strength of adsorption of the lubricating film. The load factor is, of course, an operating parameter that lies outside the realm of surface chemistry; it is the adsorption process that is of most direct significance for our present purposes.

The are two general mechanisms for the adsorption of a monomolecular film, as we have seen from earlier chapters: physical adsorption and chemisorption. The simplest type of adsorption is the physical adsorption of materials such as hydrocarbons in which the only attractive factor is dispersion force interactions. If the solid surface is relatively hydrophobic and the hydrocarbon chain is relatively long, such adsorption can be quite strong. However, the adsorbed molecules will generally lie more or less horizontal along the surface (Fig. 18.13a), which means that the thickness of the monolayer will be very small, about 0.25 nm. As a result, the effectiveness of such films will be limited and will decrease rapidly with repeated passing of the surfaces.

A more effective physically adsorbed film would involve interactions such as dipolar or hydrogen bonding interactions. Such would be the case for lubricating films of alcohols, amines, fatty acids, and fatty acid soaps, excluding the direct formation of metal soaps at the surface, which would fall under the

(a)   (b)

FIGURE 18.13. Adsorption for boundary layer lubrication can occur through either physical adsorption (*a*) or chemisorption (*b*). While physical adsorption is rapid and essentially universal, it is relatively weak and may produce thin, easily disrupted films. Chemisorption, on the other hand, is usually much stronger and may produce more structured and resistant films.

category of chemisorption, discussed below. Because dipolar and hydrogen bonding interactions are more directional than dispersion forces, the adsorbed monolayer will tend to be oriented with respect to the surface, as illustrated in Figure 18.13*b*. In this case, the adsorption is stronger and the resulting film is much thicker, approximately 0.4–0.6 nm, depending on the chain length, resulting in a significant increase in film effectiveness and durability. Comparative results for various systems are given in Table 18.1.

Clearly, the nature of the adsorbing species significantly affects its ability to function as an effective boundary lubricant. In general, one can rely on the following rules of thumb for predicting the efficacy of various potential materials:

1. For optimum effectiveness the film should be in the close-packed, condensed state (see Chapter 8). It is generally found that as lubricant

TABLE 18.1. Effectiveness of Physically Adsorbed Films as Boundary Layer Lubricants under a Normal Load of 100 g

System	Lubricant	μ_f (Sliding Speed, cm sec^{-1})
Polyethyleneterephthalate	None	0.39 (0.001)
"	Hexadecane	0.28 "
"	Oleic acid	0.25 "
"	Stearic acid	0.14 "
Steel on glass	Octanoic acid	0.18 (0.01)
"	Decanoic acid	0.13 "
"	Dodecanoic acid	0.09 "
"	Tridecanoic acid	0.06 "
"	Hexadecanoic acid	0.06 "
"	Octadecanoic acid	0.05 "
Cadmium on cadmium (load = 2 kg)	None	0.8 (0.1)
"	Cetane	0.06 "
"	Cetyl alcohol	0.4 "
"	Palmitic acid	0.07 "
"	Copper palmitate	0.05 "

transfer occurs with repeated movement (i.e., a reduction in the density of the packed film), or as the temperature of the system increases (i.e., the condensed film "melts" to become "liquid condensed") the coefficient of friction increases.

2. The coefficient of friction decreases as the chain length of the hydrocarbon (or fluorocarbon) tail increases up to about 14 carbons, after which little change is observed.

3. Although the initial coefficient will be about the same for a given lubricant on various surfaces, its "durability" will vary with the strength of the adsorption; stronger adsorption results in greater durability.

4. Related to 3, while the initial coefficients for various lubricants of the same chain length will be similar regardless of the nature of the polar head group, durability will usually increase in the order $-NH_2 < -OH < -COOH$, although some reversal of the amine–alcohol relationship may be seen in going from, say, a glass surface to a metal, depending on the reactivity of the metal.

Chemisorbed films will generally follow these rules, except that their interactions with the surface are generally stronger, resulting in the production of more durable films. The dividing line between physical adsorption and chemisorption is, in some cases, quite blurred, especially where acid groups such as carboxylates and phosphates are concerned. In cases in which salt or compound formation between surface and lubricant is possible, particularly effective boundary lubrication can result. For example, if a metal surface is basic, in the sense that it can react with a carboxylic acid to form a metal carboxylate, the resulting film will be especially durable. The same may be said about surfaces that are acidic and therefore react with amines or other basic groups.

Examples of films that overlap with the "chemical" lubrication, to be discussed below, are encountered in systems in which the bare metal is especially reactive. For example, as an asperity is worn away by friction, it may expose an area of bare metal that, in the absence of other alternatives such as oxidation by air or reaction with polar groups as discussed above, may react with other organic functionalities present, including especially elements of unsaturation (i.e., double and triple bonds, carbonyl groups). It is found, for example, that aluminum surfaces that undergo sufficient wear to produce significant areas of "clean" metal surface are lubricated much better by unsaturated hydrocarbons than by the saturated analogs. Cetene ($C_{15}H_{30}=CH_2$), for example, is found to be more effective than cetane ($C_{15}H_{31}-CH_3$). While the details of such reactions are not completely clear, there is some evidence that the clean metal surface acts as a polymerization catalyst, producing a more viscous, strongly adsorbed boundary layer film.

Similar results have been reported under conditions in which the hydrocarbon may be "cracked," producing free radicals, which then react to form polymers or react with oxygen to produce peroxides, hydroxyls, carboxylates, and other compounds, which consequently improve the efficacy of the system. In addition, it has been observed that boundary lubricants such as carboxylic acids will sometimes be much more efficient and durable lubricants for metal surfaces when the surface has a covering of at least a monomolecular layer of metal oxide. If the oxide is removed, lubrication effectiveness may fall significantly.

Obviously, under extreme conditions, any number of chemical transformations may occur in the lubricant leading to alterations in its lubricating ability. The chemistry involved is complex and difficult to study, but its existence must always be kept in mind. Other chemistries can also be very important under some circumstances, which leads to the fourth lubricating mechanism, what will be referred to as chemical lubrication.

Chemical Lubrication under Extreme Loads. So far the lubrication mechanisms discussed have involved ever thinner layers of lubricating liquid. The obvious limit to that progression is the complete absence of an external lubricant. When devices operate under extreme conditions of load, speed, temperature, and other parameters, conventional lubricants will usually begin to break down and drastic mechanisms must be employed to prevent complete seizure and failure of the machinery. One way to approach that problem has been the development of "sacrificial" lubricants, which, under extreme conditions, react with fresh metal surfaces formed by wear to produce a new inorganic chemical layer that can then be more easily sheared, thereby preventing seizure.

The conventional lubricants in this class of materials are compounds containing reactive chlorine, sulfur, or phosphorus groups, which react to produce inorganic metal chlorides, sulfides, or phosphates. Such compounds are generally included as additives in more conventional lubricants and are therefore passive until extreme conditions are reached at which time they begin to "do their thing."

The chemistry of the sacrificial lubricants can be quite complex, especially if oxygen is present. For example, systems that nominally produce metal sulfide layers may produce mixed sulfide–oxide layers with oxygen present. That may or may not be detrimental to the operating device but must be considered. Chloride-forming chemical lubricants were once popular because the resulting surface layers generally have very low shear strength, leading to better lubricating properties. However, under certain conditions of temperature and in the presence of moisture, they were found to decompose to produce hydrochloric acid, which is corrosive and certainly undesirable. Such materials have lost much of their popularity for that reason.

Because the chemical lubricants are generally included as additives in more conventional systems, certain precautions must be taken with respect to the overall composition of the mixture. Obviously, there must be no significant chemical reaction between the various components of the lubricant as formulated. In addition, one must be certain that the primary lubricant in the mixture will not dissolve or otherwise attack the chemically formed inorganic layer. If it does, both the chemical lubricant and the surface being lubricated will be rapidly depleted, leading to excessive wear of the surfaces.

While chemical lubrication is a viable mechanism for the lubrication of machinery that must operate under extreme conditions, its beneficial effects are obtained at the price of significant loss of material from the lubricated surfaces—that is, one pays for the operation of the device in terms of increased wear to the lubricated surfaces.

18.3.2. Some Final Comments on Lubrication

Before closing the book on lubrication, there are a few additional points of interest that should be mentioned briefly. Two potentially important ones from a practical standpoint are the so-called Rehbinder effect and weeping lubrication . The Rehbinder effect relates to the effects of the adsorption on the mechanical strength of materials. While there exists some uncertainty on the matter, there is significant evidence that the adsorption of surfactants or other materials onto surfaces, especially in cracks and surface flaws, can reduce the mechanical strength of the material.

It has been stated several times in other chapters that the driving force for the adsorption of surfactants is the reduction of surface energy. In solids, areas of surface flaws and cracks usually represent areas of higher-than-normal surface energies, implying that adsorption should occur in those areas more rapidly than in the more "normal" areas of the surface. If the phenomenon occurs as described, it can have a number of practical consequences—some bad, some good. On the detrimental side, the weakening of a component by an adsorbed lubricant film could obviously lead to device failure. On the positive side, however, the effect may play an important role in several processes related to lubrication. For example, the shear strength of points of contact between surfaces or oxide or other inorganic layers may be reduced, making movement of the surfaces easier and thereby reducing friction. In lubricants for metal cutting operations, the adsorbed lubricant film may weaken the metal being worked sufficiently to reduce mechanical wear of the cutting surface. And finally, in drilling operations, especially in oil exploration, the Rehbinder effect may aid significantly in reducing wear on expensive drill bits by weakening the rock structure being penetrated.

The second potentially important point to be mentioned is that of weeping lubrication, in which a lubricant is trapped inside a porous material, in surface cracks and flaws, or absorbed by a polymeric material, to be slowly extruded as needed when subjected to pressures and temperatures as a result of friction (Fig. 18.14). Such a "weeping" action can provide a mechanism for the continuous renewal of a lubricating layer under conditions in which confinement of a normal lubricant would be impractical or impossible.

Weeping lubrication has been suggested as one of the possible mechanisms by which body joints are lubricated by the synovial fluid. In that case, the

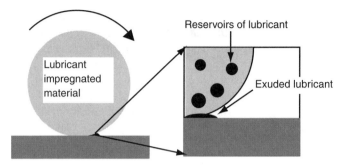

FIGURE 18.14. For some porous or absorbent materials (i.e., rubber rollers), it is possible to incorporate (impregnate) lubricant into the material so that it can be applied continuously to the interacting surfaces by pressure release, temperature changes, or simply by diffusion. Such a process may be described as "weeping" lubrication.

fluid would be trapped within the bone or cartilage structure and released "on demand." If the supply of lubricant is cut off or the porous structure of the bone and cartilage changed by some medical condition, lubrication would be lost, leading to obvious problems.

A second application of weeping lubrication is seen in the working of certain metal surfaces. If a metal is lightly abraded and lubricated prior to working, the lubricant can penetrate the flaws and cracks and be held in reserve for use during subsequent processing. For example, metals so treated can often be subjected to significant deformation by rolling and drawing without the formation of gross surface defects or scuffing. The lubricant stored in the initially abraded surface is squeezed out during processing, providing continuously renewed lubrication and preventing excessive metal–metal contact.

Lubrication is clearly an important practical topic that is continually changing and improving to meet the new demands of technology. The above discussion is obviously superficial (pardon the pun) but does introduce many of the more fundamental aspects of related problems and perhaps some solutions. While there exists a great deal of uncertainty and some disagreement on details, there are some general conclusions that can be drawn and that seem to have broad applicability.

For most conditions, the best effect is obtained when there exists a thick film of fluid between the moving surfaces affording efficient aerodynamic, hydrodynamic, or elastohydrodynamic lubrication. If the lubricating film is thinned or broken by operating conditions or system failure, additional protection is afforded by adsorbed films through boundary lubrication. Finally, under extreme conditions, protection against seizure and complete failure may be obtained as a result of chemical processes that produce "weak" oxide, sulfide, or phosphate, surface layers that can be more easily sheared that direct metal–metal contacts. For hydrodynamic and elastohydrodynamic lubrication, careful attention to the viscous properties of the lubricant, including the addition of viscosity improvers or modifiers, can significantly extend the practical operating range for many liquids.

In the case of boundary layer lubrication, in which the adsorption of monomolecular films is required, the best protection is provided by materials such as fatty acids and soaps that can adsorb strongly at the surface to form a solid condensed film. Less durable but effective protection can be obtained with polar groups such as alcohols, thiols, or amines. The least effective protection is obtained with simple hydrocarbons that adsorb more or less randomly and through dispersion forces alone. For adsorbed monomolecular films, best results are obtained when the hydrocarbon tail has at least 14 carbons. In some cases fluorinated carboxylic acids and silicones may provide a lower initial coefficient of friction, but their weaker lateral interaction sometimes results in a less durable surface film that "melts" at a lower temperature, ultimately resulting in less overall protection. If a polar lubricant can form a direct chemical bond to the surface, as in the formation of metal soaps, even better results can be expected.

At elevated temperatures and pressures, even strongly adsorbed films may be desorbed, leading to excessive surface contact. In such cases, some form of chemical lubrication must be used, because simple organic liquids may begin to degrade, oxidize, or polymerize, rapidly losing their lubricating properties. Obviously, the optimum system where extreme conditions may be expected is one in which all the bases are covered: a good hydrodynamic lubricant with viscosity modifiers, long-chain fatty acids for adsorption and boundary layer protection, and extreme pressure sacrificial additives as a last resort. Closing the subject of lubrication leads directly to the consequences of friction or the lack of lubrication: wear.

18.4. WEAR

Wear between moving surfaces is not directly a surface chemical problem. However, if the mechanisms of friction described above, namely, adhesion between points of contact followed by shearing and material transfer between surfaces, are correct, it is a direct consequence of friction, which is a surface problem. It is therefore of interest to include a few comments on the subject to close this chapter.

Unlike many practical problems, wear seems to be a complex phenomenon that has not lent itself to the formulation of useful generalizations or the generation of "laws of wear." The safest things one can say about the subject is that wear increases with time of operation, with severity of operating conditions (e.g., load, speed, temperature), and appears to be more severe with soft and brittle materials than with hard surfaces. Unfortunately, there are many exceptions to those three generalizations. In addition, the existence of chemical wear, the Rehbinder effect, and other factors complicate matters since surfaces that suffer less physical wear, may suffer more from chemical attack.

18.4.1. Physical Wear

Physical wear may be loosely defined as the removal of surface material due to adhesion–shear cycles. The process may be divided into three general regimes: light, medium or mild, and severe wear, depending on the extent of material loss or transfer and the general effect of the process on the nature of the surface produced (Fig. 18.15).

Light wear can conveniently be termed the "incidental transfer" of small amounts of material from one surface to another, or to the lubricating fluid, as a result of contact between large asperities. Light wear is an inevitable consequence of the nature of the solid state and results, in some cases, in a beneficial effect on the overall operation of a device. The "breaking in" of new machinery, for example, is partly a process of allowing the new surfaces of gears, bearings, and other components to wear off large asperities to produce a smoother surface and better "fit" between moving parts.

Medium or mild wear is usually assumed to involve the interaction of thin films, primarily oxides for metal parts, on the sliding surfaces. In this case, it is generally found that the oxide has a lower shear strength than the metal and its transfer occurs more easily (e.g., at lighter loads) than would be the

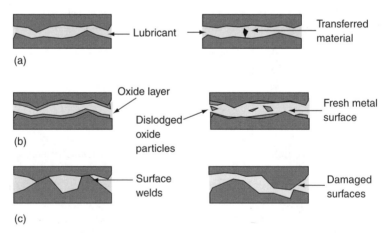

FIGURE 18.15. Wear can be loosely divided into three categories according to the amount of material lost or transferred and the probable effects of the process: (*a*) light wear implies little exchange or loss of material and little or no adverse effects on the system; (*b*) mild wear may involve significantly more exchange or loss of material, but with little observable effect on the system in the short term; (*c*) severe wear implies a major loss or exchange of surface material causing a significant change in or failure of the system involved.

case for direct metal–metal contact. Even when metal contact does occur under conditions of mild wear, those areas of contact will be small and will generally produce a reactive fresh metal surface that will rapidly oxidize, effectively returning the system to a condition of oxide contact. As a result, the material transferred will be primarily oxide, which, because of its relatively soft, brittle nature, will cause few major problems due to plowing and other abrasive action.

Severe wear is obviously the condition of most practical importance. It is normally associated with conditions of adhesion and subsequent shearing of points of contact between primary bulk materials—metals in most cases. As has been stated previously, it is observed that junctions or welds between metal surfaces often exhibit the same shear strength as the bulk material. In fact, in welds between the same materials adhesion will be strong and shear strength may be greater than that of the parent materials as a result of work hardening. In such a case, one can expect cleavage to occur at some other point in one of the two surfaces away from the original area of contact. As a result of random transfer from both of the surfaces, roughening will be relatively more severe and likely to produce operational problems more rapidly. As a result, it is generally not good practice to have critical machined parts made from the same material.

If the shear plane is at the original area of contact, little transfer of material will occur and the effect on the topography of the surfaces will be negligible. For the case for surfaces of two different materials, shearing usually occurs at a new location within the bulk of the softer of the two. In that way, unlike the case of the same material, one surface is attacked preferentially, somewhat simplifying the transfer situation.

When material transfer occurs during movement, its effect on the surfaces, the obvious wear involved, may not correspond directly to the actual amount of material transferred. For example, if material is transferred and adheres to the receiving surface, the net result may not alter significantly the topography of the surface and therefore not appear as wear. In fact, for some situations, repeated contact between surfaces may result in the passing back and forth (back transfer) of material, obscuring the true rate of wear. It is only when the transferred materials become detached that general wear becomes readily obvious. The net rate of wear, then, will depend not only on the absolute rate of material transfer but also on the degree of back transfer and complete detachment.

The relationship between the observed coefficient of friction, μ_f, and the production of wear fragments for a given system is quite complex and not subject to easy interpretation. Typical results are given in Table 18.2 for various materials interacting under a standard set of conditions. From the table one can see that although the coefficients of friction vary by a factor of 2–3, the wear rate, as defined by the fraction of contacts that result in the production of wear fragments, k, varies by a factor of 10^5. Interestingly, the system with the highest coefficient of friction, polyethylene on hard steel,

TABLE 18.2. Coefficients of Friction μ_f, and Wear Rates, k, for Representative Materials Sliding at 180 cm sec^{-1} under a Normal Load of 400 g

Interacting Surfaces	μ_f	k
Mild steel on mild steel	0.6	10^{-2}
60/40 leaded brass on hard steel	0.24	10^{-3}
Polytetrafluoroethylene on hard steel	0.18	2×10^{-5}
Stainless steel on hard steel	0.5	2×10^{-5}
Polyethylene on hard steel	0.65	10^{-7}

showed the lowest wear rate, while brass on hard steel, with μ_f of about one-third, exhibited a wear rate four orders of magnitude higher.

The reasons for such seemingly illogical results must lie in the different responses of the materials involved to the pressures and shearing forces acting in the wear process, and the mechanism by which detachment occurs. For hard materials such as steel, deformation at points of contact is expected to be plastic. With repeated passage of an area through loading and unloading cycles, it is reasonable to expect strain to be built up in the areas around and beneath the contact points, leading at some point to the situation in which surface forces (adhesion) are sufficient to break loose the fragment, or the strain will induce material fatigue leading to detachment. Brass, of course, is a rather soft, ductile material that can be gouged easily by the harder steel, with fragments being plucked off relatively frequently.

For polymers, deformation will be primarily elastic, so that strain and fatigue will be less of an immediate problem. In addition, any strains induced by contact may be annealed away by localized heating, due to the relatively low glass transition temperatures of polymeric materials. The greater wear rate of fluorinated polymers is undoubtedly related to their inherently lower cohesive strength.

18.4.2. Abrasive Wear

Once wear fragments detach from sliding surfaces, they become free to move about within the lubricating film and inflict additional damage on the surfaces by plowing and other abrasive mechanisms (Fig. 18.16). In order for significant abrasion to occur, the detached particles must be harder than the surfaces involved. That is not to say, however, that softer particles will not cause abrasive wear—it will simply be at a slower rate and often of a different nature. One can often determine the nature of abrasive particles by careful examination of the abraded surface. Spherical particles, for example, charac-teristically produce relatively smooth grooves in soft surfaces or flaking in harder materials. Sharp, irregular particles, on the other hand, will cut the surface in a more or less random pattern and generally have a higher wear rate.

(a)   (b)   (c)

FIGURE 18.16. Abrasive wear is essentially the damaging of the moving surfaces by particles deposited in the lubricant during mild or severe wear processes, or by direct surface contact due to lubricant failure. Especially during severe wear, relatively large particles may break off and abrade either or both surfaces by the plowing mechanism discussed earlier. Such abrasion will probably increase with time and may lead to system failure.

In some materials, abrasion can, over time, induce physical changes in the natures of the moving surfaces. For example, abrasion may induce changes in the structure and orientation of surface layers, or changes in crystallization, resulting in a significant alteration in the physical properties of the surface. Work hardening by abrasion, in fact, may increase resistance to further abrasion in hard materials.

The most effective way to prevent abrasive wear in many cases is to provide a harder surface for the moving parts. That can often be accomplished by some plasma deposition process or heat treatment, for example. The danger in such a process, however, is that should a fragment of the harder material become detached, it can inflict much greater damage on the surfaces than would be found for a fragment of the underlying bulk material. In some cases increasing hardness may be found to be ineffective. Often, a better approach would be to increase the elasticity of the surface so that strain and fatigue is reduced, thereby reducing the tendency for the detachment of fragments. Such an approach might be especially useful in the chemical industry, where mechanical abrasion may be accompanied by chemical attack on the surfaces. The use of an elastic, chemically resistant polymeric coating is indicated in such cases.

18.4.3. Chemical Wear

We have seen that chemical attack on sliding surfaces can be beneficial, producing distinct surface layers that serve as extreme pressure lubricants. However, under normal operating conditions, chemical reaction, especially oxide formation, is generally found to be undesirable. While oxide films may be harder than the underlying metal, they are also sometimes more brittle and become easily detached. As wear fragments, the oxides can inflict significant abrasive damage on a surface, leading to early failure. Likewise, while the use of chlorine- or sulfide-containing extreme pressure lubricants may be justified under extreme operating conditions, care must be exercised to ensure that the additive is not too reactive, since such a condition can lead to rapid corrosion of the surfaces and failure just as rapidly as the processes it was introduced to retard.

In the case of polymeric surfaces, chemical wear or degradation is a more complex problem because it involves (or may involve) a wide variety of bond-breaking and bond-making processes. Polymers, for example, may undergo oxidative degradation to produce polar surface groups ($-OH$, $-COOH$, etc.), which may increase adhesive forces at points of contact, thereby increasing friction and wear. Under repeated strain cycles, polymers may also undergo chain rupture to produce free radicals that can react further to "depolymerize" the surface, reducing shear strength and increasing wear; or lead to crosslinking, which may reduce the elasticity of the surface and, again, increase wear. Obviously, the inclusion of additives in the polymer (or the lubricant) that retard such degradation would probably be beneficial.

While the individual mechanisms of wear are fairly well understood, it is a fact of life that one mechanism seldom operates independently, and the combination of two or more will generally lead to a situation that is difficult or impossible to interpret unequivocally. While one mechanism may be entirely detrimental to the operation of a device, another may lead to an improved resistance to wear by work hardening or some other process. Processes not mentioned at all, such as alloying of the surfaces, may be important in some cases. Finally, small changes in operating conditions (load, speed, temperature, etc.) may lead to significant alteration of the predominant mechanism of wear, calling for changes in lubricant formulation.

Hopefully, even in the face of all that uncertainty, the information provided can assist the novice in the field to understand friction, lubrication, and wear processes, and apply these and other ideas to the solution of individual practical problems.

PROBLEMS

18.1. In a standard procedure for the investigation of car accidents, it is assumed that the initial velocity, v_i, of an automobile in a continuous skid can be calculated with the formula

$$v_i = (2d\mu g)^{1/2}$$

where d is the distance of the skid, μ the coefficient of friction between tires and road surface, and $g = 9.80$ m s^{-2}. Beginning with Amonton's law, derive this equation and indicate practical circumstances that may limit the accuracy of its application.

18.2. In an unusual freeway accident, a ski plane was forced to land on a snow-covered roadway and skidded into an overpass. The investigating officers from the FAA (Federal Aviation Administration), in trying to determine the landing speed of the aircraft, applied the formula used for the calculation of car speed from skid marks. Will their results be approximately correct, too high, or too low? Why?

18.3. In a process requiring moving contact between two rollers, the design engineer must select roller materials that will have a maximum coefficient of friction for a given frictional force and applied load. One roller must be made of stainless steel. The second may be of stainless steel, butyl rubber, or a viscoelastic polymer. Which material will the engineer probably select? Why?

18.4. A space shuttle experiment was designed to determine the coefficient of friction between two metal surfaces in vacuum and under an argon atmosphere. Initial results indicated a value of $\mu_f = 4.9$ under both sets of conditions. After several hours of taking data, however, the value suddenly fell to 0.7. What might have happened that could explain the sudden change in μ_f?

18.5. Police, examining the scene of an accident, observe skid marks 30 m long left by a 1200-kg car. The car skidded to a stop on a concrete highway having $\mu_f = 0.80$ between the tires and the road surface. Calculate the approximate speed of the car at the moment the brakes were applied.

18.6. Several alkyl phosphoric acids were tested as lubricants in a system. The chains employed included n-C_{10}, n-C_{12}, n-C_{14}, t-C_{14}, n-C_{16}, t-C_{18}, n-C_{20}, and n-C_{22} compounds (n refers to normal, straight-chain materials; t refers to a tertiary end group on the hydrocarbon chain). Which material or materials would you expect to produce the best lubricating effect? Why?

18.7. A high-pressure pump has a normal lubricant operating pressure of 850 kg cm^{-1} at 50°C. A system failure allows the lubricant temperature to increase to 120°C. (a) If the normal working viscosity of the lubricant is 1.0 P (poise), what will be the change in viscosity experienced during the system malfunction? (b) Assuming that the original coefficient of friction of the system is 0.2 and the lubricant pressure increases to 1050 kg cm^{-1}, calculate the new coefficient of friction.

18.8. A glass sphere is molded into an optical lens by pressing against an optically smooth silicon carbide surface under heat and pressure. In order to avoid surface irregularities and distortions due to friction between the two surfaces, a lubricant must be employed. Under the conditions of the process, what lubricating mechanism would you predict to be most effective? What characteristics should the lubricant have? Explain your conclusions.

19 Adhesion

Adhesion is an extremely important concept in both practical and theoretical terms. Unfortunately, there is no completely satisfactory definition of the term that fulfills the needs of both the theoretical surface chemist and the practicing technologist. So far in this book, the term "adhesion" has been encountered as applied in the ideal or theoretical sense—referring to the reversible thermo-dynamic process of separating unit area of two phases that originally had a common interface. That aspect of the term was defined in Chapter 2 and will not be repeated here, except where necessary for clarity. Some comments about the "reality" of that concept will be in order, however.

A more practical definition of adhesion is a state in which two bodies (usually, but not necessarily dissimilar) are held together by intimate interfacial contact in such a way that mechanical force or work can be applied across the interface without causing the two bodies to separate (Fig. 19.1). (This is, a force superior to that of the frictional forces discussed in the previous chapter.) It is the latter definition that will be of most concern in the discussion which follows, although the two concepts are, in fact, inseparable. Before entering into that discussion, it will be useful to clarify a few terms commonly encountered in the field of adhesion, but often misinterpreted.

19.1. TERMINOLOGY

"Thermodynamic adhesion" is the term that applies to the "ideal" adhesion already defined in terms of reversible work needed to separate two surfaces by overcoming the molecular interactions across the interface. "Chemical adhesion" is a term that may be applied to adhesion involving the formation of formal chemical bonds (covalent, electrostatic, or metallic) across an interface. "Mechanical adhesion" refers to the situation in which actual mechanical interlocking of microscopic asperities at the interface occurs over a significant portion of the contact area.

As a practical term, adhesion may be used with reference to the so-called failing load of a particular joint, particularly as applied to the performance of glues, cements, and other agents. "Adherence" is a term sometimes used to describe the degree of practical adhesion, while adhesive applies to a material used to join two surfaces together in an adhesive joint or weld. The "adherend" is the phase (or phases) being joined by the adhesive. A "proper"

FIGURE 19.1. In its simplest form adhesion can be defined as the phenomenon that allows a mechanical force (in excess of any frictional forces) to be applied to two contacting objects without those two objects becoming separated.

adhesive joint is a utilitarian term to describe an ideal situation in which an adhesive joint has been prepared with complete intimate contact between components (i.e., no flaws) and in the absence of intervening contaminants (e.g., moisture, dirt, oil) that might reduce the ideal strength of the system.

The following discussion will attempt to put into focus the interrelationship between the various definitions as well as illustrate some of the more important aspects of the subject. The subject is quite broad, with an extensive literature, so the coverage will be limited in scope.

19.2. THERMODYNAMIC OR IDEAL ADHESION

We have already encountered the concept of thermodynamic adhesion and its related terms such as the work of adhesion. The term is applied to a defined model system and does not take into consideration conditions before or after the formation of the interface, the presence of random flaws or defects in the system, or the bulk physical properties of the components, all of which are of primary importance in the practical application of the concept of adhesion. It is related to molecular interactions such as van der Waals, dipolar, and electrostatic forces but does not consider mechanical or chemical interactions as defined above. It is therefore not a very useful concept in terms of practical adhesion problems, but it serves as a good theoretical tool and to indicate a maximum force or work that a given interface may be expected to transmit before failure (i.e., separation) occurs.

Because, in theory at least, the concept of ideal or thermodynamic adhesion applies equally well to liquid and solid phases, it is of interest to see how a calculation of such an ideal value compares with reality. The complete expression for the work of adhesion between two phases with each phase completely saturated by the other, denoted by A(B) and B(A) is

$$W_{A(B)B(A)} = \sigma_{A(B)} + \sigma_{B(A)} - \sigma_{AB} = W_{AB} - \pi_{A(B)} - \pi_{B(A)} \qquad (19.1)$$

where the phase in parentheses saturates the one it follows. The maximum force required to separate unit area of interface, the ideal adhesive strength of the interface, F_{ad}, can be approximated by

TABLE 19.1. Ideal and Real Cohesive (Tensile) Strengths of Some Common Materials ($r_0 = 0.4$ nm)

Material	Cohesive (tensile) strength, MPa	
	Ideal	Real
Polyethylene (molded)	180	382
Polystyrene (molded)	210	692
Aramid yarn	7,900	2,760
Drawn steel	9,800	1,960
Graphite whisker	100,000	24,000

$$F_{ad} = \frac{1.03 \ W_{A(B)B(A)}}{r_o} \tag{19.2}$$

where r_o is the equilibrium distance of separation, usually on the order of a few molecular diameters (0.2–0.5 nm). Referring to the calculation for the system benzene–water given in Chapter 8, the work of adhesion is $W_{A(B)B(A)} = 56$ mJ m^{-2}. If one assumes that the major portion of that work will be done over a distance $r_o = 0.4$ nm, then the ideal force required to separate the two phases would be on the order of 1.44×10^5 kg m^{-2}, well above the actual strength of most adhesive joints. As a more practical example, for two polymers that interact by dispersion forces alone, a typical value for W_{AB} will be 100 mJ m^{-2} leading to a maximum ideal adhesive strength of about 2.6×10^5 kg m^{-2}, which, again, is several orders of magnitude greater than practically obtainable adhesive strengths. Similar calculations can be applied for cohesive strengths. A few comparisons of ideal and practical cohesive strengths for some materials are given in Table 19.1.

On the basis of the ideal calculations similar to the above, it has been suggested that for practical adhesives, if good wetting of the surfaces to be joined can be achieved, dispersion forces alone should be sufficient to ensure a strong adhesive bond. Unfortunately, theory and reality do not always agree. The ideal calculations are, of course, based on the concept of thermodynamically reversible separation processes, while the fact of life is that such conditions are almost never attained. In fact, fracture processes are invariably accompanied by irreversible viscoelastic processes that dissipate energy and complicate the analysis of the situation. In addition, and perhaps more importantly, real adhesive joints will contain flaws that will greatly reduce the practical strength of the system. Some of those points will be discussed more fully below.

19.3. PRACTICAL ADHESION

The brief analysis of ideal adhesion given above indicates that, under the best of circumstances, one might expect to be able to attain very strong adhesive joints with materials interacting through the universal dispersion energy, with no need to invoke stronger molecular interactions, mechanical interlocking, or chemical bond formation. Experience, unfortunately, generally indicates otherwise. A simple analysis of the situation indicates that the adhesive strength between two dissimilar materials should be greater than the cohesive strength of the weaker of the two, leading to cohesive failure rather than adhesive failure. In some instances, particularly in the adhesive component of friction (see Chapter 18) that concept seems to be borne out since shearing of the contact area between surfaces normally occurs in the weaker material. In adhesive joints, however, the area of contact between surfaces is several orders of magnitude greater than that in friction, and the question of the completeness of interfacial contact becomes important.

A liquid spreading over a rough surface (the normal situation) can easily trap air in depressions in the surfaces, leading to the formation of a composite surface (Fig. 19.2). That will be the case whether the spreading liquid has a small or large contact angle, θ, although a small advancing angle, θ_A, will obviously improve matters. When a portion of the composite involves air–adhesive interfaces, the actual area of adhesive contact is greatly reduced. In addition, the three phase boundaries thus formed represent excellent sites for the initiation of cracks and flaws in the system. The net result—a significantly weakened joint. Similar effects can be seen if the entrapped material is water, oils, or other materials with significantly lower adhesional interactions or cohesion strength.

In practice, it is usually found that the actual adhesive strength of a joint prepared with a "good" (i.e., wetting) adhesive will be at least one order of magnitude less than the ideal value. Poorly wetting systems would be expected to perform correspondingly less effectively. The primary reason for the discrepancy between ideal and real adhesive strength appears to lie in the almost invariable presence of bubbles, cracks, and flaws that are associated with the interfacial zone (Fig. 19.3). When stress is applied to a joint, it tends to concentrate at such flaws, so that the "local" stress is significantly greater than the average value. When the local stress exceeds the local strength (already lessened by the presence of the composite interface, e.g.), failure occurs.

Gas bubbles trapped by spreading adhesive

FIGURE 19.2. Since all surfaces have a certain degree of roughness, it is common that an adhesive applied to such surfaces will entrap air bubbles, reducing the area of contact of adhesive with the surfaces and reducing the effectiveness of the bond or weld.

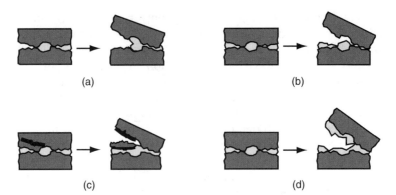

(a)   (b)

(c)   (d)

FIGURE 19.3. In what is generally classified as "adhesive failure," breakage may occur at various locations including entrapped bubbles (*a*), at the interface ("true" adhesive failure) (*b*), in the substrate (cohesive failure) (*c*), or in the adhesive (also cohesive failure) (*d*).

The fracture of practical adhesive joints involves two primary processes—cohesive or adhesive failure at or near the joint and work (reversible and irreversible) involved in plastic, elastic, or viscoelastic deformation of one or all of the components of the joint—one of the two solid surfaces or the adhesive (Fig. 19.4). As indicated in the preceding chapter on friction, cohesive failure of the weaker of two solids in contact is common. The same can be said for normal adhesive joints, in that actual adhesive failure (i.e., exactly at the interface) is less common that cohesive failure, of, for example, the adhesive material, near the interface. What, then, are the necessary conditions for obtaining "good" adhesion between two surfaces?

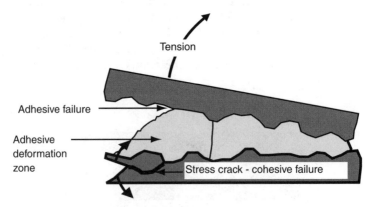

FIGURE 19.4. When a tension or shear strain is placed on a joint, the energy may be dissipated by several mechanisms including adhesive and cohesive failure at various points, as already mentioned, but also by the plastic, elastic, or viscoelastic stretching of the adhesive and/or one or both substrates.

19.4. SOME CONDITIONS FOR "GOOD" ADHESION

Any workable theory of adhesion must take into consideration all of the possible aspects of energy transfer across an adhesive joint, not only molecular forces or ideal adhesive strengths, but also the presence, number, and size of flaws, various energy dissipation processes, and irreversible fracture processes. It has already been demonstrated that in an ideal situation, forces of molecular interaction should be sufficient to produce a very strong adhesive joint, assuming perfect contact across the interface, but that reality lies far from the ideal.

For reference purposes, Table 19.2 lists the various attractive molecular forces that may operate across an interface along with the approximate range of strengths they cover. Since van der Waals forces, for example, fall off rapidly with distance of separation by r^{-3}, for such forces to be effective the interacting surfaces must be as close as possible, typically 0.2–0.5 nm. Beyond that distance, such interactions will be quite weak and the ability of the joint to transmit any applied stress will be accordingly reduced. Clearly, intimate molecular contact between phases is a necessary condition for good adhesion—necessary but not sufficient!

It is a fact of life that in practical adhesion problems, materials properties are often as important as interfacial forces. It has been stated that a significant portion of the fracture energy of a joint is dissipated in various deformation processes. Obviously, the nature of the joint interface in terms of physical "mixing" is an important aspect of the overall problem.

The intermolecular forces listed in Table 19.2 are common to all liquids and solids, depending on chemical composition, yet liquids have significantly lower mechanical strength than comparable solids. In solids, the intermolecular distances are generally smaller and are reinforced by the forces stemming from the more ordered structure (crystal lattice, etc.). In polymers, which constitute the majority of adhesives, mechanical strength (e.g., the ability to

TABLE 19.2. Values of Attractive Molecular Interactions at Interfaces

Type of Interaction	Approximate Energy Range (kcal mol^{-1})
Van der Waals	
Dispersion	5
Dipole–dipole	0–10
Dipole-induced dipole	0–0.5
Hydrogen bonding	0–40
Chemical bonds	
Covalent bonds	15–170
Ionic bonds	140–250
Metallic bonds	27–83

transmit stress without failure) is also a function of molecular weight. Below a certain value, strength falls off rapidly with molecular weight. In such a system, the intermolecular forces between chains are essentially unchanged, so that the loss in strength must be a result of some more physical phenomenon, in this case so-called molecular entanglement. That is, as the polymer molecules become longer, they become more "wrapped up" in their neighbors (Fig. 19.5). For the polymer to fail under stress, the entangled chains must slide past one another and disentangle, a process that requires a significant amount of energy. Molecular entanglement, therefore, serves as an excellent process for the dissipation of the stress forces, and provides a good mechanism to back up intermolecular forces for producing good adhesion.

This dissipation process depends directly on the amount of chain entanglement and the nature of the forces acting between chains. In a bulk polymer, the degree of entanglement is a direct result of the nature of the polymer (molecular weight, branching, etc.) and its method of preparation (e.g., cast from the melt, from solvent, spun). At an interface, entanglement is more problematical—the process (for polymeric surfaces, in any case) may be assisted, for instance, by the use of a solvent that swells the adherend surface, allowing interpenetration of adhesive and adherend; the application of heat, which increases the mobility of the polymer chains; or the use of a monomeric

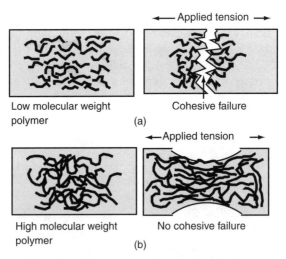

FIGURE 19.5. For polymeric adhesives (and polymers in general) molecular weight can have an important effect on the cohesive strength of the material. For low-molecular-weight polymers (*a*), there is relatively weak interaction among adjacent chains; movement of one chain past another is easy and the tensile or shear strength is low. In high-molecular-weight materials (*b*), chain interactions are greatly increased, they may become tangled, and the material exhibits a much greater strength: stretching and "necking" as shear is applied, but resisting a much greater force before failing.

system that is polymerized after applications. In any case, if entanglement can be increased, the strength of the bond at the interface will be improved.

If entanglement does occur, it should be obvious that one can no longer talk about a sharp interface, in the classic sense, but must consider an interfacial zone, the structure of which will be a primary determinant of the strength of the joint. For simplicity, one may consider two types of interfaces—sharp, in which the primary component of strength derives from classical intermolecular attractive forces, and *diffuse,* in which entanglement plays a significant role. These can be further classified by the relative strengths of the molecular interactions to give four general classes of adhesive behavior (Fig. 19.6).

The first, and simplest, class (Fig. 19.6*a*) can be described as having a sharp interface and weak intermolecular interactions. An example might be a joint between a nonpolar polymer (e.g., polyethylene) and a polar polymer (e.g., polyvinyl alcohol–polyvinyl acetate copolymer). In such a system, the only molecular interaction is that due to dispersion forces, with little tendency for chain entanglement due to the inherent incompatibility of the two polymers. The mechanical strength of the resulting joint derives solely from the dispersion forces, which will not be able to inhibit the movement or slippage of the joint significantly. A joint of very poor strength is the result.

The second class (Fig. 19.6*b*) is a joint with a sharp interface, but with significant specific chemical interactions between the two phases. For example, a polymer containing groups capable of significant hydrogen bonding or acid–base interactions ($-OH$, $-NH-$, or $-COOH$) can interact strongly with, say, a metal or metal oxide surface (M^{n+} or $-M-O-M-$), even though no significant entanglement is possible. The resulting joint would have significant mechanical strength because of the larger magnitude of the interactive forces, even without the assistance of entanglement.

In the third class (Fig. 19.6*c*), there is significant interpenetration of the adhesive into the surfaces to be bound. The interpenetration may be at the

FIGURE 19.6. It is usually possible to estimate the probable strength of an adhesive joint based on the types of interactions present between adhesive and adherend: (*a*) Dispersion forces alone usually produce joints of limited strength; (*b*) the presence of polar interactions will usually improve the situation significantly; (*c*) penetration of adhesive polymer chains into the adherend surface also adds greatly to the potential strength of the joint; (*d*) physical interlocking, coupled with any or all of the other mechanisms, usually insures the strongest practical joint.

molecular level, in the case that the adhesive polymer and polymeric adherend are mutually miscible, or more at the microscopic bulk level, as for porous substrates. If the interface is diffuse and significant entanglement occurs, a strong joint may be expected, regardless of the nature of the intermolecular forces acting between elements. In such a case, the entanglement of entire polymer molecules is not necessary for significant strength to be developed. Entanglement may be considered a sufficient condition for good adhesion, even in the absence of strong intermolecular interactions.

The fourth class (Fig. 19.6*d*) would be that involving some form of physical or mechanical interlocking of the two surfaces. Such a situation might be encountered when a molten polymer or pre-polymer mixture is applied to a rough surface under condition where it can flow into the solid surface irregularities.

If one considers the action necessary to displace the two surfaces in a system with a sharp boundary, it can be seen that for the first case, the molecular forces being overcome are not only small (relative to specific interactions) but are essentially limited to a two-dimensional action across the interface (i.e., a plane approximately parallel to the two surfaces being bonded) as illustrated in Figure 19.7*a*. If entanglement occurs, the same small forces will be acting in three dimensions (Fig.19.7*b*)—that is, each entangled molecule will have nearest neighbors on all sides, which will mean—roughly speaking—that the interactions will be multiplied many times, depending on the degree of interpenetration. If stronger specific interactions are present (mechanical interlocking), or if chemical bonds are formed, so much the better!

From the preceding discussion, then, it appears that the optimum conditions for good adhesion include a diffuse interfacial zone and/or strong specific intermolecular interactions between phases. One may add to that (in the opinion of some) the existence of a direct physical interlocking between surfaces. With all the best-designed practical systems, however, adhesive joints

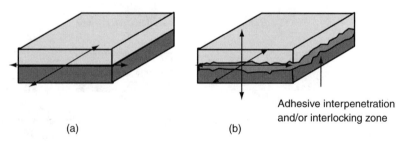

Adhesive interpenetration
and/or interlocking zone

(a)   (b)

FIGURE 19.7. For an adhesive joint with a smooth boundary (*a*) the tension on the joint can be dissipated only in a plane parallel to the joint, otherwise joint failure will be immediate. In a joint where interpenetration or interlocking occur (*b*), the tension may be dissipated in the additional direction perpendicular to the joint. The added "option" results in a stronger joint.

never attain the strength one would predict on the basis of theory. The following section addresses the question of why life can be so difficult.

19.5. ADHESIVE FAILURE

When an adhesive bond fails under a small applied stress, it is commonly described as a "weak" bond or "poor" adhesion. In fact, such a description may be misleading since failure may have occurred at the interface, near the interface within one of the phases comprising the system, or well away from the interface. These scenarios are illustrated in Figure 19.3. It is only the first case that one can accurately describe as being a result of poor adhesion. The other two are more properly called "failures" of the bulk materials, that is, failure of cohesion, which is not the same thing. However, usage dictates that failure of any kind be termed "adhesion failure." Of course, in an actual joint, failure may be due to a combination of all three processes.

When failure occurs exactly at the joint interface or well away from it (say, more than 100 nm), then identification of the locus of failure is generally a simple process. It is when the failure occurs within 10–100 nm of the interface that identification becomes a problem.

19.5.1. Importance of Failure Identification

Correct identification of the locus of failure can be of great practical and theoretical importance. If one can determine that failure occurred cohesively near the actual interface, it can be inferred that improvement in bond strength can be obtained by increasing the cohesive strength of the ruptured material without worrying about the nature of the interactions at the interface (i.e., molecular attraction or entanglement). If the failure is found to occur at the interface, on the other hand, it will clearly be necessary to change the chemical nature of the components to increase the intermolecular attractive forces— introduce more specific interactions, form chemical bonds, and/or increase interpenetration of the phases. If the locus of failure is not correctly identified, a great deal of time, energy, and money may be wasted solving problems that do not exist! All of that depends, of course, on the fact that the bond in question was actually between the phases expected (i.e., a "proper" joint), and that some cohesively weak contaminant (moisture, oil, air, dirt, etc.) is not the primary cause of failure.

On the basis of the simple calculations of ideal adhesive bond strength given earlier, it has been suggested that bond failure in a "proper" adhesive joint will seldom occur at the interface. Instead, failure will occur in a weak boundary layer near the true interface, or within the weaker of the two bonded phases. Modern experimental techniques and theoretical considerations, however, indicate that all three possibilities for failure do, in fact, occur, depending on the given situation.

For a system with a sharp interface and weak intermolecular interaction, as in the nonpolar–polar polymer system mentioned above, failure exactly at the interface is a distinct possibility (or even probability). Thus, interfacial separation may be expected when interfacial strength is weaker than the bulk strength of the bonded materials. As we have seen, if the intermolecular interactions across the interface are more specific (including chemical bond formation) or if significant interpenetration of polymer chains occurs, rupture at the interface becomes less likely. In some cases, the locus of failure may depend on the rate at which stress is applied; rapid application leads to cohesive failure and very slow application, tending more toward true adhesive failure (since slow application of stress gives more time for the entangled molecules to "slide past" one another).

A great deal of theory has grown up around the subject of failure in adhesive joints. It would be prohibitive to attempt to cover the subject here; however, the practical importance of understanding the topic cannot be overemphasized. One has only to consider the large number of critical structural joints employed in modern construction (e.g., of airplanes) to see how important the subject has become.

19.5.2. The Role of Joint Flaws in Adhesive Failure

To this point the discussion has focused primarily on so-called "proper" or ideal adhesive joints, assuming intimate contact between components and the absence of flaws and contaminants. In the real world, such conditions are difficult to attain, so that the question of practical joint failure may not be concerned so much with intermolecular forces and entanglement, but with the mechanics of stress propagation in the system. That subject, like so many related to practical applications of surface chemistry, is very extensive and beyond the scope of this book. However, the basic principles involved are such that a few words may serve a useful purpose.

A typical flaw, for purposes of the present discussion, would be an entrapped bubble of air or other contaminant that is itself relatively weak cohesively (Fig. 19.8). In the presence of such flaws, little or no energy can be transmitted through the flaw so that the stress becomes concentrated at the junctions of the flaws with the interface or in one of the bulk materials. The "local" stress, therefore, is greater than the average value and is more likely to exceed the adhesive or cohesive strength of the system near the flaw. As a result of that situation, the applied stress induces the formation of a crack that continues to propagate along the line of least resistance (with continued application of stress) until joint failure results.

Cracks or other such flaws present in the adherend surface may also serve as foci for crack propagation under stress leading to apparent adhesive failure. In joints or welds that undergo cyclic mechanical (bending) or thermal (hot–cold–hot cycling) stress may also develop stress cracks at or near a joint or

Trapped air

Dirt and oil
contaminants

(a)

Contaminants and
flaws in adhesive

(b)

Contaminants and
flaws in substrate

(c)

FIGURE 19.8. In summary, adhesive joint failure may be caused by a number of factors some not connected with the actual adhesive bond. A common source of problems is the presence of trapped air or other contaminants that serve as loci for failure at the interface (*a*). Perhaps less common, but of significance, is the presence of contaminants or flaws in the bulk of the adhesive (*b*) or in the substrate (*c*) that weaken the physical properties of that phase and lead to joint failure.

weld, again leading to apparent adhesive failure when in fact the problem lies within the adherend alone.

Because there are so many geometries of adhesive joints encountered, and so many types of stress applied (tension, shear, torsion, thermal, etc.), the analysis of a given system must be tailored to meet the specific application. The processes of experimental design and data analysis, therefore, become quite complicated. It should also be kept in mind that flaws such as those often implicated in adhesive failure can also lead to apparent cohesive failure in the bulk material.

PROBLEMS

19.1. Calculate the surface tension that a liquid should have so that the work of adhesion to a surface with $\sigma_c = 20$ mN cm^{-1} will be a maximum. What will be the contact angle?

19.2. The statement has been made that the work of adhesion between two dissimilar substances should be larger than the work of cohesion of the weaker one. Demonstrate whether this should always be correct or show circumstances in which it will not be so.

19.3. A surgeon wishes to repair a hernia with a biodegradable adhesive instead of normal sutures. The tension of the skin at the lesion is 290 N m^{-1}. If the total surface area to be bound is 4 cm^2, what must be the minimum adhesive strength of the bond to ensure a safe closure?

19.4. For bonding two smooth glass surfaces, one might select a hot-melt polyethylene adhesive, an aqueous acrylic latex adhesive, or two-component epoxy adhesive. Which material would you expect to produce the strongest bond? Why?

19.5. In the photographic industry it is necessary to coat hydrophilic gelatin emulsions onto essentially nonionic acetate or polyester surfaces. Direct application of the aqueous phase to the surface often results in a system with poor adhesion—that is, the dried gelatin phase peels away from the polymer surface. Suggest the reason for the observed result.

19.6. In order to overcome the adhesion problem described in Problem 19.6, it has been found useful to apply a thin coating of an acrylic polymer containing small amounts of free acrylic acid prior to the application of the gelatin emulsion. Suggest an explanation for any observed improvement in the resulting adhesion.

19.7. In some cases, an improvement similar to that in Problem 19.7 can be obtained by treating the hydrophobic surface with electron beams, a high-intensity short-wavelength light (corona discharge), or with a corrosive etching chemical. Suggest the mechanism operating to improve adhesion in those cases?

19.8. Roughing a surface can often produce an improvement in adhesion. Suggest a mechanism by which such improvement might be explained.

19.9. The use of surface "roughing" to improve adhesion may produced the opposite effect if not properly applied. Give two problems that may arise from surface roughing that could result in poor adhesion.

Bibliography

GENERAL READINGS

Adamson, A. W., *Physical Chemistry of Surfaces,* 4th ed., Wiley-Interscience: New York 1984.

Aveyard, R., Haydon, D. A., *An Introduction to the Principles of Surface Chemistry,* Cambridge University Press, Cambridge, 1973.

Everett, D. H., *Basic Principles of Colloid Science,* Royal Society of Chemistry Paperbacks, Royal Society of Chemistry, London, 1988.

Evans, D. F., Wennerstrom, H., *The Colloidal Domain,* VCH Publishers, New York, 1994.

Hiemenez, P. C., *Principles of Colloids and Surface Chemistry,* 2nd ed., Marcel Dekker, New York, 1986.

Hunter, R. J., *Foundations of Colloid Science,* Oxford University Press, New York, 1987.

Israelachvili, J., *Intermolecular and Surface Forces,* 2nd ed., Academic Press, San Diego, 1991.

Jaycock, M. J., Parfitt, G. D., *Chemistry of Interfaces,* Ellis Horwood, Chichester, U.K., 1981.

Kruyt, H. R., *Colloid Science,* Vol. 1, Elsevier, New York, 1952.

Shaw, D. J., *Introduction to Colloid and Surface Chemistry,* 3rd ed., Butterworths: London, 1980.

Tanford, C., *The Hydrophobic Effect. Formation of Micelles and Biological Membranes,* 2nd ed., Wiley, New York, 1980.

Vold, M. J., Vold, R. D., *Colloid and Interface Chemistry,* Addison-Wesley, Reading, MA, 1983.

CHAPTER READINGS

Chapter 1
van Olphen, H., Mysels, K. J., *Physical Chemistry: Enriching Topics from Colloid and Surface Science,* Theorex, La Jolla, CA 1975.

Chapter 2
Adamson, A. W., *Physical Chemistry of Surfaces,* 4th ed., Wiley-Interscience, New York, 1984, Chapters 2 and 7.

Aveyard, R., Haydon, D. A., *An Introduction to the Principles of Surface Chemistry,* Cambridge University Press, Cambridge, 1973, Chapters 3, 5, and 6.

Shaw, D. J., *Introduction to Colloid and Surface Chemistry,* 3rd ed., Butterworths: London, 1980, Chapters 4–6.

Vold, M. J., Vold, R. D., *Colloid and Interface Chemistry,* Addison-Wesley, Reading, MA, 1983, Chapters 1, 3, and 4.

Chapter 3

Myers, D. Y., *Surfactant Science and Technology,* 2nd ed., VCH Publishers, New York, 1992, Chapter 2.

Rosen, M. J., *Surfactants and Interfacial Phenomena,* Wiley-Interscience, New York, 1976.

Rieger, M. M., Rhein, L. D., eds., *Surfactants in Cosmetics,* Marcel Dekker, New York, 1997, Chapter 1.

Chapter 4

Adamson, A. W., *Physical Chemistry of Surfaces,* 4th ed. Wiley-Interscience, New York, 1984, Chapter 6.

Israelachvili, J., *Intermolecular and Surface Forces,* 2nd ed., Academic Press, San Diego, 1991, Chapters 2–6.

Mahanty, J., Ninham, B. W., *Dispersion Forces,* Academic Press, New York, 1976.

Rieger, M. M., Rhein, L. D., eds., *Surfactants in Cosmetics,* 2nd ed., Marcel Dekker, New York, 1997, Chapter 5.

Vold, M. J., Vold, R. D., *Colloid and Interface Chemistry,* Addison-Wesley, Reading, MA, 1983, Chapter 7.

Chapter 5

Adamson, A. W., *Physical Chemistry of Surfaces,* 4th ed., Wiley-Interscience, New York, 1984, Chapter 4.

Aveyard, R., Haydon, D. A., *An Introduction to the Principles of Surface Chemistry,* Cambridge University Press, Cambridge, 1973, Chapter 2.

Israelachvili, J., *Intermolecular and Surface Forces,* 2nd ed., Academic Press, San Diego, 1991, Chapter 3.

Rieger, M. M., Rhein, L. D., eds., *Surfactants in Cosmetics,* 2nd ed. Marcel Dekker, New York, 1997, Chapter 5.

Vold, M. J., Vold, R. D., *Colloid and Interface Chemistry,* Addison-Wesley, Inc.: Reading, MA, 1983, Chapter 7.

Chapter 6

Adamson, A. W., *Physical Chemistry of Surfaces,* 4th ed., Wiley-Interscience, New York, 1984, Chapter 1.

Aveyard, R., Haydon, D. A., *An Introduction to the Principles of Surface Chemistry,* Cambridge University Press, Cambridge, 1973, Chapter 3.

Evans, D. F., Wennerstrom, H., *The Colloidal Domain,* VCH Publishers, New York, 1994, Chapter 2.

Vold, M. J., Vold, R. D., *Colloid and Interface Chemistry,* Addison-Wesley, Reading, MA, 1983, Chapter 1.

Chapter 7
Adamson, A. W., *Physical Chemistry of Surfaces,* 4th ed., Wiley-Interscience, New York, 1984, Chapter 5.
Aveyard, R., Haydon, D. A., *An Introduction to the Principles of Surface Chemistry,* Cambridge University Press, Cambridge, 1973, Chapter 5.
Shaw, D. J., *Introduction to Colloid and Surface Chemistry,* 3rd ed., Butterworths: London, 1980, Chapter 5.
Vold, M. J., Vold, R. D., *Colloid and Interface Chemistry,* Addison-Wesley, Reading, MA, 1983, Chapter 3.
Zettlemoyer, A. C., ed., *Nucleation,* Marcel Dekker, New York, 1969.

Chapter 8
Adamson, A. W., *Physical Chemistry of Surfaces,* 4th ed., Wiley-Interscience, New York, 1984, Chapter 12.
Aveyard, R., Haydon, D. A., *An Introduction to the Principles of Surface Chemistry,* Cambridge University Press, Cambridge, 1973, Chapter 3.
Shaw, D. J., *Introduction to Colloid and Surface Chemistry,* 3rd ed., Butterworths, London, 1980, Chapter 4.
Vold, M. J., Vold, R. D., *Colloid and Interface Chemistry,* Addison-Wesley, Reading, MA, 1983, Chapter 1.

Chapter 9
Adamson, A. W., *Physical Chemistry of Surfaces,* 4th ed., Wiley-Interscience, New York 1984, Chapters 8 and 12–14.
Aveyard, R., Haydon, D. A., *An Introduction to the Principles of Surface Chemistry,* Cambridge University Press, Cambridge, 1973, Chapters 5 and 6.
Shaw, D. J., *Introduction to Colloid and Surface Chemistry,* 3rd ed., Butterworths: London, 1980, Chapters 5 and 6.
Vold, M. J., Vold, R. D., *Colloid and Interface Chemistry,* Addison-Wesley, Reading, MA, 1983, Chapters 2–4.

Chapter 10
Adamson, A. W., *Physical Chemistry of Surfaces,* 4th ed., Wiley-Interscience, New York, 1984, Chapter 4.
Evans, D. F., Wennerstrom, H., *The Colloidal Domain,* VCH Publishers, New York, 1994, Chapter 5.
Shaw D. J., *Introduction to Colloid and Surface Chemistry,* 3rd ed., Butterworths, London, 1980, Chapters 1 and 8
Vold, M. J., Vold, R. D., *Colloid and Interface Chemistry,* Addison-Wesley, Reading, MA, 1983, Chapters 6–10.

Chapter 11
Adamson, A. W., *Physical Chemistry of Surfaces,* 4th ed., Wiley-Interscience, New York, 1984, Chapter 11.

Becher, P., *Emulsions: Theory and Practice,* 2nd ed., Reinhold, New York, 1965.

Becher, P., ed., *Encyclopedia of Emulsion Technology,* Vols. 1–4, Marcel Dekker, New York, 1985.

Davies, J. T., Rideal, E. K., *Interfacial Phenomena,* Academic Press, New York, 1961.

Evans, D. F., Wennerstrom, H., *The Colloidal Domain,* VCH Publishers, New York, 1994, Chapter 11.

Griffith, W. C., *J. Soc. Cosmetic Chemists* **5,** 249 (1954).

Rieger, M. M., Rhein, L. D., eds., *Surfactants in Cosmetics,* 2nd ed., Marcel Dekker, New York, 1997.

Shaw, D. J., *Introduction to Colloid and Surface Chemistry,* 3rd ed., Butterworths: London, 1980, Chapter 10.

Vold, M. J., Vold, R. D., *Colloid and Interface Chemistry,* Addison-Wesley, Reading, MA, 1983, Chapter 12.

Chapter 12

Adamson, A. W., *Physical Chemistry of Surfaces,* 4th ed., Wiley-Interscience, New York, 1984, Chapter 11.

Berkman S., Egloff, G., *Emulsions and Foams,* Reinhold, New York, 1961.

Bikerman, J. J., *Foam,* Springer-Verlag, New York, 1973.

Little, R. C., *J. Colloid Interface Sci.* **65,** 587 (1978).

Shaw, D. J., *Introduction to Colloid and Surface Chemistry,* 3rd ed., Butterworths, London, 1980, Chapter 10.

Vold, M. J., Vold, R. D., *Colloid and Interface Chemistry,* Addison-Wesley, Reading, MA, 1983, Chapter 12.

Chapter 13

Evans, D. F., Wennerstrom, H., *The Colloidal Domain,* VCH Publishers, New York, 1994, Chapter 9.

Vold, M. J., Vold, R. D., *Colloid and Interface Chemistry,* Addison-Wesley, Reading, MA, 1983, Chapter 11.

Chapter 14

Evans, D. F., Wennerstrom, H., *The Colloidal Domain,* VCH Publishers, New York, 1994, Chapter 7.

Vold, M. J., Vold, R. D., *Colloid and Interface Chemistry,* Addison-Wesley, Reading, MA, 1983, Chapters 13–15.

Chapter 15

Evans, D. F., Wennerstrom, H., *The Colloidal Domain,* VCH Publishers, New York 1994, Chapter 4.

Israelachvili, J., *Intermolecular and Surface Forces,* 2nd ed., Academic Press, San Diego, 1991, Chapters 16 and 17.

Rosen, M. J., *Surfactants and Interfacial Phenomena,* Wiley-Interscience, New York, 1976.

Shinoda, K., Nakawaga, T., Tamamushi, B., Isemura, T., *Colloidal Surfactants, Some Physico-Chemical Properties,* Academic Press, New York, 1963.

Vold, M. J., Vold, R. D., *Colloid and Interface Chemistry,* Addison-Wesley, Reading. MA, 1983, Chapter 18.

Chapter 16

Elworthy, P. H., Florence, A. T., Macfarlane, C. B., *Solubilization by Surface-Active Agents,* Chapman and Hall, London, 1968.

Evans, D. F., Wennerstrom, H., *The Colloidal Domain,* VCH Publishers, New York, 1994, Chapters 4 and 11.

Fendler, J. H., Fendler, E. J., *Catalysis in Micellar and Macromolecular Systems,* Academic Press, New York, 1975.

Prince, L. M., ed., *Microemulsions Theory and Practice,* Academic Press, New York 1977.

Vold, M. J., Vold, R. D., *Colloid and Interface Chemistry,* Addison-Wesley, Reading, MA, 1983, Chapter 19.

Chapter 17

Adamson, A. W., *Physical Chemistry of Surfaces,* 4th ed., Wiley-Interscience, New York 1984, Chapter 10.

Aveyard, R., Haydon, D. A., *An Introduction to the Principles of Surface Chemistry,* Cambridge University Press, Cambridge, 1973, Chapter 6.

Shaw, D. J., *Introduction to Colloid and Surface Chemistry,* 3rd ed., Butterworths: London, 1980, Chapter 6.

Vold, M. J., Vold, R. D., *Colloid and Interface Chemistry,* Addison-Wesley, Reading, MA, 1983, Chapter 14.

Chapter 18

Adamson, A. W., *Physical Chemistry of Surfaces,* 4th ed., Wiley-Interscience, New York 1984, Chapter 9.

Girifalco, L. A., Good, R. J., *J. Phys. Chem.* **61,** 904 (1957).

Fox, H. W., Zisman, W. A., *J. Colloid Sci.* **5,** 514 (1950).

Chapter 19

Adamson, A. W., *Physical Chemistry of Surfaces,* 4th ed., Wiley-Interscience, New York, 1984, Chapter 9.

Bickerman, J. J., *The Science of Adhesive Joints,* Academic Press, New York, 1961.

Houwink, R., Salomon, D., eds., *Adhesion and Adhesives,* Elsevier, New York, 1965.

Kaelble, D. H., *Physical Chemistry of Adhesion,* Wiley-Interscience, New York, 1971.

INDEX